T0331728

Coal, Steam and Ships

Crosbie Smith explores the trials and tribulations of first-generation Victorian mail steamship lines, their passengers, proprietors and the public. Eye-witness accounts show in rich detail how these enterprises engineered their ships, constructed empire-wide systems of steam navigation and won or lost public confidence in the process. Controlling recalcitrant elements within and around steamship systems, however, presented constant challenges to company managers as they attempted to build trust and confidence. Managers thus wrestled to control shipbuilding and marine engine making, coal consumption, quality and supply, shipboard discipline, religious readings, relations with the Admiralty and government, anxious proprietors and the media – especially following a disaster or accident. Emphasizing interconnections between maritime history, the history of engineering and Victorian culture, Smith's innovative history of early ocean steamships reveals the fraught uncertainties of Victorian life on the seas.

Crosbie Smith was Professor of History of Science at the University of Kent until he retired in 2014 to concentrate on research. Two of his books have won the History of Science Society's Pfizer Award, *Science of Energy* and *Energy and Empire*, which he co-wrote with Norton Wise.

SCIENCE IN HISTORY

Series Editors

Simon J. Schaffer, University of Cambridge

James A. Secord, University of Cambridge

Science in History is a major series of ambitious books on the history of the sciences from the mid-eighteenth century through the mid-twentieth century, highlighting work that interprets the sciences from perspectives drawn from across the discipline of history. The focus on the major epoch of global economic, industrial and social transformations is intended to encourage the use of sophisticated historical models to make sense of the ways in which the sciences have developed and changed. The series encourages the exploration of a wide range of scientific traditions and the interrelations between them. It particularly welcomes work that takes seriously the material practices of the sciences and is broad in geographical scope.

Coal, Steam and Ships

Engineering, Enterprise and Empire on the Nineteenth-Century Seas

Crosbie Smith

University of Kent

Shaftesbury Road, Cambridge CB2 8EA, United Kingdom

One Liberty Plaza, 20th Floor, New York, NY 10006, USA

477 Williamstown Road, Port Melbourne, VIC 3207, Australia

314–321, 3rd Floor, Plot 3, Splendor Forum, Jasola District Centre, New Delhi – 110025, India

103 Penang Road, #05–06/07, Visioncrest Commercial, Singapore 238467

Cambridge University Press is part of Cambridge University Press & Assessment, a department of the University of Cambridge.

We share the University's mission to contribute to society through the pursuit of education, learning and research at the highest international levels of excellence.

www.cambridge.org
Information on this title: www.cambridge.org/9781107196728

DOI: 10.1017/9781108164894

First published 2018

A catalogue record for this publication is available from the British Library

ISBN 978-1-107-19672-8 Hardback
ISBN 978-1-316-64747-9 Paperback

For
The Crew

The sea, with such a storm as his bare head
In hell-black night endured, would have buoyed up
And quenched the stellèd fires.

The Tragedy of King Lear (Act 3 Scene 7)

Contents

Figures

Tables

Acknowledgements

The research foundations for this book derive from the AHRC-funded 'Ocean Steamship: A Cultural History of Victorian Maritime Power'. The project originated in an informal conversation between Ian Higginson and me as to promising fields for future research in the cultural history of science and technology. It initially focused on Alfred Holt's celebrated 'Blue Funnel' line of steamers deploying high-pressure double-cylinder (compound) engines that complemented their owner's moral commitment to minimizing waste in all its forms. The aims of the larger project, however, included an undertaking to evaluate the claim that Scottish cultural networks, practices and values played a central role in promoting and securing the trustworthiness of British ocean steam navigation. This theme was to feature prominently in a series of papers on the Holts, Cunard, RMSP and PSNC published during the subsequent years of the research programme.

Because the 'Ocean Steamship' project has been so fundamental to the writing of this book, I am exceptionally grateful to Ian Higginson, Phil Wolstenholme, Anne Scott, Trish Hatton and Don Leggett, all of whom served on the Kent team at various times and in different ways. During early visits to Liverpool's rich archives, for example, Ian and Phil mined a number of remarkable archival gems, including Alfred Holt's personal diary and a sermon by his Unitarian minister on 'The Doctrine of Waste'. Both Anne and Trish later worked tirelessly on tracing and extending the maritime networks of ship-owners and shipbuilders while also providing many of the historical insights around which the project flourished. Co-applicants Ben Marsden and Will Ashworth supplied invaluable advice and support. Ben played the major role in developing the methodological grounding for the programme, culminating in his and my *Engineering Empires* (Palgrave, 2005), while Will guided the team unerringly in relation to its engagement with Merseyside historical contexts. Don's Greenwich-based researches in the unpublished papers of William Schaw Lindsay opened up fresh perspectives into ocean steamship history in the period, not least because Lindsay knew personally many of the early ship-owners, their families and their communities.

The programme brought us academic support from a wide range of other scholars with strong and diverse interests in the fields of maritime history, history of science and history of technology. The maritime historians included the late Robin Craig (whose cliff-side house at St Margaret's Bay stood as the most wonderful private maritime library imaginable), the late Joe Clarke (with his inimitable dedication to the history of marine engineering in the north-east), and the late Mike Stammers (with his great command of deep-sea and coastal sailing ship history in the period). This book, as will be seen from the Introduction and subsequent chapters, owes an immense debt to the late Tom Hughes's inspirational systems approach within the history of technology, as well as to Steven Shapin's ground-breaking historical studies on matters of trust, credibility and truth. As always, I thank Norton Wise for the insights gained and the skills honed during our memorable collaboration on *Energy and Empire: A Biographical Study of Lord Kelvin* (Cambridge, 1989).

Over the years, I have also benefited immeasurably from fruitful discussions with Ludmilla Jordanova, Robert Fox, John Brooke, Geoffrey Cantor, John McAleer, Mike Hatton, David Lambert, Roy Fenton, John Hume, Ray Thompson, Phil Shore, Graeme Gooday, Jon Agar, Jenny Rampling, Aileen Fyfe, Ian Wilson, Megan Barbrook, Andrew Lambert, Lucy DeLap, Anyaa Anim-Addo, Yakup Bektas, Will Pettigrew, Pratik Chakrabarti, Peter and Christine Froome, Sam Hamilton and the late Alison Winter.

For facilitating excellent and varied opportunities to present project papers at seminars, conferences and symposia, I am especially indebted to Sir David Cannadine in his capacity as Institute of Historical Research Director and organizer of the Annual Anglo-American Conference (2001) at which we delivered our initial findings, Robert Lee at Liverpool University, Robert Blyth and Nigel Rigby at the National Maritime Museum, John Clarke at St Andrews University, David Livingstone at Queen's University Belfast and Charles Withers at Edinburgh University. Successive editors of *Technology and Culture*, John Staudenmaier and Suzanne Moon, have also been unfailingly supportive. University of Kent seminar presentations and informal meetings over coffee owed much to the comments and insights offered by colleagues, notably Charlotte Sleigh, John Clarke, Grayson Ditchfield, Kenneth Fincham, Jackie Waller, Stefan Goebel, Philip Boobbyer, David Ormrod and Hugh Cunningham.

As the project matured, I was very fortunate in having as successive doctoral students Don Leggett and Oliver Carpenter. Don's thesis, together with a goodly proportion of his manifold publications, offered an important complementary study to the original 'Ocean Steam Ship' project by developing a cultural history of the Victorian Navy. His book *Shaping the Royal Navy: Technology, Authority and Naval Architecture, c. 1830–1906* (Manchester University Press, 2015), promises to shape the field for years to come. Don also kindly undertook to read critically a draft of this book. Oliver's less glamorous subject, north-east steam tramp shipping, showed how modest Methodist trampship-owners, the Runcimans, could rise to high socio-political status in twentieth-century national life. Both Don and Oliver played significant roles in shaping my History Special Subject at Kent. Over three sessions, 'The Ocean Liner' proved that not all undergraduate history needed to replicate and extend well-trodden school history. Much of the content of this book, moreover, was forged and re-forged in the intense seminar discussions. I am grateful to many of the protagonists for their contributions, and especially to Edward Gillin and Stephen Courtney whose dialectic has helped a synthesis emerge in the following chapters. Tim Marshall's insights were also invaluable. Specific acknowledgements will be found in the footnotes.

An invitation to deliver the eighteenth Rausing Lecture in the History of Science (hosted by the Department of History and Philosophy of Science) at the University of Cambridge came as a wonderful opportunity to present to a wider audience several of the methodological themes now integrated in this book. My special thanks go to Simon Schaffer whose insights and comments before and after the lecture were, as ever, inspiring, multifaceted and challenging. I chose the lecture title 'Coal, Steam and Ships' to mirror Turner's famous painting 'Rain, Steam and Speed: Great Western Railway'. Turner's painting can be viewed both as a depiction of the violent destruction of the English landscape *and* as a representation of a new icon of human progress, but it also subtly merges the natural and the industrial: the rain that appears through the mist is also the provider of the water that becomes the steam which generates the speed that epitomizes a self-styled 'progressive age'. It is, in short, both a representation and a reminder of the *interconnectedness* of natural and human-built systems of power.[1]

The lecture, however, also contained the deliberately provocative sub-title 'Economic Historians *versus* Historians of Technology?' I posed the question primarily in order to highlight, through steamship case studies, the danger of uncritically adopting, as self-evident, patterns of inevitable 'improvement' and universal diffusion. Such interpretations often presuppose anachronistic views of technology in which, for example, 'more efficient' steam engines have a real existence at some time in the future. Above all, to display overconfidence in numerical data, and thereby to reduce other historical perspectives to merely decorative roles, is to fail to learn the lessons that a mature cultural history of technology teaches its practitioners.

My thanks go to other members of the audience who, formally and informally, offered comments and criticisms, especially Jim Secord, Hasok Chang, Liba Taub, Jim Bennett, Peter Bowler, Charlotte Sleigh, Edward Gillin and

[1] See Marsden and Smith (2005: 157–8).

the late John Forrester. With the support of Jim Secord, Simon Schaffer and Hasok Chang, I applied for and secured a Visiting Fellowship at Clare Hall (2013–14) which gave me a unique space (as well as time) in which to work up the book draft. I also owe much to Clare Hall's President, David Ibbetson, and all the members of the College community, for an unforgettable seven months of stimulating discussion and fellowship.

As someone brought up on the shores of Ireland's County Down, I have come to appreciate a very different kind of fellowship – what Joseph Conrad aptly termed 'the fellowship of the craft'. Introduced to small boats at an early age, I came to appreciate the shared values and experiences of seafarers of all ages. Conversations with them, I believe, have immensely deepened and enriched my readings of those nineteenth-century maritime texts that form the backbone of this study. In particular, I thank Captain William McCrea (retired coaster skipper and former master of the MV *Saint Aidan*), Captain Jim Fullerton (former senior master of the Belfast to Liverpool ferries *Saint Colum* and *Ulster Prince*), Adrian Adair (skipper of the fishing vessel *Aurora*), Valerie and Roderick Monson (of the ketch *Valfreyia*), Michael Griffiths (formerly of the ketch *Maplin Bird*), Chris Thompson (fishing vessel *Dawnlight*), Josh Porter (yacht *KMIT* and fishing vessel *Rosebank*), Emma Morgan (who served as *Rosebank* mate) and Matt Thompson (naval architect, musician and mariner).

Archivists and staff at Cambridge's University Library (especially Rare Books), Glasgow's Mitchell Library, University Archives, University Special Collections and Riverside Museum (especially Martin Bellamy), Liverpool's Record Office, Merseyside Maritime Museum Library and Sidney Jones Library, Halifax's Nova Scotia Archives, Edinburgh's New College Library, Newcastle's Tyne and Wear Archives, Greenwich's Caird Library, UCL Library Archives and the British Library all deserve my wholehearted thanks and appreciation for their patience. A delightful meeting with Bob Forrester in the old Caird alerted me to the late Stuart Nicol's through work on Royal Mail Lines history, based on the UCL and NMM archives.

Last but far from least, I thank Lucy Rhymer, Ian McIver, the editorial and production staff at Cambridge University Press (especially Abirami Ulaganathan), the Press's two anonymous readers and the series editors for Science in History (Simon Schaffer and Jim Secord) for their patience, dedicated hard work and enthusiastic commitment to the publication of this book. I am also much indebted to the excellent work of the Press's copy editor Diane Chandler.

In the opening months of 1960, the author's father introduced his ten-year-old son to the hitherto mysterious world of Belfast shipbuilders Harland & Wolff, whose industrial empire also extended to sites in Liverpool, Glasgow,

Southampton and London. In the Thompson graving dock (now promoted to tourists as the 'Titanic Dock') sat P&O's liner *Chusan* (1950), a stalwart of the Company's long-established mail and passenger services to India, Hong Kong and Japan. Fitting out at the nearby Thompson Wharf (no longer extant) lay Royal Mail Line's new passenger-cargo liner *Aragon*, second of a trio of large refrigerated vessels for its South Atlantic service to Rio de Janeiro and Buenos Aires. In the Musgrave Channel (now the site of the Building Dock), the largest crude-oil tanker yet constructed in the yards, *William Wheelwright*, bore the initials 'PSNC' on the rounded stem-head. On the slipways was P&O's *Canberra*, promoted as the largest British passenger liner since the Clyde-built *Queen Elizabeth* (1940) and designed to revolutionize travel to Australia. So too was the cargo-liner *Port Alfred*, six months away from launching for Cunard Steamship Company's subsidiary known as the Commonwealth & Dominion or Port Line to the Antipodes. It was one of the last times that ships of P&O, Cunard, Royal Mail and Pacific Steam could be seen together under construction in one place. Now, over half a century later, this book offers an account of the founding years of those same mail steamship lines that came to reify Britain's maritime empire in global space.

Abbreviations

ACKC	A. C. Kirk Collection
AGM	Annual General Meeting
AHD	Alfred Holt Diary
AHP	Alfred Holt Papers
AR	*Amazon* Report
ASNC	Aberdeen Steam Navigation Company
BAAS	British Association for the Advancement of Science
BoT	Board of Trade
CDSP	City of Dublin Steam Packet Company
CP	Cunard Papers
DM	Daily Minutes
EIC	East India Co.
EISNC	East India Steam Navigation Company
fp	facing page
FP	Forbes Papers
FRS	Fellow of the Royal Society of London
GHD	George Holt Diary
GM	General Meeting
GMT	Glasgow Museum of Transport (now Riverside Museum)
GPS	Glasgow Philosophical Society
GSNC	General Steam Navigation Company
GWR	Great Western Railway
GWSS	Great Western Steam Ship Company
GUA	Glasgow University Archives
HAPAG	Hamburg Amerikanische Packetfahrt Aktien Gesellschaft
HCO	House of Commons Order
HM	Her Majesty's
HMS	Her Majesty's Ship
hp	horsepower
HSBC	Hong Kong and Shanghai Banking Corporation
ICE	Institution of Civil Engineers
IES	Institution of Engineers in Scotland

IESS	Institution of Engineers and Shipbuilders in Scotland
ihp	indicated horsepower
ILN	*Illustrated London News*
IN	Indian Navy (EIC)
INA	[Royal] Institution of Naval Architects
KC	Kelvin (William Thomson) Collection
lb	pounds weight
LFP	Laird Family Papers
L&MR	Liverpool and Manchester Railway
LNRS	Liverpool Nautical Research Society
LRO	Liverpool Record Office
LUL	Sidney Jones Library, University of Liverpool
MD	Managing Director
MM	*Mechanics Magazine*
MMM	Merseyside Maritime Museum
NAR	*North American Review*
NC	Napier Collection GMT
NDL	Norddeutscher Lloyd
NM	*Nautical Magazine*
NMM	National Maritime Museum
NP	Napier Papers GUA
NSA	Nova Scotia Archives
NYC	Northern Yacht Club
OA	Ocean [Steamship Company] Archives
ODNB	Oxford Dictionary of National Biography
OSS	Ocean Steamship Company
PO	Post Office
P&O	Peninsular and Oriental Steam Navigation Company
PP	George Peacock Papers
psi	pounds per square inch (steam pressure)
PSNC	Pacific Steam Navigation Company
RM	*The Railway Magazine*
RMSP	Royal Mail Steam Packet Company
RN	Royal Navy
rpm	revolutions per minute
RSE	Royal Society of Edinburgh
RYS	Royal Yacht Squadron
SC	Select Committee of the House of Commons
SMG	*Shipping and Mercantile Gazette*
SP	Scotts of Greenock Papers
SPCK	Society for Promoting Christian Knowledge

SS	Steamship
TIES	*Transactions of the Institution of Engineers in Scotland*
TIESS	*Transactions of the Institution of Engineers and Shipbuilders in Scotland*
TWA	Tyne & Wear Archives, Newcastle
ULC	University Library Cambridge
ULG	University Library Glasgow
WSL	William Schaw Lindsay papers

Introduction: Coal, Steam and Ships

> These are not times in which Steam Navigation Companies, of all others, can find safety in concealment and mystery; too much agitation prevails among Shareholders, and it will be found that the interest of both Directors and Shareholders will best be consulted by candid publicity.
>
> The Railway Magazine *takes steamship lines to task for lack of openness (1843)*[1]

Writing a credible history of ocean steam navigation is a challenging enterprise. Central to the methodology of this book is the inspirational work of the late Thomas P. Hughes on the history of large technological systems. Hughes' own historical focus lay with electrical power generation and distribution.[2] But his approach was never intended to be restricted to a single field of study. Nor did it seek to impose a rigid deterministic framework on its subject matter. Indeed, a principal strength is its malleability, enabling it to be reshaped to capture the histories of different technologies in different periods.

Four features of the Hughesian methodology are especially relevant to unlocking the secrets of ocean steam navigation. First, Hughes' approach gives priority to human agents. Technological systems are human inventions, shaped by choices, values, contingencies and, not least, fallibility. Second, the methodology focuses on the interconnectedness of the diverse components (material and social) that constitute a complex system. Third, his analysis concerns large systems that occupy geographical space. Spatially located, they are invented, designed, shaped by, constructed and operated within contexts that are often at once material, moral, spiritual, political, social and/or economic. And fourth, he recognizes that the projectors and managers of large engineering systems seek to maximize order and minimize disorder through a deliberate strategy of attempting to control recalcitrant elements that threaten to disrupt or destroy the system. Maximum control, he infers, often means that the elements in question have been incorporated into the system. As we shall now see through illustrative examples drawn from the principal chapters in this book, all four

[1] *The Railway Magazine* (*RM*) (1843: 206) (Editor).
[2] Hughes (1983).

interrelated features provide an invaluable toolkit for understanding the history of ocean steamships in the early Victorian period.[3]

'Never Known or Thought of in the Trial of the Steam-Engine': Human Agency at Work

'The men who worked out problems in diagrams and algebraics at home were not seen, and their names were never known or thought of in the trial of the steam-engine', the Glasgow clergyman Norman Macleod told 250 distinguished guests seated at lunch aboard the newest and largest Cunard mail steamer *Persia* in January 1856 (Chapter 15). 'So, perhaps, the names of ministers might never be thought of; but if they made the men who made the steam-engine, – if they made them more sober, more honest, more faithful, and more trustworthy, – then, perhaps, the clergy had more to do with that occasion than the world thought'. Amid the mutual congratulations, Rev Macleod's speech alone drew attention to the unnamed and unacknowledged engineers rather than simply to the leading projectors responsible for the ship.[4] In so doing, he highlighted the often forgotten but integral part played by the moral and spiritual capital of human agents in shaping a material system.

Rendering an individual human agent central to any narrative can readily slide into a story of a heroic entrepreneur, individual pioneer or inventor of genius. For example, Hyde's *Cunard* (1975) trades in 'forceful characters', chief among them Samuel Cunard who 'had the foresight of genius coupled with the gift of choosing men of ability as his associates'. Fox's *Ocean Railway* (2003) places Cunard centre stage: '[i]t was all there from the start', Fox declares. 'And at the centre of it all, driving and organizing, the elusive figure of Samuel Cunard and the great transatlantic line he founded'. In other accounts, it is the heroic, pioneering representation of Isambard Kingdom Brunel who overtakes Cunard in the narratives of pioneering geniuses. Conversely, Harcourt's *Flagships of Imperialism* (2006) retrospectively presents the Peninsula and Oriental Steam Navigation Company's (P&O) celebrated co-founder Arthur Anderson as 'crude and showy'.[5]

In *Heroes of Invention* (2007), Christine MacLeod offers a cultural history of the diverse ways in which engineers and inventors across industrializing Britain acquired heroic status. Once set in specific historical contexts, the making of a heroic genius can be seen to serve a variety of local and national agendas: civic pride, national identity and prestige, and the interests of a later

[3] Hughes (1987: 51–82); Hughes (1994: 100–13); Hughes (2004).

[4] *Liverpool Courier*, 16 January 1856; Smith (2014: 76); Shapin (1989, 1994: 355–407) (invisible technicians).

[5] Hyde (1975: 323); Fox (2003: xviii) (Cunard); Falkus (1990: 83–4) (Brunel's 'visionary engineering genius'); Harcourt (2006:160) (Anderson).

generation of engineers' intent on securing their professional standing, for example. James Watt or Brunel might have enjoyed comparatively little iconic status in their own lifetimes, but through the efforts of later generations of engineers they were transformed into national, as well as local, heroes through statues, portraits, biographies and all manner of civic commemorative events well reported in the press of the time.[6]

Heroes of Invention reminds us that heroic status is not intrinsic to a human agent but has to be constructed within social and other contexts. In the long run, ship-owners and shipbuilders as 'pioneering' or 'heroic' individuals have received a generous share of hagiographical literature, even if portraits were largely for in-house display and public statues were less common than in the railway world. At the beginning of ocean steam navigation, however, projectors had no guarantees of immortality, still less gifts of prophecy, but there was every prospect of ridicule or obscurity.

From the mid-1830s, for example, New England merchant Junius Smith aimed to raise upwards of 1 million pounds capital from London and New York investors to launch a fleet of eight large steamers. Craving the accolade of being the first projector of a transatlantic steam shipping line, he imagined himself receiving a knighthood from Queen Victoria. To that end he chose the name *Victoria* (later renamed *British Queen* while on the stocks) for his first vessel. But as engineering and financial problems mounted, the grand ambition foundered amid mistrust and tragedy (Chapters 4 and 6).[7]

The Royal Mail Steam Packet Company (RMSP) projector James McQueen also had mighty ambitions. With a track record in journalism, geography and Caribbean trade, this lowland Scot made it his goal to persuade metropolitan bankers and merchants to launch a large-scale mail steamship system. Awarded a state contract initially worth £240,000 per annum, the line lost ships and money. Directors identified McQueen as the scapegoat. Because of his supposed miscalculations of construction costs, coal consumption and route schedules and for his open authorship of press articles exposing the shortcomings of the line, he was erased from the Company's collective memory (Chapters 7 and 8). A century on, he had been redeemed as a heroic founder for the purposes of the Company's centenary history, especially in the wake of the financial scandal centred on its recent chairman Lord Kylsant.[8]

While wary of elevating its projectors as individual heroes, steamship companies took self-fashioning and public image very seriously. Following the Honourable East India Company (EIC) practice, both P&O and RMSP often used the term 'Court' to refer to their Boards of Directors. The Court

[6] MacLeod (2007: esp. 1–26).
[7] Pond (1927: 186–8) (knighthood), 266–79 (death).
[8] Green and Moss (1982: 141–3).

expected due deference from its officials, servants and proprietors. In the case of RMSP, Scottish landed gentlemen with close ties to northern British aristocrats initially dominated the Court. Attacks from the radical *RM* generated hostile reactions from company directors in both the RMSP and P&O. Company reputations were vulnerable to press exposés. Shareholders became restless, the more so if the competence of captains or officials was called into question. And once confidence began to retreat, investors took fright, share values declined, insurance on ships became more expensive, passenger numbers fell and changes of Government could threaten an inquiry into mail contracts and raise the possibility of their non-renewal (Chapters 8–10, 13).

A Company Secretary or managing director (MD) often wielded autocratic power both inside and outside a large steamship company. Particularly in cases where a Court of Directors remained aloof from day-to-day affairs, the MD acted rather like a courtier who knew his place but in practice exercised much authority. He not only managed the practical running of the business, but also protected confidential information regarding matters both of finance and discipline. Wayward masters and officers too had reason to fear him. For almost fifteen years, for example, Captain Edward Chappell, RN, managed RMSP's troubled system of ocean mail steamers. 'No complaint can reach the aristocratic board of the Royal Mail Company, but through Captain Chappell', a detractor observed in the *RM*. 'The noble lords, the directors, are as difficult of approach as his Celestial Majesty' (Chapters 8–10). Yet in the annals of the Company, Chappell's name is barely acknowledged.[9]

The Pacific Steam Navigation Company (PSNC) MD William Just used all his diplomatic skills to carry through a fraught transition from simple to compound engines in co-operation with Glasgow engine makers Randolph, Elder and Company. Serving in the post as guardian of confidence in the line for thirty years, Just saw PSNC expand from a mere two steamers to the largest steam navigation company in the world (Chapters 17 and 19).[10]

The historians of P&O have generally ignored James Allan in favour of the Company's public figures such as Anderson or Thomas Sutherland. Allan began his shipping career as a clerk in the Irish office of Richard Bourne's Dublin & London Steam Packet Company. When Bourne sold out to a rival, Allan transferred to the London office of the then Peninsular Steam Navigation Company (trading to the Iberian Peninsula) while still in his mid-twenties (Chapter 12).[11] First as secretary, then as Assistant Manager from 1847 and finally as one of the key MDs from the early 1850s, this most trusted of servants was with the P&O Company almost to the day of his death in September 1874

[9] Bushell (1939: 110).
[10] Wardle (1940: 73, 93, 132, 149).
[11] Harcourt (2006: 38).

(Chapters 14 and 18). Allan's only character flaw, in the verdict of merchant shipping historian William Schaw Lindsay, 'consisted in believing all other men to be as upright as himself'.[12]

In our period, character (rather than qualification) was often regarded as the 'stock in trade' of an individual who occupied a position of skill but was not necessarily a person of gentrified or aristocratic pedigree. This attribution was especially the case for engineers in general and for marine engineers in particular.[13] The RMSP's engineering superintendent, George Mills, carried heavy responsibilities that covered both the construction of new steamers and the maintenance and repair of vessels in the mail service. At a fraught meeting of directors and proprietors in 1854, shareholders expressed their concerns over weak financial returns, the poor performance of many of the newer steamers and the public perceptions of a company prone to physical and financial shipwreck. Several voices found the perfect scapegoat in Mills. But in appointing Mills to such a pivotal position, the directors' own wisdom and judgement was effectively under attack. They therefore rallied to the defence of Mills' character (Chapter 10).

David Elder, on the other hand, sustained his stock in trade as a trustworthy engineer throughout his life. '[H]e sought no public fame, and it is doubtful whether one in a hundred of those who, in all parts of the world, have seen and admired his works have ever heard of David Elder', *Engineering*'s anonymous obituary stated in 1866. Elder's works were indeed many. Over a period of four decades, this unassuming works manager for Robert Napier had overseen the design and construction, in every practical detail, of the side-lever marine engines that powered many of the Clyde-built river boats, coastal and cross-channel steamers and Cunard mail steamers for which the firm was soon famous. Omitted from the press reports of the *Persia*'s trial trip (Figure 0.1), Elder was the prime example of Revd Macleod's unseen and unnamed engineers who, self-taught and with no privileged family pedigree, played crucial roles in realizing the promises of the projectors of ocean steam navigation (Chapter 2).[14]

'How Beautiful a Machine is this Magnificent Ship'!: The Steamship as System

When the Revd Macleod crossed the Atlantic in 1845 aboard the first-generation Cunard mail steamer *Acadia*, he stood in awe of the steam engines that David

[12] Lindsay (1874–6: vol. 4, 379–81n).

[13] Marsden and Smith (2005: 8) (Babbage on character and confidence). I thank Ben Marsden for this insight.

[14] Smith (2014: 80–4) (Elder's engineering practices).

Figure 0.1. Cunard's crack iron paddle steamer *Persia* (1856) built by Robert Napier

Elder had designed. 'What a wonderful sight it is on a dark and stormy night to gaze down and see those great furnaces roaring and raging', he wrote. '... And then to see that majestic engine with its great shafts and polished rods moving so regularly night and day, and driving on this huge mass with irresistible force against the waves and storms of the Atlantic'. Macleod had here witnessed for himself the very interconnectedness of furnaces, boilers, side-lever engines, crank shaft, paddlewheels and hull, all working together to produce regularity in crossing the hostile ocean. But neither did he ignore the integral role of the humble human actors in the guise of the firemen 'laughing and joking' as they fed coal into the hungry furnaces. And when he reflected on the human engine-builder's intellect and skill, he saw a pointer to God: '[i]f the work glorifies the intellect of the human workman, what a work is man himself'![15]

In a less definite, yet compatible vein, the author of RMSP's official *Guide to the Madeiras* (1844) recognized the god-like qualities of the human designer of the steamship: '[h]ow beautiful a machine is this magnificent ship, and how like a god is man who can create such a machine, so complete, so perfectly applicable to his purposes'! A good steamship was a well-designed machine, an engineering system whose constituents parts – hull, masts and sails, engines, boilers, furnaces, steering, accommodation, master, officers and deck and engine-room crew – formed part of a single, orderly entity. Take away, alter or damage any of these constituents and the vessel risked losing that wholeness.[16]

[15] Macleod (1876: vol. 1, 238); Marsden and Smith (2005: 249). Contrast Burgess (2016: 25, 39) (steamships entailing a disconnection from religious belief).

[16] Osborne (1844: 12–13). I thank Stephen Courtney for this reference.

Steamships, however, were not the only machines of the ocean in which every component functioned as part of an integrated whole. Through a gale, wrote author and former master mariner Joseph Conrad from the vantage point of 1906, 'the silent machinery of a sailing ship would catch not only the power, but the wild and exulting voice of the world's soul ...'. In contrast to steam, the machinery of a vessel under sail does 'its work in perfect silence and with a motionless grace, that seems to hide a capricious and not always governable power, taking nothing away from the material stores of the earth'. Conrad also spoke for many fellow-seafarers in his late novel *The Rescue* (1920), begun in the 1890s and set in the eastern seas of the 1860s, when he describes Captain Lingard's view of his brig: '[t]o him she was as full of life as the great world ... She – the craft – had all the qualities of a living thing: speed, obedience, trustworthiness, endurance, beauty, the capacity to do and to suffer – all but life'.[17]

Changes to hull-, mast- and engine-systems could radically alter a ship's performance. The character and consequences of such changes were often contested. Screw propulsion, for example, was not self-evidently superior to paddlewheels, the celebrated *Rattler* demonstrations notwithstanding (Chapter 14).[18] Contemporary reports of Brunel's *Great Britain* in her original role on the Atlantic often centred not on the benefits, but on the fragility of the screw. Passengers tended to praise her qualities as a sailing vessel, qualities made all the more noteworthy by the frequency with which the propeller lost blades and ceased to function for the remaining duration of the voyage (Chapter 6).

With passenger requirements integral to the design and operation of an ocean mail steamship, voyagers often complained of the discomforts of screw vessels. Having survived a January crossing to Boston aboard Cunard's first paddle-steamer *Britannia* in 1842 (Chapter 5), Charles Dickens returned to the United States in 1867 on the same line's much larger iron screw steamer *Cuba*. 'These screws are tremendous ships for carrying on, and for rolling, and their vibration is rather distressing', he told his son. Writing to his dentist a couple of months later, he recounted how divine service had taken place 'with a heavy sea on'. Two big stewards attempted to bring the officiating minister to the reading desk but were thwarted by the ship's rolling and pitching. As a result, the 'extremely modest' reverend gentleman had to embrace 'with both arms as if it were his wife' the mast in the middle of the saloon. And all the while, Dickens observed with a tone of delighted irony concerning the line's policy of insisting on Church of England services (Chapter 15), 'the congregation were breaking up into sects and sliding away'.[19]

[17] Conrad (1906: 38); Conrad (1950: 20–1).
[18] Lambert (1999: 40–52) (*Rattler* trials).
[19] Storey (1999–2002: vol. 11, 494–5; vol. 12, 24–5).

The large iron screw steamers of P&O were similarly prone to rolling in ocean waves and also exhibited steering problems in confined waters (Chapter 14). Reverend Macleod took passage to India on the *Rangoon* in 1867. One Sunday on the Red Sea he stood preaching for nearly an hour on the rolling deck, and became so exhausted with the heat and exertion that 'Old Indians ministered to me, and poured iced water over my head, and gave me some to drink with a little brandy in it, which quite restored me'.[20] The sources do not record how the assembled company received the brandy-assisted sermon.

Satirically minded passengers could view the ocean steamer as a self-contained system isolated from a wider humanity. A return passage from England to New York on the *Great Western* inspired Judge Haliburton to write some twenty-eight satirical letters, each under a different persona, including 'From the Professor of Steam and Astronomy (otherwise called the clerk) to the [GWSS] Directors'. At the end of this *Letter-Bag of the Great Western*, Haliburton vividly captured the notion of a passenger steamer as a self-contained microcosm that reflected the life and lives of the larger world. 'One hundred and ten passengers, taken indiscriminately from the mass of their fellow beings, are a fair "average sample" of their species', he affirmed. '[T]he vessel that carries them is a little world, and life in a steamer is a good sample of life in "the great world"'. Thus there are 'the same complaints, the same restlessness, and the same air of perverse dissatisfaction in their letters, as we meet with on land'. It was, he claimed, 'the power of steam' that was at work both in the Atlantic crossing and in life: 'although the scene is varied by calms, fair breezes, and storms, still the great machine is in continual progress'.[21]

'The Railway Train *Minus* the Longitudinal Pair of Metal Rails': Systems of Ocean Steam Navigation

'We have been able to traverse wild regions that were previously almost impenetrable, and to bear the fruits of civilization to parts of the globe where they would otherwise have remained unknown for many years to come', French engineer–economist Sadi Carnot observed of steam-powered vessels in 1824. 'There is a sense in which steam navigation brings the most widely separated nations closer together'. On the arrival of the first Cunard mail steamer in Boston sixteen years later, the Unitarian Revd Gannett shared Carnot's optimistic Enlightenment vision. 'This great and wide sea', he told his congregation, '... at first sight appears to be a barrier ... to the intercourse of the nations who live on opposite shores'. But the arrival of the *Britannia* showed that God

[20] Macleod (1876: vol. 2, 261–2).
[21] [Haliburton] (1840: 184–5). Burgess (2016: 147–9, 249) discusses ships as self-contained worlds in relation to social class in the later nineteenth century.

intended the ocean 'to be the highway of the nations, on which they might pass and repass and never crowd its ample space'.[22]

In the late 1830s Scottish journalist and geographer James McQueen constructed a 'General Plan' to initiate, throughout the western and eastern oceans of the world, lines of mail steamers which he termed 'mail coaches of the ocean' (Chapter 7). In contrast to McQueen's imagery, Cunard told friends of his belief that 'steamers, over a route of thousands of miles in length, might start and arrive at their destination with a punctuality not differing greatly from that of railway trains'. In order to achieve this result, 'the ships should be thoroughly built and thoroughly well manned, and their course laid down with the greatest accuracy'. The analogy required the steamship 'to be the railway train *minus* the longitudinal pair of metal rails'. Half in jest, however, he suggested that such rails were required only on 'ugly, uneven land' but not on the 'beautiful level sea'.[23]

While the authenticity of Cunard's verbal remarks cannot be verified, they are consistent with contemporary assumptions that steamship systems would deliver that regularity and punctuality that sailing packets did not. Mail lines were perennially concerned to protect their reputation and finances by maintaining punctuality, often at high cost. '[T]he matter of punctuality', the chairman told RMSP shareholders in 1868 with respect to persuading the Government to renew its large mail contract, '... is only attained at a very enormous outlay [on fuel], and to secure it forms the most considerable charge upon our Government subsidy'. Higher speeds in such cases were dictated by state demands for faster communication and greater punctuality, rather than simply by public demands for more rapid passages or by company desires to show off their latest steamers.[24]

Quite a few of the men involved with steamship affairs had been weaned on railway engineering and finance. Most often cited is Brunel's representation of the GWSS project as a maritime extension to New York of the Great Western Railway (GWR) between London and Bristol.[25] Junius Smith obtained the services of Isaac Solly as first chairman of his Atlantic steamship company. Solly's chairmanship of the Great Birmingham and London Railway Company gave him valuable credibility, not least because railway companies were especially adept at raising large amounts of capital for their joint-stock projects (Chapter 3). Dionysius Lardner's knowledge of railway engineering almost certainly prompted him to introduce the term 'locomotive duty' (the number of miles one ton of coal per horsepower took a locomotive or steamship) into his

[22] Carnot (1986: 62); Gannett (1840: 7–8).
[23] Hodder (1890: 193–4).
[24] RMSP GM (28 October 1868: 12). Compare Burgess (2016: 23–5) (speed).
[25] Griffiths et al. (1999: 15); Buchanan (2002: 57); Fox (2003: 70). Fox's *The Ocean Railway* has little to say on the analogy. For insightful perspectives on railway experiences see Schivelbusch (1986: esp. 1–15).

evaluation of the prospects for viable Atlantic steam navigation (Chapter 3). Henry Booth, promoter of the Liverpool and Manchester Railway (L&MR), contributed a rigorous paper on steamship design to the Liverpool Polytechnic Society in the mid-1840s.[26] And Alfred Holt acquired his engineering skills on the same celebrated railway (Chapter 17).

The validity of an analogy between steamship and railway systems became a hotly contested topic at RMSP's general meetings (GMs) in the mid-1850s. A proprietor, among many such shareholders with interests in railways, put the question '[w]hy should this Company be dealt with differently from a Railway Company'? The acting chairman, Captain Shepherd of the EIC, vigorously asserted that 'there is the greatest possible contrast between a Railway and a fleet of ships under [mail] contract for a service of five or six years only'. A railway company with its line already constructed has 'under ordinary circumstances a monopoly of that line in perpetuity'. It could, for instance, calculate the traffic, apportion its expenditure to its receipts and estimate what its district was likely to require were the population to go on increasing. Then if 'things go on well they know that they will be justified in calculating upon a continuous and a profitable trade'. It was, he declared, not so with RMSP. The shipping company more nearly resembled 'a party who had the lease of a line for six years'. Its 'magnificent property' of mail steamers was therefore at best akin to 'the whole of the rolling stock sufficient to perform the service of the line for that period but which may afterwards be thrown upon your [the proprietors'] hands and become comparatively valueless'. Mail steamship company tracks remained little more than lines on Admiralty charts.[27]

The large-scale geographical system of the RMSP had been premised on a fleet of identical mail steamers in order to ensure regularity of operation. In the ideal system, the dozen or more mail steamers would be found at definite intervals from one another along the tracks they followed. Changing any one of these steamers for one of a different size and power changed the system. The two main building programmes, completed in 1842 and in 1852 to coincide with each new mail contract, consisted respectively of classes of fourteen and five vessels. Replacing these fleets or even updating in terms of size and speed was thus an extremely expensive business (around £500,000 in the second programme). And as the Chairman told the proprietors in 1856, 'the ship built in 1852 upon the most approved plan and fitted with all the improvements in her Engines and Machinery known to Science, may be altogether superseded by the vessel built in 1853 or 1854'.[28]

As part of increasingly complex systems, railway companies possessed locomotives, rolling stock, stations, maintenance depots and even hotels. Steamship

[26] Booth (1844–6: 24–31); Marsden and Smith (2005: 100–1).
[27] RMSP GM (10 April 1856:13–16).
[28] RMSP GM (10 April 1856: 16–17).

lines also not only owned 'rolling stock', but often invested in considerable property on shore, including head offices in London or Liverpool, offices and agencies in principal ports, hotels, repair facilities and feeder vessels bringing passengers and freight from more remote places. Lindsay recognized the central, system-building role that P&O's Allan played in constructing physical networks. 'To establish agencies at the leading ports of India and China, open depots for coals, erect docks and factories for the repairs of their ships, to bring the whole into systematic and harmonious working order, and above all, to keep [maintain] agencies remote from each other and far from home, under proper control', he explained, 'required a master mind of no common order, the more so that the system he organized was then entirely new'.[29]

Government patronage was also crucial to the stability of a mail steamship system (Chapters 5 and 6). As someone very close to Cunard's success in obtaining and sustaining the contract for the North American mails, Haliburton (under the guise of anonymous 'author') wrote fulsomely in 1840 that 'the [British] Government is always distinguished by the same earnest desire to patronize, as it is to protect the colonies, who have experienced nothing at the hands of the English, but unexampled kindness, untiring forbearance, and unbounded liberality'. The recent grant of some £55,000 per year 'for the purpose of affording us [in Nova Scotia] the advantage of a communication by steam with the mother country, ... was not made grudgingly, or boastingly, or as an experiment, but as early as it was safe to be done ...'. Speaking perhaps with forked tongue, he further observed that this act 'leaves us in doubt whether to admire most the munificence of the gift, or the power and wealth of the donors'.[30]

Integral to steamship systems in our period were privileged sites of skill, geographically located: shipbuilding yards, marine engine and boilermakers' workshops, maintenance and repair depots at home and overseas, coal mines, coal shipment sites and coaling stations, the Admiralty with its hydrographic skills, and a range of further specialist skills from engineers and men of science.[31] The Cunard Company had intimate ties with Clyde shipbuilding and marine engineering through the Napier network: hulls from John Wood and other lower Clyde wooden shipbuilders and later through Napier's iron shipbuilding establishment at Govan, but always using David Elder's engines right up to the last of the North Atlantic paddle-steamer contracts around 1860. Moreover, major repairs, including boiler replacements, were done at Napier's Glasgow works (Chapter 2).[32]

[29] Lindsay (1874–6: vol. 4, 379–81n).
[30] [Haliburton] (1840: 185).
[31] See esp. Schaffer (2007: 279–305). Schaffer uses the apt phrase 'geographies of skill'.
[32] [Anon.] (1864: 505–7); Smith et al. (2003a: 455–6).

The RMSP had special links with William Pitcher on the Thames, both for wooden hull construction and for maintenance. For their compound-engined vessels, they privileged John Elder in Glasgow (Chapter 18). Similarly, P&O developed strong links with Tod & MacGregor in Glasgow, but later renewed ties with Cairds at the end of our period (Chapters 14, 17 and 18). After a difficult relationship with Napier's, the PSNC found in Elder's firm the key to its prosperity in the late 1850s (Chapter 16).

Cunard, RMSP and P&O established ties with South Wales' mines to secure the best quality steam coal. The PSNC attempted to exploit local Chilean coal, but found little elsewhere on the Pacific Coast. Through brokers such as Lindsay, RMSP, P&O and PSNC adopted the method of contracting with small independent sailing vessels to ship British coal to stations overseas (Chapter 15). Sailing ship-owners often made a fortune from this new trade in British coal. On the strength of their earnings, Lindsay built up a smart fleet of auxiliary steamers (sailing vessels fitted with very low-powered steam engines for deployment in the approaches to port or in calms) in the early 1850s, and Liverpool's Lamport & Holt (hitherto committed to small sailing vessels) launched their prestigious cargo services with iron screw steamers from the early 1860s (Chapter 17).

'She has Gone, and Left no Track upon the Pathless Sea': The Control Imperative

Mail steamship-owners faced the multiple difficulties of controlling recalcitrant elements (physical, political, financial and social) threatening the stability of ocean steam navigation systems. An early case was that of Junius Smith's transatlantic project. Reluctant to involve contractors within his company, he increasingly lost control of the construction programme. Delays multiplied and trust between owner and engine builder fell dramatically as Smith sought a scapegoat. Opposition from the GWSS further threatened to destroy Smith's ambitions for a monopoly on Atlantic steam. Then both ship-owners failed to secure Government patronage, and with it the prospect of an all-important mail contract. Damaging rumours had been circulating in the press for some time that Smith's two vessels in service by 1840 were underpowered, structurally weak and potentially unstable. He fought hard to quell the reports but when in 1841 the *President* went missing with everyone aboard, the battle was lost (Chapter 6).

In contrast, engine-builder Robert Napier, who had long warned of the risks to confidence of accidents with steam at sea, acted in close collaboration with Cunard's Company. The Line's public practices also offered passengers reassurance. Carefully controlled Sunday services, an absence of Sunday departures and a policy of not vaunting its safety record all combined to counter suggestions of tempting Providence. Masters and officers would be known to be of the best character. Discipline among the deck and engine room crew would be strict. Even when the *Columbia* ran ashore in dense fog at the southern tip of Nova Scotia and became a total loss, the Company gained rather than lost prestige from press reports of the calm and controlled rescue of all the passengers. And there were ample authoritative voyagers, among them Dickens, William Makepeace Thackeray and Mark Twain, with the literary skills to convey to the reading public a profound sense of a systematic competence, lack of ostentation and professional control amid the most terrifying seas. It was, remarked Twain, 'rather safer to be in their vessels than on shore'[33] (Chapters 5, 6 and 15).

As joint-stock chartered companies based in London, RMSP and P&O were especially vulnerable to attack from the radical financial press. The safety record of RMSP was unenviable but the larger more stable P&O was far from immune from criticism and resorted to legal methods to suppress perceived aggression by the *RM*. The *Magazine*'s manager and editor was John Herapath, a self-taught mathematician whose unorthodox papers on the molecular behaviour of gases had been rejected by the Royal Society of London. Often at loggerheads with such metropolitan establishments, Herapath was to prove the greatest bane of joint-stock railway and steam shipping enterprises during the 1840s.[34] Favourite themes included uninhibited criticisms of companies that promised more than they could deliver to shareholders and the wider public, that attempted to exploit monopoly powers for the benefit of an aristocratic or gentlemanly coterie and that compared unfavourably with private partnerships constituted of a few individual, often family, members.[35]

The *RM*'s editor, for instance, wrote in 1843 of RMSP's shareholders being 'led on with the scheming of men [the directors] who have failed in everything and succeeded in nothing' (Chapter 8). That same year an anonymous correspondent drew the editor's attention to 'the jobbing that has characterised this [P&O] Company throughout' (Chapter 13). As large joint-stock enterprises, each with the privilege of a Royal Charter that gave protection to proprietors against unlimited liability and each with generous contracts to convey Her Majesty's Royal Mails across half the globe, both RMSP and P&O had much to lose through adverse press coverage.[36]

Even Cunard's transatlantic line of steamers, often held up as exemplary on account both of its noteworthy safety record and its status as a partnership rather than a joint-stock company failed to escape unscathed from the *Magazine*'s

[33] Smith (2011: 297–315) (literary passengers); Salamo and Smith (1997: vol. 5, 578) (Twain); Babcock (1931: 142).
[34] 'Herapath, John (1790–1868)', *ODNB*.
[35] See also Taylor (2006: 125–57) (wider contexts).
[36] *RM* (1843: 327) (Editor), 121 ('A. B.').

sarcasm. In reference to a recent £20,000 per annum enhancement of its North Atlantic mail contract, for example, another anonymous correspondent spoke of Mr Cunard in 1841 again exerting 'his power over the Government in favour of his Scotch friends [that is, his partners George Burns and David MacIver] ... the very day before the Whigs left Downing Street'.[37] Here was a less-than-subtle hint that a company, promoted as highly respectable, worthy and safe, owed not a little of its continuing existence to Government patronage from the party most associated, in the public estimation at least, with a spirit of Victorian free enterprise and competitiveness (Chapter 6).

As we shall see in the following chapters, control of recalcitrant elements, both internal and external, of steamship systems presented constant challenges to company managers as they attempted to build confidence into their lines of steamers. Control of shipbuilding and marine engine making, control of shipboard practices and discipline, control of the media (especially following a shipwreck or other accident), control of relations with the Admiralty and so with Government patronage, control of the body of anxious proprietors in the case of joint-stock companies and control of navigation especially when threatened with derangement through the presence of iron in engines and, in due course, hulls – all of these strands, and more, appear throughout the following chapters.

Adapting a Hughsian systems approach has historiographical consequences. First, functioning as a toolkit rather than as a theoretical or scientistic model, it avoids rigid rational categories derived from other periods and instead offers contextual flexibility. Second, a systems approach emphasizes spatial, and not simply temporal, analyses that resonate with other turns in the history of technology. In an inspirational article originally presented to a symposium on 'Big Pictures' in the history of science, John Pickstone offered strong pointers away from traditional linear, developmental narratives. He suggested instead that a reconstruction of 'Big Picture' conceptualizations could and should be achieved through spatial breadth rather than temporal length.[38] The consequences for this book are significant. By focussing on a comparatively short period of up to three or four decades in the early to mid-Victorian era, by not limiting the selection of material to a single company extended over its entire existence, by highlighting the interconnections between or among actors and their concerns and by dividing the chapters into four broad spatial groups, something of that desired 'Big Picture' can be attained.

[37] *RM* (1843: 58) ('Fair Play').
[38] Pickstone (1993: 433–58).

Part I

North Atlantic Steam

1 'Trust in the Promises of God'

The Moral and Spiritual Credibility of Steam Navigation

> Dependence on the providence, and trust in the promises of God, are duties which must be acknowledged by all those who believe in a Providence ... How wonderful is the power and knowledge which can regulate the universe and direct the secret thoughts of the human race, which can so connect the changes in the different parts of the material world, the very winds which blow, with the purposes of the heart of man, as in every instance to bring to pass that which is wise and proper.
>
> *Dr John Burns, Glasgow University Professor of Surgery and brother of steamship-owner George Burns, reflects on the principles of Christian philosophy (1828)*[1]

Summary

From the 1820s, projectors promised systems of coastal and cross-channel steam navigation that would convey passengers, mail and high-value freight more rapidly, regularly, reliably and safely than their sailing counterparts. Claiming for their countryman Henry Bell the accolade of putting to work the first sea-going steamer, *Comet*, in 1812, Scottish ship-owners initially promoted the linking of their industrializing port towns not only with Liverpool and London in the south, but also with Belfast and Dublin in the west. Among them were the Burns and the MacIver families, who would soon play a central role in a new transatlantic project. With steam navigation open to accusations that it was tempting Providence by defying the winds of heaven, projectors recognized the importance of aligning the new material systems with moral and spiritual cultures. Confidence in steamships could rise in harmony, or fall in discord, with a contemporary, earnestly professed faith in an all-powerful, redeeming God.

1.1 'Amid the Gloom of This World's Depravity': A Very Presbyterian Economy

The eighteenth-century Scottish Enlightenment focused on Glasgow and Edinburgh tended to soften the harsher doctrines of Scotland's Reformation founded on Calvinist readings of human depravity and eternal punishment. At the heart of the ancient town of Glasgow there stood in close proximity

[1] Burns (1828: 279, 282–3); Smith and Scott (2007: 471–96, esp. 475–80).

the medieval cathedral of St Mungo and the commercial High Street which descended past the university and down towards the River Clyde. Threatened in the sixteenth century with dismemberment by zealous reformers, the cathedral survived the fate of other Catholic foundations, thanks to the resistance of Glasgow's artisans whose ancestors had wrought the edifice. Under Presbyterian governance that disavowed hierarchy, however, the building now functioned as three separate churches, formed from a twofold division of the nave and from the crypt. It was to the lower site, known as 'The Barony', that the Revd John Burns was called as minister in 1774. A new Barony Church opened in 1801 and the crypt resorted to its traditional role as a burying-place within the Cathedral.[2]

The religion of John and Elizabeth Burns was a Protestantism that ran like a powerful tide through every facet of their lives, material as well as spiritual. Several years before his ordination as a minister of the kirk, Burns had prayed to an omnipotent God 'who settest bounds unto the spacious sea, and who art the Absolute Governor of the whole universe, of all things and creatures in heaven and on earth'. In contrast to 'the righteous ways of God', human beings existed in a 'state of sin and misery'. Boasting of its excesses, its vanities and its pursuit of false gods, this 'city of man' was above all characterized by instability and impermanence. Yet such was the providential nature of God that He had appointed means, through Christ, 'for bringing men ... to a state of holiness and happiness'. Even then, however, the 'instability' of the human heart meant that the individual Christian required at all times divine succour attained through prayer.[3]

The family's evangelical Christianity, based on biblical authority and grounded on a belief in the need for individual salvation, cut across denominational differences. This perspective enabled the Burnses to freely attend Episcopalian (Anglican) services – anathema to earlier generations of Scottish Presbyterians – for the purposes of listening to English evangelicals.[4] The eldest son, also John, professed the same beliefs but followed a medical vocation that took him eventually to the new chair of surgery at the nearby university, itself a medieval foundation with powerful Christian underpinnings.[5]

The youngest son, George, shared his father's commitment to evangelical Presbyterianism. Serving in his youth as treasurer of the 'Penny-a-Week North-West District [Bible] Society', he delighted in listening to speakers preaching against Unitarians who might question the fundamentals of traditional belief, including the special divinity of Christ. But whereas many sons of Glasgow's merchants forsook the trade of their fathers for the ministry of the kirk,

[2] Hodder (1890: 24–33).
[3] Hodder (1890: 23–4).
[4] Hodder (1890: 24–33, 36).
[5] Hodder (1890: 53, 59).

George's career took him in the opposite direction. Already well versed in the political economy of the cotton and produce trades, he married Jane Cleland in 1822, a daughter of Glasgow's leading statistician, Dr James Cleland.[6]

George's marriage into the Cleland household brought him close to the centre of Glasgow's commercial networks. Cleland played a leading civic role in the growing industrial town, serving as superintendent of public works between 1814 and 1834, and overseeing construction of agricultural markets, churches and Clyde bridges, as well as the introduction of standardized weights and measures.[7] He addressed the British Association for the Advancement of Science's (BAAS) statistical section in 1835 on a politically controversial system of relief for the poor introduced by his clerical ally, the Revd Thomas Chalmers. Three years later Cleland lobbied for the BAAS to hold its annual meeting in Glasgow for the first time.[8]

Various editions of his pamphlet *Statistical Facts Descriptive of the Former and Present State of Glasgow* secured Cleland's reputation as a social statistician concerned to promote civic cultures of improvement and progress through the science of political economy. In addition to data on population, manufacturing and Clyde shipping (especially steam), the work contrasted past and present states of the city with respect to health, wealth, morality and property. Against 'powerful influence', Cleland claimed to have been a key supporter of the *Comet*, and thus of steam navigation on the Clyde.[9] His publications on Glasgow's growing trade perfectly complemented the mercantile vocation of his son-in-law.[10]

The Revd Chalmers arrived in 1815 as minister of Glasgow's Tron Church near the foot of the High Street. One of his first acts was to deliver a series of weekday 'Astronomical Sermons' that placed humankind in relation to God's creation of the vast universe of stars and planets. Attending every lecture on astronomy, George Burns 'was struck to find that many of [the congregation] ... were the most unlikely he would have expected to see – rich and poor, learned and illiterate, religious and profane, all had flocked together to the church that day'. Burns quickly became part of a small inner circle within the Tron and later within St John's parish, where Chalmers inaugurated his system of Christian political economy for the poor of the parish. On a foundation of godliness, Chalmers believed, even the poorest of communities, industrial as well as rural, could be made self-reliant and independent of institutionalized philanthropy.[11]

Chalmers' obituarist in the *North British Review* (1847) claimed that the late minister 'converted the populous and plebeian parish of St. John's into an

[6] Hodder (1890: 73–4, 93–8, 137).
[7] 'Cleland, James (1770–1840)', *ODNB*.
[8] Morrell and Thackray (1981: 204, 293); Hodder (1890: 115).
[9] Cleland (1840: 33).
[10] Hodder (1890: 65, 122).
[11] Hodder (1890: 77–83).

isolated district – with an elder and a deacon to every family, and a Sabbath school for every child – and had wellnigh banished pauperism from within its borders'. The parish thus stood as 'a reproachful oasis ... shaming the wastes around it' while Chalmers hoped and prayed 'that its order and beauty would have said to other ministers and their elders, "Go ye and do likewise"'.[12] In his biography of George Burns, Edwin Hodder noted that throughout his life Burns 'ever looked back with affectionate interest on those times when he was privileged to work with the gifted and noble man who wrought so great a reformation in Glasgow, who made Evangelicalism popular, who raised men's views of Christianity, and who dispersed much of the infidelity which, like a dark cloud, hung over Britain'.[13]

As with the Revd Burns, Chalmers placed Providence at the centre of a nuanced and scholarly Presbyterian theology. He invoked a venerable Christian tradition distinguishing God's absolute power to create or destroy from His ordained power that acted to sustain the laws of nature as uniformities. 'We admit that His creative energy originated all, and that His sustaining providence upholds all', he wrote in one of his sermons.[14] Unlike the Deity of more extreme evangelicals, God did not arbitrarily act by way of special punishments. Instead, transgressing nature's laws would bring upon human beings retribution from within the divinely maintained natural and moral order. Recognition of this truth was a prelude to conversion and thus to salvation, rendering possible a moral regeneration of society.[15]

Although Chalmers firmly believed in the divinely ordained constancy of nature's laws, he argued that the arrangements of nature, and especially the structures of human society, were inherently unstable. As a consequence of the Fall both human beings and nature had passed from a state of perfection to one of sin and decay. 'Nature contains within itself the rudiments of decay ... Every thing around us should impress the mutability of human affairs', he told his congregation in one sermon. 'For death is at work upon all ages ... A gust of wind may overturn the vessel, and lay the unwary passenger in a watery grave'. Such instabilities and imperfections in nature and society – especially the unpredictable, contingent character of individual death – underscored the urgent need among all human beings for repentance and salvation. And the same sort of instability in society emphasized the value of an unyielding faith during the severe moral trials wrought by both natural and commercial storms.[16]

[12] [Anon.] (1847: 560–74, on 565).
[13] Hilton (1991: 58); Hodder (1890: 103–4, 118–19).
[14] Chalmers (1836–42: vol. 4, 387–8). The classic study of voluntarist theology is Oakley (1961: 433–57). See also Smith (1979: 62).
[15] Hilton (1991: esp. 13–15, 64–7).
[16] Chalmers (1836–42: vol. 2, 371–2; vol. 7, 266–7, 272–3); Hilton (1991: 147–8); Smith (1979: 60–6).

Chalmers believed that these natural tendencies could only be stabilized by spiritual power. Human individuals and communities could not by themselves act to prevent disorder. Under the redemptive force of Christ, however, the imperfect material, moral and spiritual conditions of human beings could be improved. That kind of argument underlay Chalmers' practical schemes in Glasgow's poorest parishes. A minister drawing his spiritual inspiration from the gospels and acting as an instrument of the Divine Will could reshape the lives of a church community. In such a way, the condition of the community would be brought from (in Revd Burns' words) a state of sin and misery to one of holiness and happiness.

G. R. Searle's investigation of Chalmers' Christian political economy contains two insights of particular relevance to Scottish ship-owning practice in the period. First, Chalmers argued in his *Application of Christianity to the Commerce and Ordinary Affairs of Life* (1820) that 'an affection for riches, beyond what Christianity prescribes, is not essential to any extension of commerce that is at all valuable or legitimate', and that an excess of 'the spirit of enterprise beyond the moderation of the New Testament' produces 'overdone speculation' and consequent commercial calamity. And second, Chalmers represented trade between honourable merchants as a more worthy testimony to the character of Britain than 'the renown of her victories, or the wisdom of her councils'. Indeed, 'amid the gloom of this world's depravity', he could think of nothing 'finer than such an act of homage from one human being to another' than 'the credit which one man puts in another, though separated by oceans and continents'.[17]

Chalmers' sermons relating to the sea provide clear illustrations of providential control over nature. Taking as his text the famous verses from Psalm 107 that begins 'They that go down to the sea in ships', he focused on the possible means by which God could answer the cries for help of the storm-tossed mariners whose plight the passage described. First, the omnipotent God might intervene directly in the visible world and suspend a law of nature. Such would be the case in divine miracles. Chalmers, however, argued that in the biblical passage God neither raised nor subdued the tempest by miraculous intervention, but did so through the natural agency of wind.[18]

His reasoning here drew implicitly on the works of the eighteenth-century Glasgow University professor of moral philosophy, Thomas Reid. Arguing

[17] Searle (1998: 8–15, 25–6); Chalmers (1820: v–vi), 30 (also quoted in Searle); Taylor (2006: 22, 31).

[18] Psalm 107: 23–30; Chalmers (1836–42: vol. 7, 234–62, esp. 234–36; vol. 2, 314–58); Smith (1979: 62–3). Chalmers was delivering a discourse 'On the consistency between the efficacy of prayer – and the uniformity of nature'. The discourse was reworked as a section 'On the doctrine of a Special Providence and the efficacy of prayer' for his 'Bridgewater Treatise' (Chalmers 1836–42: vol. 2, 314–58).

that 'the rules of navigation never navigated a ship', Reid claimed by analogy that the laws of nature by themselves had no power. Just as the navigation of the ship required the active agency of the captain, so nature's rules depended for their efficacy upon direct divine agency or upon the action of instruments under God's direction. Human beings, made in the image of God, had been endowed with active powers that, limited though they were in comparison with the deity's, differentiated human beings from all other creatures on earth and supplied the power to act for good or ill.[19] Chalmers developed the theme of human beings as key instruments in a sermon on the shipwreck of St Paul in Acts. Paul knew that it was 'God's will that they should be saved by the exertions of the sailors – that they were the *instruments*'.[20]

Very similar evangelical perspectives informed the life and work of George Burns' brother John. These views found expression in his *Principles of Christian Philosophy* (1828). 'The belief in a Providence', Dr Burns affirmed, 'enables us to prosecute our course with steadfastness, to live without harassing anxiety, regarding the present or the future; to give ourselves up to the management of a wise and holy God with confidence and hope'. Like Chalmers, he held that Providence, as a divine agency, acted through nature's laws in the material world and that, from a human perspective, success depended on working with – and not in defiance of – those laws. But there were no guarantees: 'A man may indeed be successful who has no title from his diligence and skill to expect it; and on the other hand may fail, although he employ the utmost human prudence and exertion, God seeing it fit thus to demonstrate his superintendence over the affairs of men'. In the end, the Christian believer had to trust that 'even our most bitter afflictions were decreed by the tender mercy of the Lord'.[21] According to his biographer, George Burns 'was wont to say that if he wished to give expression to his own views on Christian life generally ... he could not do better than repeat the words of his brother in this work'. Providence, as we shall now see, pervaded not only his (and his family's) whole personal life, but also assigned moral and spiritual credibility to their 'business in great waters'.[22]

1.2 'Hurried to Their Eternal Audit': The Tempting of Providence

For the early nineteenth-century travelling public, the sea represented a place for unrelenting suffering. Extreme evangelicals exploited these fears in their endeavours to convince humankind of the sin-laden and transitory nature of

[19] Reid (1846–63: vol. 2, 509–697, esp. 527); Smith and Wise (1989: 86–7); Laudan (1970: 103–31).
[20] Chalmers (1836–42: vol. 9, 155–9) (my italics).
[21] Burns (1828: 281–7, 295).
[22] Psalm (107: 23); Hodder (1890: 167, 144–5).

the visible world. A Calvinist legacy, grounded on a belief in human depravity, pervaded nineteenth-century Scotland, especially in seafaring and rural communities. Chief among manifestations of depravity was the sin of pride. Promoters of steamships were vulnerable to accusations of hubris. A contributor to the United Presbyterian Church's *Scottish Christian Journal* reported hearing a fellow witness warn that the 'power will soon be taken out of God's hands' as they watched a steamboat depart from Glasgow's Broomielaw, the local quay most familiar to passengers embarking on coastal passages.[23] The remarks were prompted by the knowledge that the vessel, heading into the teeth of a gale within the confines of the narrow river, was doing what no sailing vessel could have even attempted.

Popular forebodings that it was 'flying in the face of Providence to encourage [steam at sea]' were widespread. John Scott Russell, for example, recalled that in 1816 the steamer *Glasgow* had departed on a brief sea passage described by 'friends' of the vessel's crew as 'a tempting of Providence'.[24] When, therefore, George Burns contemplated a move into coastal steamships, he could expect accusations of impiety, pride and overconfidence. Already possessing first-hand knowledge of cotton manufacturing and trading, Burns in 1818 had set up a firm of general merchants (mainly in the produce business) with his older brother James. George, however, was no stranger to early Clyde steamers, for he had witnessed the *Comet*'s departure from the Broomielaw in 1812. He also patronized Henry Bell's Baths at Helensburgh, to which resort the *Comet* conveyed Glasgow's genteel mercantile citizens. Around 1824 the Burns brothers entered the coasting trade between Glasgow and Liverpool as Glasgow agents for the Liverpool-based firm of Matthie and Theakstone, which ran a small fleet of sailing vessels between the two ports.[25]

Burns had won the agency in the face of strong competition by demonstrating to Hugh Matthie an unwillingness to ingratiate himself with any person. At their meeting, Matthie stated that the rival firm's bid was supported 'by a round-robin of recommendations from the most influential people'. But his parting shot to Burns was that he looked 'to personal fitness as of the first importance ... it will be given to the best and most capable man I can get'. Having demonstrated face-to-face trust, Burns obtained the agency. Shortly afterwards, G. & J. Burns also acquired Theakstone's 50 per cent share in the fleet.[26]

The Burnses' earliest involvement with steam navigation, however, was a precarious one. Built to serve a passenger trade between Glasgow and

[23] [Anon.] (1849: 156–8, on 156); Smith and Scott (2007: 474–5).
[24] Hodder (1890: 154); Russell (1861: 142); Bruce (1888: 40–5).
[25] Hodder (1890: 122–4, 145).
[26] Hodder (1890: 145–9).

Figure 1.1. Greenock's Custom House Quay with early Clyde steamboats

Galloway in south-west Scotland, the wooden paddle steamer *Ayr* (part-owned by the Burnses) was only two weeks old when it ran down and sank the second *Comet* off Gourock, near Greenock (Figure 1.1), with the loss of about seventy lives on 21 October 1825.[27]

The melancholy disaster fulfilled the forebodings of Scotland's Calvinists. One anonymous pamphleteer quickly highlighted 'the fate of the *Comet* as a signal instance of the uncertainty of life' and, echoing Psalm 107, 'the constant peril which besets those who "go down to the sea in ships"'. And while the *Edinburgh Observer* concluded that as a result of the disaster it would 'require a considerable length of time to restore public confidence in steam navigation', the *Edinburgh Weekly Journal* lamented in uncompromising evangelical tones the tragedy of 'so many immortal creatures ... in a few brief seconds, hurried to their eternal audit'.[28]

In a series of even more severe sermons addressed to Glasgow's improvident youth, another preacher, Abraham Perrey, used the terrible event to impress upon his audience, not its singular character, but an instance of the innumerable 'calamities the supreme Ruler of our destiny is inflicting on our kind'. He urged his listeners to reflect on the passengers spending the previous night 'if not in sinful dissipation, yet, undoubtedly in all the thoughtless levity of a most frivolous amusement ... on the very threshold of eternity'. That night was then

[27] Hodder (1890: 154).

[28] [Anon.] (1825). This pamphlet published extracts from contemporary newspaper accounts, including the *Edinburgh Observer* and *Edinburgh Weekly Journal*. I thank Anne Scott for this material.

the prelude to a divine visitation when 'many felt the displeasure of an angry God', but no longer had 'the privilege of attending the means of grace'. Failure to turn to God before meeting with such a sudden termination of human life would indeed entail a far worse fate than the mere physical extinction of the *Comet*'s passengers. Cutting off the impenitent, the Supreme Judge 'consigns them to everlasting woe', there to meet with 'those deadly pangs ... which are to be the eternal portion of all who incessantly revolve in that lake of fire and brimstone, which is the second death'.[29]

While Perrey represented the *Comet* disaster as the direct result of the actions of an angry Deity intent on delivering yet another warning and punishment to depraved humanity, the *Edinburgh Weekly Journal* placed the blame for the disaster firmly on 'those very persons in whom the confidence of the unhappy sufferers was naturally and necessarily reposed'. It therefore made a clear distinction between tempest – in which 'the voice and the hands of the Deity are reverentially traced, and [in which] the soul falls prostrate before His chastening rebuke' – and human agency characterized by negligence and folly.[30]

Much of the press blame for the collision and subsequent loss of life fell upon the *Ayr*. The *Scotsman* sarcastically pointed to the advantage that steamers had over sailing vessels in their capacity to escape danger through 'their self-directing power, as in the case of the *Ayr*'. The pamphleteer indicted the steamer for ramming the *Comet* and then leaving the vessel to sink without attempting to rescue any of those in the water: 'The captain of the *Ayr* panicked [sic] and sailed for Gourock ... Survivors claimed he even ran some of them down!'[31]

The disaster was a powerful lesson to George Burns as he contemplated entering steamship ownership on a larger scale. At stake was the moral and spiritual credibility of the venture. 'Personal fitness' of masters and the experience of ship-owners now became the guiding moral principles. Confident that the old Presbyterianism with its presumption of inevitable disaster and divine punishment no longer held sway, his moderate evangelicalism emphasized the importance of combining (in Chalmers' phrases) 'the wisdom of experience' with a 'sense of deepest piety'. These values could persuade a fearful public to place their trust and confidence in a Christian ship-owner working humbly in accordance with the natural and moral laws of God.[32]

Cleland had noted of mid-eighteenth-century Glasgow that 'the people were particularly strict in their observance of the Sabbath'. There were 'families

[29] Perrey (1826: 6, 45, 61, 70, 80). I again thank Anne Scott for this material. The 'lake of fire' reference is to Revelation 19:20 and 20:14. For an excellent North American comparison, see Kurihara (2011: 1–16).

[30] [Anon.] (1825).

[31] [Anon.] (1825).

[32] Burns (1828: 286–7); Chalmers (1836–42: vol. 7, 258–9).

who did not sweep or dust the house, did not make the beds, or allow any food to be cooked', and some who 'opened only as much of the shutters of their windows as would serve to enable the inmates to move up and down, or an individual to sit at the opening to read'.[33] It was a culture of Sabbatarianism that long persisted. When, in 1826, George Burns secured the agency for a line of steamers between Belfast and Glasgow, he threatened to withdraw unless the company rescinded its decision that the steamers sail on Sundays in defiance of the fourth commandment to keep the Sabbath day holy. Not for the first or last time would evangelical Christianity shape Burns' shipping and ship-owning practices.[34]

At the end of the 1820s, Burns won over doubters – including his brother James and his senior partner Matthie – to replace most of the sailing vessels engaged in the Liverpool to Glasgow trade. Seeking to build confidence into his fleet, he ordered the new steamers from John Wood of Port Glasgow and Robert Steele of neighbouring Greenock, both shipbuilders with high reputations and long experience in the construction of strong, sea-kindly and elegant wooden hulls. Indeed, Wood's extensive experience with steamer hulls extended back to the first *Comet*. Matthie's opposition gave way to support and he even proposed, 'as a compliment to Burns', that the first new steamer, completed early in 1829, be named the *Doctor* after his brother John, 'who was then one of the most popular men in Glasgow'. George, however, decided on the more modest names of *Glasgow* and *Liverpool*. So impressed were passengers by the maiden voyage in late July 1830 of the latter that they inserted in the press an appreciation of the vessel and master.[35]

With the Burnses now fully in control of management, they announced Friday sailings, not with the primary aim of defying popular superstition, but in order to maintain the principle of avoiding Sunday working wherever possible. Concerned with business more than divine law, Matthie pointed out that canal freight arrived in Liverpool from inland sites on Saturdays and suggested, sarcastically, that George Burns, if he thought it necessary, might provide chaplains to allow for Saturday departures. In his reply, Burns said that he 'thought very well of the suggestion about providing chaplains, and that he and his brother would pay the entire expense of the experiment'. Once again, Burns defied popular superstition, this time regarding the presence of clergymen on board or around ships. Broomielaw spectators at first jeered the

[33] Cleland (1837: 23).

[34] *Exodus* (20: 9–11); Hodder (1890: 152–5).

[35] Hodder (1890: 156–7). Duckworth and Langmuir (1977: 3, 131) claim that the *Liverpool*'s sister ship *Glasgow* was built by Scotts of Greenock, but list the builders as Cairds of Greenock. Cairds, however, did not enter shipbuilding until the mid-1840s (Chapters 7 and 14). Russell (1861: 147) includes the ship in his list of hulls built by the Wood brothers. It is thus more likely that Cairds supplied the machinery and Wood the hull.

practice, and wits even suggested that the master of one ship was '[s]ailing in a steam chapel'. Yet it was a practice that seemingly remained until the Church of Scotland split into Free and established kirks at the Disruption (1843), after which ministers were in high demand.[36] Less concern for Sabbath observance among later Clyde pleasure steamer operators called forth the condemnation of the established presbytery of Glasgow: 'that which you have created amounts to a virtual abrogation of the divine ordinance and practically excludes it from the statute-book of Heaven'.[37]

Burns regarded his role in steamship ventures as a Divine 'calling', paralleling the 'call' of ministers to their vocation or to a particular parish. Searle has noted that this view of a mercantile life was characteristic of evangelical perspectives in the period.[38] While travelling overland to London in 1832, Burns wrote thus to his wife Jane after a Sabbath spent attending three Church of England services throughout the day:

> with all my deep unworthiness, were it not that the gospel of Christ and His grace afford me support, what would become of me under the fits of depression that at times weigh me down! My lot in life is now fixed, and doubtless by His unerring wisdom. And He has hitherto helped me and delivered me wonderfully out of difficulties, and enabled me to persevere in the struggle. I have still many strong temptations to contend against, but trust that God will help me through. At present, although it is sorely against my nature, *I am engaged in the occupation in which I am called* in discharging a Christian duty for you, and for our dear children and myself.[39]

In late summer 1831 the House of Commons debated a private member's bill to regulate the speed of steam vessels on the Thames. Alderman Wood argued that steamboat owners seemed indifferent to the loss of life, especially among the myriad small boats and lighters at work on the river, occasioned by the rapidity of those steam vessels and by their propensity to race one another for the arches of London's bridges. It was a controversial measure. Opponents argued variously that 'a rapid rate [among steamboats] was their chief excellence' or that the bill would 'materially affect steam navigation in general on rivers'. One MP reminded the House of a similar but subsequently rejected bill brought in after the loss of life of the second *Comet*. Instead, the Clyde

[36] Hodder (1890: 157–8). Led by Chalmers, the Free Church was so-called because its members upheld the 'democratic' right of a parish congregation to choose – over that of wealthy patrons to determine – the appointment of its minister. In general, Free Kirk clergy were stricter in interpretation of the scriptures than their Church of Scotland brethren, though the doctrinal distinctions were often less than clear-cut.

[37] Paton (1853: 3). I thank Anne Scott for this reference.

[38] Searle (1998: 25).

[39] Quoted in Hodder (1890: 172). My italics.

authorities had introduced local regulations 'since which no accident whatever had occurred'.[40]

A Select Committee reported in mid-October. It noted in particular that legislative enactments designed to prevent steamboat calamities 'cannot prove detrimental to the proprietors of steam vessels, in any degree to counterbalance the security and satisfaction they are likely to afford the public; but that on the contrary it is probable they will eventually prove advantageous to the steamboat proprietors themselves, from the increased confidence which will thereby be created in this mode of conveyance'.[41] All seemed set fair for the passage of Alderman Wood's bill in the early summer of the following year.

Burns made it his mission to see the bill defeated. To that end he won some very well-connected allies, most notably the moderate evangelical Viscount Sandon (Dudley Ryder), recently elected an MP for Liverpool with support from the Gladstone dynasty. Sandon had married into the family of the first Marquess of Bute whose seat was on the Isle of Bute on the Clyde, and whose members knew well the Clyde steamer fleet that served its principal town of Rothesay. Sandon had also been one of the Lords of the Admiralty. Although the precise grounds for Burns' opposition are not recorded, there can be little doubt that he and his supporters saw it as a threat from narrow vested interests such as the lighter men and watermen to subvert and inhibit the new steamship era through unwarranted state intervention.[42]

Burns' mission was also inseparable from his evangelical agenda: his steamship vocation was a divinely sanctioned mission, and as such had to be fought for at all costs. He thus represented his role in opposing the legislation as integral to the Christian battle against the forces of darkness that threatened to restrict the advance of steam. His approach mirrored that of Chalmers' rejection of institutionalized philanthropy at St John's in favour of a voluntary system of self-reliance based on Christian practice. As Burns reported to his wife Jane, he was 'endeavouring in the strength of Christ to fight hard in this department of the Christian warfare. It is the hardest struggle in which I was ever engaged, but in some shape or other we must encounter the enemy'.[43]

On the Sabbath aboard his new steamer *Liverpool* moored in the Thames, Burns arranged for divine service for the benefit of the thirty-strong crew and for other shipping in the vicinity. 'It was delightful to hear the voice of praise raised on the bosom of the Thames', he told Jane. 'I dare say the surrounding crowds of shipping were surprised'. Like Chalmers's St John's parish, the steamer served as 'a reproachful oasis' of spirituality 'shaming the wastes

[40] *Hansard's Parliamentary Debates* (1831): 289–91.
[41] SC (1831: 3–6, esp. 5).
[42] 'Ryder, Dudley, second earl of Harrowby (1798–1882)', *ODNB*.
[43] Quoted in Hodder (1890: 175–6).

around it', and acted as an exemplar to other Christian evangelicals to go and do likewise. The powerful imagery rendered 'church' and 'ship' interchangeable. As fallible human constructs, both were all too prone to error and disaster. Without strong Christian leadership to sustain the community in times of trouble, church and ship would disintegrate as every member fought for individual survival.[44]

Wood's bill met a strange fate at the end of July 1832 in the hands of the Hon. Member for Glasgow, Mr Dixon. For reasons unknown, Alderman Wood 'persisted in quitting the House, although apparently not at all indisposed'. Having given advance notice to Wood, Dixon moved that the 'Bill be read a second time that day [in] six months'. Denying that he had taken any unfair advantage of the absence of the principal supporter of the bill, he had effectively defeated the measure and left the Alderman aggrieved at having had the measure 'taken ... out of the hands of the person who introduced it'.[45] Glasgow interests had clearly won the day.

Trust in Providence and the implementation of Divine will in matters of steamship practice meant that G. & J. Burns won increasing levels of confidence from a travelling public. In fulfilling promises to convey passengers and freight safely and reliably between destinations, the line became the embodiment of what a trustworthy steamship company could and should be. As Burns' biographer noted, with 'splendid steamers, good captains, an excellent system of business, and a wide influence, the Glasgow Company carried everything before it'.[46]

Early vexation with his Irish Sea venture, however, came with the establishment of a rival Liverpool to Glasgow service by fellow-Scot David MacIver based in Liverpool. In September 1830 the celebratory opening of the Liverpool and Manchester Railway (L&MR), several years in the making, promised an orderly and progressive world. The engineers, father and son George and Robert Stephenson, represented the triumph of the new skills of steam locomotion, civil engineering and system building that could overcome all of nature's obstacles. Present and future generations of passengers and shippers would, unlike their predecessors, henceforth enjoy regularity, reliability and economy. Only the accidental death of William Huskisson MP marred a memorable demonstration of the railway's perfect working.[47]

Determined both to match the progressive vision of the L&MR and to break the Burnses' control of the Glasgow to Liverpool line of steam navigation, MacIver set up the City of Glasgow Steam Packet Company in 1831 with capital from a wealthy cotton broker, James Donaldson. Supported by

[44] Quoted in Hodder (1890: 174–5). For an example of 'kirk as ship' see [Anon.] (1847: 566).
[45] Hansard (1832: 930–1).
[46] Hodder (1890: 159). On the issue of trust linked to 'experience', see Winter (1994: 69–98).
[47] Marsden and Smith (2005: 136–45).

the engineering skill of Robert Napier and David Elder (Chapter 2), MacIver himself 'vowed that he would, if possible, drive the Burnses off the seas'. He later confessed that he 'had travelled in the *City of Glasgow* backwards and forwards between Liverpool and Glasgow, going down himself into the engine-room to superintend the firing of the furnaces, in order that he might leave nothing undone' in the effort to conquer the rival firm.

Between 1832 and 1835 three new steamers, all built by Wood and with engines by Robert Napier, joined the City of Glasgow Company. Emphasizing the close personal and financial interconnections, the first two MacIver steamers were named *John Wood* after the hull builder and *Vulcan* after Napier's engine works. Cleland later reported that the last of the trio, the second *City of Glasgow*, made the passage from Greenock to Liverpool in less than eighteen hours in 1835, compared to Burns' steamers' average in 1831 of about twenty-four hours. But MacIver's emphasis on speed failed to match the Burnses' reputation and profits. Rather than exploit the weakness of the rival company, however, George Burns persuaded MacIver to agree to combine the fleets on a division of revenue ratio of two-fifths (MacIver) to three-fifths (Burns). The arrangement was honoured and Burns undertook to win for Christ the wayward soul of his erstwhile tormentor in the years that remained before David MacIver's death in 1845. Meanwhile, the Burns brothers, MacIver and his brother Charles, Donaldson, Wood and Napier forged among themselves a powerful new community of trust embodied in the transatlantic steam navigation project popularly known both as Cunard's line of steamers and as Burns & MacIver (the principal managers) (Chapters 5 and 6).[48]

During two decades of Burns & MacIver's construction of an interwoven system of coastal steam navigation extending from the Western Isles to the Mersey Estuary, the company sustained a reputation for reliability, safety and sound management. Burns and his associates had explicitly abstained from any tempting of Providence, and thus consolidated their moral and spiritual credibility. The best builders and engine makers supplied the steamers, the owners ensured a high standard of maintenance and dedicated masters and crews manned the vessels as they plied their various tracks. But on a calm summer night in June 1850, the iron paddle steamer *Orion* (Figure 1.2), completed by Cairds of Greenock (Chapters 7 and 14) four years previously as the progressive pride of the Burns & MacIver Irish Sea fleet, struck a rock close to Portpatrick on the south-west coast of Scotland.

The steamer was on a routine passage from Liverpool to the Clyde with a large number of passengers, many of them bound for the annual BAAS

[48] Duckworth and Langmuir (1977: 99–101, 188–9); Hodder (1890: 160–3, 244–5); Cleland (1840: 14).

Figure 1.2. G. & J. Burns' iron paddle steamer *Orion* (1846)

meeting to be held that year in Edinburgh. So great was the damage that, despite watertight bulkheads, the vessel quickly foundered. Fifty lives were lost, including Dr John Burns. Here indeed was a sore trial, seemingly inflicted on those who had never knowingly tempted Providence. 'We went into another room and humbled ourselves before God', the Revd Gribble recollected to one of George's two sons, James Cleland Burns, with respect to the Burns household at the time. '[We acknowledged] our sins, submitting to His dispensations, and imploring His mercy to console the relations of those who had been drowned ... the catastrophe was met by your honoured parents in the sweet spirit of submission to God'.[49]

Ship-owning could be credibly presented to the public as a noble business operating within a profoundly moral and spiritual universe of divine laws that left no space for chance or luck. In Chalmers' reading, human beings had the power to act either as instruments of God within His framework of moral and natural laws, or as tools for evil. On the one hand, human 'success' was an indication of good work and skilful exertion, of an accumulation of moral and

[49] Hodder (1890: 276–9); *The Times*, 20 June and 2 September 1850. The *Orion* disaster is considered from a number of cultural perspectives in Anne Scott, 'The wreck of the *Orion*: Reading steam-ship wrecks in nineteenth-century Britain', unpublished paper presented at the British Society for the History of Science Annual Conference, University of Kent, Canterbury, July 2006. I also thank Rachel Flynn for *The Times* references.

spiritual credibility and capital; by abiding with divine laws, human beings could thus properly share in, and even imitate, the Creator's capacity for good and wise government. On the other hand, human 'failure' implied a tempting of Providence by challenging or ignoring God's laws. Such failure entailed, as in business, an accumulation of moral and spiritual debt that could only be redeemed by salvation through Christ. Yet in recognition of the omnipotence of God compared to the limited powers of human beings, 'success' remained wholly contingent upon divine choice. The God of Chalmers and of Burns would not inflict arbitrary punishments on evil-doers, but He might choose to impose adversity and suffering as purposeful trials upon His people. Disasters at sea might ultimately befall even the best-found vessels as reminders of their fundamental dependency upon Divine governance.

2 'The Character of Fine Workmanship'

Making Clydeside's Marine Engineering Reputation

> Glasgow also, in its commercial relations, trades with every quarter of the globe, and its merchants deal in the various products of every country. It hence appears that one branch of manufactures or trade may be dull while another may be prosperous; and, accordingly, Glasgow does not feel any of those universal depressions which so frequently occur in places limited to one or two branches of manufacture or commerce.
>
> *Statistician John Strang identifies the economic strengths of the rising Victorian city of Glasgow (1850)*[1]

Summary

Robert Napier and his self-effacing works manager David Elder built confidence first into estuary and coastal steamers and later into early transatlantic steamships. Napier's engine-building and boiler-making workshops were the privileged sites for the practice of a range of engineering and shipbuilding skills. Craftsmen who acquired their practical knowledge while in his employ frequently established neighbouring sites of skill on the Clyde and beyond. Under Elder's direction, Napier's Vulcan, Lancefield and Govan works constructed and integrated the components – steam engines, boilers and iron hulls – that constituted steamships as human-built systems which in turn made the firm's reputation for reliable and durable vessels.

2.1 'Seldom Idle': Constructing Steamships

In the early 1840s, *The Penny Magazine* described Glasgow's Broomielaw as a quay 'studded with ships and boats as closely as it is possible to be without injury'. Here were moored passenger steamers 'of a splendid character' serving Liverpool, Dublin and Belfast. Others 'of a smaller class' operated to Dumbarton, Greenock, Helensburgh, Rothesay and lesser piers on the Firth of Clyde. And sailing vessels, brought up river by steam tug, arrived from 'every part of the globe' to berth at deep-water quays in the process of being extended further and further downstream. Indeed, little over half a century had passed

[1] Strang (1850: 162); Morrell and Thackray (1981: 202–22) (discussing Strang's role in attracting the BAAS to Glasgow for the first time in 1840).

since engineers had begun a long, slow process of disciplining Glasgow's shallow river into a waterway for sea-going vessels.[2]

At Washington Street, just clear of the commercial quays on the north bank, stood Robert Napier's recently established Vulcan Works constructing marine steam engines on an increasingly large scale. Close-by was his cousin David's Lancefield premises, including Lancefield House and a tidal basin for installing steam engines in ships' wooden hulls built at one of the many yards sited in the traditional shipbuilding towns of Greenock, Port Glasgow and Dumbarton near the mouth of the River Clyde. By the time *The Penny Magazine* published its account, however, Robert Napier had not only added the Lancefield properties to his expanding engineering empire, but also begun to create an iron shipbuilding yard across the Clyde at Govan. As the magazine informed its readers, '[a] short steamboat trip down the Clyde from Glasgow to Greenock affords the means of observing the nature of the improvements which have been wrought'. Narratives of progress were the order of the day.[3]

Although Glasgow's 'busy and closely-filled harbour' contrasted with the narrow river further downstream, there too civil engineers were remaking undisciplined spaces into a navigation system: 'though narrow, it is deep and regular: every yard of it for many miles is carefully watched, the bed dredged, the banks brought to a regular slope and faced with stones, useless inlets and creeks filled up, projecting points pared away and all obstructions sedulously guarded against'. As the Clyde at length widened towards the Firth, 'bell-shaped structures are placed at intervals to mark the progress of the tide'. Here stone training banks directed the stream to scour its human-built channel and to leave aside the former meanderings among the mud shallows of the south shore. On the north side, the rugged landmark of Dumbarton Rock, guardian of its ancient seaport and shipbuilding town, marked the slow broadening of the river. And finally, on the south side, the adjacent towns of Port Glasgow and Greenock, both centres of high-quality wooden shipbuilding, but both in slow decline as major seaports due to the rise of Glasgow Harbour, heralded the waters of the Firth of Clyde and the highway to the open sea.[4]

Born in Dumbarton, Robert Napier inherited a tradition of iron working. His grandfather, also Robert, 'followed the calling' of blacksmith and married Jean Denny of that town. Three of their sons also became blacksmiths, two (John and James) in Dumbarton and one (Robert) in Inveraray as smith to the Duke of Argyll domiciled in the neighbouring castle. James married, in 1789, Jean Ewing from the hamlet of Rosneath on the Gareloch. They had six sons, of whom Robert (born in 1791) was the eldest. While his father and uncle

[2] [Anon.] (1843: 379); Shields (1949: 12–23); Riddell (1979: 1–51); Smith and Wise (1989: 21–2).
[3] [Anon.] (1843: 379).
[4] Strang (1850: 162).

operated a foundry with two steam engines (one a Newcomen type working a boring mill for finishing cannons), the expectation was that the eldest son would train for the kirk. Robert, however, persuaded his father to allow him to follow mechanical pursuits.[5]

Napier's whole life was nevertheless lived in close proximity to the Church of Scotland. Not only did his brother Peter fulfil the promise of a minister of the kirk within the family, but Robert was also a good friend of Chalmers. En route to take up the chair of moral philosophy at St Andrews in November 1823, for instance, Chalmers wrote to apologize for not having had enough time upon leaving St John's parish to thank Napier for the use of a 'child's coach' for his little daughter. Napier regularly attended the modest Church of Scotland meeting house at Row (Rhu) that was under the patronage of the Dukes of Argyll. From 1831 the Revd Laurie Fogo, 'full of religious enthusiasm and evangelical fervour' and 'earnest and evangelical as a preacher', occupied the pulpit. He was also Napier's lifelong friend. Napier contributed £800 towards a new, much larger kirk on the Row site in 1851 and two years later funded a memorial with a bronze statue on Henry Bell's grave close to the front entrance to the church.[6]

Cousin David (son of John) had already entered the field of marine engineering, having constructed the boiler for Bell's *Comet* (1812) and engined the *Rob Roy* (1818) that became celebrated as the first cross-channel steamer. At Camlachie, not far east of Glasgow's High Street, David established a foundry and used the nearby burn to conduct simple experiments on different hull forms. As he himself explained, 'on perusing the works of [Charles] Bossut, a celebrated French mathematician, on the resistance of fluids, I began to have serious doubts of the full [i.e., bluff] bow being the proper shape, to prove which I set about a series of experiments, for which a large mill-dam that bounded the premises at Camlachie afforded great facilities'. Erecting at the water's edge a high framework supporting a drum for winding up a weight, he attached the other end of the line to an 'experimental' block of wood with dimensions proportional to those of the proposed *Rob Roy*. Carefully noting 'the time the weight took to descend', he 'continued fining the bow [of the block] as long as there was any perceptible increase to the speed, always taking care to put the block into the scales ... so as to keep the weight of the block the same'. Once satisfied, he delivered the 'block or model' to the shipbuilder (his Dumbarton kinsman William Denny) 'to take off his lines for building the *Rob Roy*'.[7]

[5] Napier (1904: 1–9).

[6] Napier (1904, 166–81, 179–80); Maugham (1896: 61, 63) (on Fogo's evangelicalism); Duncan (2002) (Rhu Church and its clergy); Osborne (1995: 235–6) (Bell's grave was unmarked for sixteen years).

[7] Bell (1912: 22–3); Napier (1904: 18–27); Ferreiro (2007: 163–8) (Bossut).

Figure 2.1. David Elder, self-effacing works manager for Robert Napier

According to Scott Russell's *Steam and Steam Navigation* (1841), David also chose to take passages on a sailing packet between Glasgow and Belfast at 'the stormy period of the year'. Constantly positioned at the bows of the vessel, he watched as the seas broke on the ship's side. At intervals he would persistently inquire of the captain if conditions could yet be considered rough. When the seas began to sweep the vessel from end to end he repeated his question, to which the master replied that he could not 'remember to have faced a worse night in the whole of his experience'. Delighted, Napier went below to his cabin muttering 'I think I can manage if that be all'.[8] This very hands-on craftsman-engineer duly established his engine works at Lancefield and leased the Camlachie foundry in 1821 to Robert, who selected a former millwright, David Elder, as his works manager (Figure 2.1).

Born in 1785 at Little Seggie near Kinross in Fife, Elder hailed from a family of country joiners, wheelwrights and house builders. His grandfather's papers apparently exhibited 'a remarkable proficiency in mathematics and the principles of mechanics'. His father, a 'country wright', had dissented from the established Church of Scotland during a period of intense religious controversy among Presbyterians, and as a result his son was not admitted to the local parish school. Undaunted, he studied Simson's *Euclid* and an unnamed work on algebra translated from French. While his 'seniors would be devoutly employed at a tent-preaching', he would instead be studying practical hydraulics at the sites of local waterwheels.[9]

[8] Russell (1841: 246–7). Also quoted in Bell (1912: 32–3).

[9] Napier (1866: 92–3); Smith (2014: 80–4) (on David Elder's engineering practices). 'Tent-preaching' refers to a Presbyterian practice of erecting a covered wooden pulpit outside a church

By 1804 Elder was working on the construction of buildings in Edinburgh's Charlotte Square at the heart of the fashionable New Town. Outside working hours, he read texts borrowed from a bookstall at the rate of one penny per night. After marriage to Grace Gilroy of Paisley in 1812, he moved to that Renfrewshire textile town where he undertook the design of a large thread mill. In 1817 he planned and directed the construction in Glasgow of several large cotton mills, alum works, thread mills and portions of Tennant's chemical works. Thereby well versed in the mechanical system-building of mills and mill machinery, he joined Napier's firm and designed his first pair of side-lever marine engines which he installed in the river steamboat *Leven* for service between Glasgow and Dumbarton. At this stage, the stock of workshop tools was limited to some ten- to fourteen-inch turning lathes, a horizontal boring mill and a smaller vertical boring machine.[10]

Elder introduced unorthodox practices into Glasgow marine-engine building, hitherto the preserve of millwrights familiar with iron working. Finding that their notions of 'accurate workmanship' fell short of his own standards, he selected instead cartwrights and house-joiners to superintend the various departments and to take charge of important new tools. His preference for craftsmen skilled in woodworking rather than iron-making rested on his belief that the former 'would carry their ideas of close fitting joints of wooden structures along with them, when they would be called upon to construct iron ones'. The new tools also embodied an overarching belief in the minimization of 'waste' and the maximization of useful work in all matters of engine construction.[11]

One such tool was a paring machine (known by the thwarted millwrights as the 'devil') that performed 'work with an expedition and accuracy which no hands could execute'. Elder originally designed the machine to trim thin strips of metal during the finishing stages from the connecting rods of the new side-lever engines. From a surviving diagram, the system appears to have consisted of a vertical wheel fitted with a large number of paring blades protruding on each side of the circumference. As the wheel turned, it advanced towards the inner end of a U-shaped metal strap clamped in a horizontal position. The strap could thus be pared mechanically to make a perfect fit within the system of connecting rods.

Elder put the machine to other works, 'formerly unthought of, on account of the difficulty and expense attending their execution'. It was, therefore, 'seldom idle'. Other machines similarly reduced waste. A modification to an

for overflows, especially at communion times when 'preachings' could take place all day long and over several days. See Hodder (1890: 28).

[10] Napier (1866: 93–4); [Anon.] (1866: 103).

[11] Napier (1866: 94–5). See also Wise (1989–90).

ordinary turning-lathe 'secured the greatest strength with the smallest quantity of material in his side-rods, cross-heads, &c'. His punching or piercing 'engine' reduced loss by friction to a minimum when the pressure was greatest. In every machine, James Napier later noted, Elder seemed to possess a knowledge of the forces 'as a sort of instinct, and for which he provided in the most scientific manner'. For example, he prevented the joints of engine frames from performing any work by shaking and vibrating. In so doing he 'economised the available power of the engine, by having every joint metal to metal, and as much bearing surface, and as closely fitted as it was in the power of the men and tools at the time to execute'.[12]

Robert Napier acquired the Washington Street site, soon known as the Vulcan Works, in the late 1820s to cater for his growing marine engine business. He and his manager designed and constructed the spaces within the site in order to maximize their economic value, and soon equipped it with heavy machine tools. They therefore aimed to make engine construction a systematic process.[13]

Invited into the Vulcan Works, *The Penny Magazine* represented this new world to its readers (Figure 2.2). The visitor entered 'an open or quadrangle or space bounded on all sides by the buildings wherein the manufacturing operations are carried on'. The scene resembled an anatomical theatre in which dissection had been replaced by creation and assembly. The arm of a large crane could sweep over most of the area, picking up heavy weights from different quarters as required and permitting the assembly of 'an enormous steam-engine, about forty feet high, intended for the hot-air blast-furnace of an iron work[s]'. Each of the surrounding buildings served a different purpose and trade. On one side stood the 'founding or casting house', with a furnace for melting iron and a sandy pit for making the castings; in another part of the works the boilers and funnels took shape amid the thunder of metallic riveting.[14]

Within the engineering shops, the 'character of fine workmanship' was apparent. '[H]ighly finished machines' worked on wrought iron. The lowermost shop handled the boring of cylinders and the fitting and adjustment of pistons. A lathe could turn wheels up to eighteen feet in diameter. On the higher floors, 'workshops contain benches, tools and lathes, for preparing and finishing the smaller and finer pieces for steam-engines'. Also above-stairs were store-rooms for the existing patterns as well as pattern-making shops, crucial to the practices of moulding specific sizes of casting in the foundry across the

[12] Napier (1866: 95–7). It is important, however, to note that Napier read Elder's life through the lens of a science of energy constructed by his friends Macquorn Rankine and William Thomson, professors of engineering and natural philosophy respectively at the University of Glasgow (Smith 1998a: 150–66).

[13] Napier (1904: 38, 48–9).

[14] [Anon.] (1843: 382).

Figure 2.2. Napier's Vulcan Works with large boiler under construction (1843)

quadrangle. Finally, the forge or 'smithery' had been equipped not only with forge-fires and anvils, but with a small steam-engine to power 'a continuous blast of air directed to any particular forge at pleasure, by moving a lever connected with a valve in the blast pipe'. The Vulcan Works was indeed a Napier smithery on a large scale, systematic and highly integrated.[15]

In the mid-1830s, Napier leased the Lancefield site from his cousin. By the *Magazine*'s visit, Napier and Elder had laid out the works in ways very like the older Vulcan Works. But Lancefield was now engaged more in shipbuilding activities than in marine engine-building. 'We observed many of the rough elements for an iron steam-boat lying about the works': the magazine reported, 'iron bars for the ribs, in one place; huge castings and bent pieces for the stem

[15] [Anon.] (1843: 382). See also Fox (2003: 90) who claims that '[t]he foundry sprawled across a large quadrangle on Washington Street'. The *Magazine's* point was that the site was anything but a sprawling one.

and stern, in another; quadrangular pieces from which the keel was to be built up, in a third'.[16]

Between the works and the Clyde was Napier's Dock for the fitting out of steamships. On the west side of the dock stood an extensive ship's carpenters' yard, and on the north side a shed to facilitate the transfer of engines and boilers from the works to the vessels lying alongside where 'we may often see huge boilers, many tons in weight, suspended in mid-air, and gradually descending to their place of reception'. There and at the neighbouring engine works and foundries, many of the best-known Clyde builders and engineers acquired their skills prior to setting up on their own account. Just to the east of the dock, Hydepark Quay offered further accommodation to steamers under construction. On the opposite bank of the Clyde, the recently established iron shipbuilding yard of Tod and MacGregor (former employees of David Napier), as well as an engine-building works of Thomas Wingate, testified to the promise of these new industries.[17]

Like an island of green in a black industrial ocean, Napier's detached residence, gardens and bowling green at Lancefield House stood adjacent to his works. It was there that he hosted his Nova Scotian visitor, Samuel Cunard, in 1839 (Chapter 5). The neighbourhood was scarcely appropriate for a gentleman's residence, filled as it was by another foundry, an iron forge, a cotton spinning works, a chemical grinding works, a pottery and a property of the Glasgow Gas Light Company. But it was representative of what City Chamberlin and statistician Strang, in his paper presented to the BAAS in 1850, termed the city's 'economic cosmopolitanism', upon which the sustainability of Glasgow's wealth and power rested. From the very doorstep of Napier's residence, therefore, prospective customers could witness for themselves a microcosm of the sheer diversity of industrial production in the city, many of whose products would be directed into the realization of that new and highly complex engineering system known as an ocean steamship.

Successive statistical accounts, often presented to the BAAS, confirmed Glasgow's economic and industrial rise: the population increased from almost 84,000 in 1800 to over 270,000 in 1840 and nearly 370,000 in 1850; a tenth of Britain's iron was produced in the Clyde region in 1835 and 123 cotton mills in 1845 constituted the staple trade of the city and its neighbouring communities. But more telling than statistics was the remaking of geographical space through human agency. Proud of their ancient cathedral and university, and

16 [Anon.] (1843: 384).

17 [Anon.] (1843: 384); 'Map of Lancefield District 1840', Mitchell Library, Glasgow. Abbott (1851: 721–34) offers a similarly detailed and illuminating account of one of the major New York engine-building sites. While many of the practices may reflect connections and exchanges between New and Old World firms over the previous decade or more, the layout of the works appears much less compact and less driven by economy than Napier's.

growing wealthy as merchants in tobacco and sugar from the New World, these citizens envisioned an artificial waterway connecting their town to the ends of the earth. Their promises of improvement made a slow journey to fulfilment, but by the 1830s sea-going vessels with a draft of fourteen feet could reach the Broomielaw. And at mid-century, the harbour handled nearly 400,000 tons of sailing vessels and 900,000 tons of steamers per annum, significantly showing a large preponderance of steam over sail at a time when the British register recorded, on the contrary, an overwhelming dominance of sail over steam.[18]

Thanks in large part to the reputation for the sound engineering skills of his manager, in the early 1840s Napier established an iron shipbuilding site at Govan, on the south side of the river downstream of Glasgow. An 1842 map of the area marks the 'Site of Mr Robert Napier's new engine works', but the engine works remained firmly at the Vulcan and Lancefield sites. A year later, *The Penny Magazine* observed that there were

> no huge trunks of oak lying in heaps in the yard; no 'converting' sheds, where the business of sawing and shaping the timbers is carried on; no 'steaming tank' for steaming the planks preparatory to bending round the contour of the ship; no trenail-cutters [for producing the large number of wooden rods that would secure the planks to the frames]; no oakum-spinners [for the preparation of the fibrous caulking to be inserted between the planks that would render the hull watertight].

Instead there were 'workers in iron, with their forges, their powerful punching-engines, their enormous cutting-machines, which will sever a bar of iron as if it were a ribbon, their anvils, their hammers, their rivet-furnaces'. Lying around were 'sheets of iron from the rolling-mill, and bars from the drawing mill, with guide-pieces for bending them to the required form'.[19] Napier and his team had thus constructed a highly specialized site, distinguished from traditional wooden shipbuilders by the concentration of skilled workers crafting both iron bars for the ships' ribs and iron plates for the skin.[20]

The *Magazine* also challenged the apparent similarities between the construction of iron and timbered hulls. In both practices, the draughtsman 'selects a large flat surface of flooring [the mould-floor], and on this he chalks a series of lines destined to show the exact size and shape of the various "ribs" of the ship'. With timber frames, the 'breadth and thickness of every individual piece at every part of its length' had to be indicated. 'Mould-boards' of thin wood then had to be prepared 'in exact conformity with the chalk lines on the mould-floor' to act as guides for the sawyers in cutting the massive timbers.

[18] Smith (1998b: 119–20).
[19] [Anon.] (1843: 380).
[20] [Anon.] (1891: facing p. 5) (1842 map of the Clyde at Govan); [Anon.] (1843: 380). See esp. Thiesen (2006: 1–15) ('The origin of practical shipbuilding methods').

With iron frames, however, 'the trouble is saved, since the contour or cross-section of every rib is given to it while yet in the state of an iron-bar' and 'thin rods only are necessary, sufficiently firm to give curvature to the several ribs'. Thus 'nothing in the appearance of the skeleton of an iron vessel strikes a stranger as being more remarkable than the extreme slightness of the ribs and keel, compared with the bulky timbers of a wooden vessel'. Indeed, for an iron vessel 'of considerable dimensions', the three-quarter-inch thick frames might measure only three inches on each of two sides. Construction of a very small iron sailing vessel in the yard required little more than three weeks from keel-laying to sailing order, illustrating 'the extraordinary rapidity with which iron vessels can now be constructed'.[21]

2.2 'Of the Best Materials and Workmanship': Selling Steamers

'Possessed of superior taste', the journal *Engineering* asserted, 'Mr Elder succeeded in giving a new character to this class of work, and Mr Napier's factory was soon filled with engines for a number of steamboats, not only for the Clyde but for service between Glasgow, Belfast, Londonderry and Liverpool, and between Aberdeen and London, and Dundee and London'.[22] Among these coastal passenger steamers were two – *Robert Napier* (1832) and *Isabella Napier* (1835) – for the North-West of Ireland Steam Packet Company, in which the Napiers clearly had a stake. The ships' regular but often stormy passages from the Clyde to the Atlantic ports on the rugged north of Ireland were enough to test any vessel. One especially striking portrait of the *Isabella Napier* shows the steamer defying the dangers of a lee shore. Between 1832 and 1840, Elder also fulfilled contracts for five steamers for David MacIver's City of Glasgow Steam Packet Company (Chapter 1), including the prestigious 705-ton *Commodore*.[23]

Concerned with building the reputation of the firm, Napier and Elder had the satisfaction of witnessing two of their earliest steamers, *Clarence* and *Helensburgh*, winning first and second places in a steamboat contest staged in 1827 by the Northern Yacht Club (NYC) whose clubhouse was at Row. The race took place between Rothesay Bay and the north end of the Great Cumbrae in the Firth of Clyde. The hull of the eighty-one-ton *Helensburgh*, like the *Rob Roy*, had been built by William Denny at Dumbarton two years earlier for a Glasgow and Garelochhead service. Elder had fitted the steamer with one

[21] Thiesen (2006: 1–15). Compare Thiesen (2006: 80–112) ('Building iron ships in a wooden shipbuilding culture'). He notes that wooden shipbuilding itself tended to become more industrialized in this early period of sea-going iron ships.

[22] [Anon.] (1866: 103).

[23] Napier (1904: fp. 60) (Isabella Napier portrait); Smith and Scott (2007: 482) (portrait reproduced); Duckworth and Langmuir (1977: 181, 188–9).

Figure 2.3. Assheton Smith's final Napier-built steam yacht *Serpent* (1851)

52-hp side-lever engine. The older *Clarence* also ran on the route. According to one witness several 'commanders of the [Glasgow-Garelochhead fleet] were desperately addicted to racing their boats against those of their rivals'.[24]

This spectacular triumph, widely publicized, attracted the interest of a 'powerful English gentleman' and member of the prestigious Royal Yacht Squadron (RYS), Thomas Assheton Smith of Tedworth in Hampshire. Accused of the ungentlemanly intention of seeking the admission of a steam vessel into the RYS for alleged business purposes, the indignant Smith withdrew his name from the Club. He then summoned Napier to his residence before placing an order worth over £20,000 for a steam yacht, the *Menai*, with hull built by Wood and twin side-lever engines fitted by Elder. Such was Smith's trust in Napier that he never visited the yacht during construction, and indeed went on to order eight more steam yachts from Napier up to 1851 (Figure 2.3).[25]

[24] Macleod (1883: 71–2); Napier (1904: 37–8).
[25] Napier (1904: 37–47). See also Fox (2003: 174–6).

Smith's wealth flowed from a Caernarvonshire estate (Vaynol in north-west Wales) that he inherited from his father in 1828. Although the estate covered 37,000 acres (including half of Snowdon, Wales' highest mountain), he owed his prosperity to the mountain slate quarries at Dinorwig near Llanberis and the simultaneous development of Port Dinorwig, with eventual accommodation for 120 coastal sailing vessels engaged in carrying the slate far and wide to growing industrial towns and cities. From 700 men in 1832, employment in the quarries increased to 2,400 by 1858. As a wealthy gentrified patron and establishment figure, Smith operated within a sporting culture (especially that of fox hunting).[26] With his steam yachts, he facilitated the kind of practical demonstration which, if promises were fulfilled, brought prestige to the marine engineers but which, if promises were unfulfilled, carried few risks compared to similar 'experiments' with commercial vessels, where unforeseen accidents could ruin the reputation of everyone concerned. David Napier suffered just such a disaster in 1835 when the boiler of his steamer *Earl Grey* exploded, killing at least ten persons, while lying at Greenock awaiting trials against one of his cousin's steamers. He moved to London soon after.[27]

Robert, however, remained resolutely on the Clyde, where his and Elder's marine engines began to drive the firm's reputation further south. That reputation, according to Napier's biographer, benefited more from the completion and successful operation of the steamers *Dundee* and *Perth* – and later the *London* – for the Dundee, Perth and London Shipping Company than any other work he ever did. These ships, linking Scottish east-coast ports and the British capital city, showed publicly that Elder's engines could steam continuously for more than twenty-four hours. Most cross-channel runs were of a much shorter distance and duration, and most coastal passages almost certainly required intermediate calls for coal. David MacIver put up the financial security for his friend Napier who afterwards admitted that he 'lost a good deal of money' by the two vessels in part due to 'the very low prices' he had quoted. But immediate profit and loss was less important than raising public confidence. On the arrival of the *Dundee* in London early in April 1834, the Company's chairman, George Duncan, MP, reported to Napier – himself a shareholder – that the captain was 'highly pleased with her operations' and that all the passengers 'seemed delighted with their voyage and arrangements'. While on the Thames the ship and her complement also 'came in for a severe ordeal of criticism, out of which they emerged triumphant and universally admired. Large numbers of people flocked to see them ... and they became one of the sights of London'.[28]

[26] 'Smith, Thomas Assheton (1776–1858)', ODNB.

[27] Napier (1904: 24); *Nautical Magazine* 8 (1839): 774 ('Abstract of the Parliamentary Report on steam vessel accidents').

[28] Napier (1904: 49–55, 59).

In his answers to the Post Office Commissioners in 1836, Napier cited the *Dundee* and *Perth* as exemplars of steam navigation. They had 'plied regularly from Dundee to London for about eighteen months, and I am informed they have not cost the company eighteen pounds for repairs to the engines or boilers during that period'. They had run summer and winter, departing on fixed days, and laden 'generally with heavy cargoes'. Data from their log-books showed that over a ten-week period in 1834 their average speed was eleven and three-quarter miles per hour. And, he emphasized, 'they still continue to keep up their character for speed, and good sea-vessels'. Duncan, as owner and MP, had already endorsed the quality of the steamers. Not only were they built in a 'first-rate manner ... superior to any other vessels', he told the Commission, '[but] have now run since April 1834, and they have never been an hour off the passage since that time'. As we shall see, the performance of these steamers in speed and economy acted as a standard for the projection of some key ocean mail services in the late 1830s (Chapters 7 and 11).[29]

In 1833 Napier worked in a consultative capacity for a London correspondent, Patrick Wallace, on the feasibility of launching a regular steamship service between Liverpool and New York. His sober evaluation bore many of the hallmarks of David Elder. 'I have not the smallest doubt upon my mind but the speculation in a short time (if well managed) will be one of the most lucrative in the country', Napier told Wallace, 'provided always that the company set out on a proper plan – by having first rate vessels fully fitted for the trade in every department'. 'Fully fitted' meant not merely 'fitted' out, but properly adapted and having the requisite 'fitness' for every task demanded by the ocean.[30] Cost and running expenses of the first two vessels, he emphasized, were as nothing 'compared with having them so efficient as to set all opposition at defiance, and to give entire confidence to the public in all their arrangements and appointments, cost what it may at first; for upon this depends entirely the success, nay, the very existence, of the Company'.[31]

Napier recommended vessels of 'not less than 800 tons, probably more, & propelled with two engines of not less than 150 horse power each, or 300 horse power in whole, in order to ensure good passages in almost any kind of weather'. In addition to design, construction of the vessels and machinery 'should all be of the best materials and workmanship'. He estimated the cost of each vessel – two being proposed initially – at £34,000. He also set forth

[29] PO (1836: 263–6).

[30] Robert Napier to Patrick Wallace, 3 April 1833, DC90/2/4/11, NP. This cautious draft version differs from the more assertive version printed in Napier (1904: 102–13). In addition to this 1833 letter there is evidence in an undated document in NC that Napier had already taken notes of the 'Dimensions &c. of the Steam Ship *Savannah* of Savannah ... from Cleland's *Rise and Progress* of the City of Glasgow. Published in 1820'.

[31] Napier (1904: 104–5).

annual operating costs, including coal consumption based on six round trips at £4,500, insurance at £3,400, upkeep on engines and boilers at £1,000, renewal of sails and rigging at £1,000, wages at £2,240, and victualling at £872. Total costs came to £18,128. Revenue in goods and passengers (at £30 each per single crossing with annual numbers based on statistics – most probably drawn from Cleland's statistical works – for the previous two years from the Clyde to the United States and British North America) he estimated at £29,100 in total, making a 'clean profit per annum' of some £10,972 per vessel, contrasting with the caution with which he had commenced his evaluation.[32]

Crucially, Napier urged Wallace to impress upon the 'minds of his friends the great necessity of using every precaution that can be thought of to guard against accident on such a long passage'. Should accidents happen, however, the ship's company ought to 'be prepared as much as possible with a remedy for every common accident that is at all likely to take place with the engines or boilers'. To this end, he advised having 'duplicates of such parts of the machine as are more likely to go wrong – also to have a good workshop on board with a complete set of tools'. Furthermore, in a move that took as given close ties between the running of the ships and the engine builders, Napier declared that 'all the men connected with the engine department' should be 'regular tradesmen as engineers and boiler makers & the whole of that department Superintended & directed by a Master Engineer (under the Captaincy)'.[33]

Although the project did not materialize as such, the estimates are very similar to those advanced in Junius Smith's later proposal for a system of regular Atlantic steam navigation (Chapter 3). Meanwhile, Napier and Elder continued to build the reputation of the Glasgow firm for accurate and reliable marine steam engine systems of superior economy both in terms of operation and maintenance. Assheton Smith, 'on intimate terms with the Duke of Wellington and other members of the aristocracy', provided a gateway for Napier to the East India Company. One of Smith's aristocratic hunting friends, Lord George Bentinck, was manorial lord of an Ayrshire estate on which stood the Muirkirk Iron Works. In 1834 Bentinck and Napier entered into a gentlemanly correspondence – characterized by mutual trust – concerning Napier's purchase of shares in the rather unpromising enterprise.[34]

Lord George's father, William Cavendish-Bentinck (fourth Duke of Portland), indulged his own aristocratic enthusiasm for fast sailing yachts by constructing vessels at his Troon Harbour shipbuilding yard on the Firth of Clyde to William Symonds' radical designs that combined large beam with unusual sharpness of bow and stern. Symonds' appointment as Admiralty Chief Surveyor (1832–47)

[32] Robert Napier to Patrick Wallace, 3 April 1833, DC90/2/4/11, NP.
[33] Robert Napier to Patrick Wallace, 3 April 1833, DC90/2/4/11, NP.
[34] Lord George Bentinck to Robert Napier, 5 March 1834, in Napier (1904: 96–9).

in place of both traditional dockyard men and School of Naval Architecture graduates owed much to the Duke's patronage and the politics of Whig reform in the early 1830s.[35]

The Duke's brother, Lord William Bentinck, had served as Governor-General of India between 1828 and 1835. Long associated with 'Old Corruption' in the language of its critics, the Honourable East India Company (EIC) began, in an era of Whig governments espousing rhetoric of progress, to reinvent itself as reform-minded, especially under Bentinck's leadership. Modern scholarship, however, challenges the traditional view that middle-class Utilitarians successfully infiltrated, and ultimately replaced with rational government, 'Old Corruption' patronage and nepotism. Instead, aristocratic Whigs such as Charles Wood aimed primarily to prevent patronage falling into the control of middle-class radicals, a perspective consistent with the actions of Whig statesmen in relation to mail contracts (Chapters 5 and 6).[36] Integral to this new age of 'improvement', the traditional East Indiamen, those large sailing ships carrying the valuable freight, mails and passengers to and from India round the Cape of Good Hope, would give way to steamers conveying the mails from Bombay to Suez.[37]

In 1835 Napier received the contract for one of two projected EIC steamers. Assheton Smith, according to Napier's own account, volunteered to become his security for this (and later Government contracts), which involved heavy financial commitments. Mutual trust underpinned the execution of the entire project. '[T]hey have given me my own way with the vessel, trusting to my honour in everything', Napier told George Duncan with reference to the EIC. 'The surveyor has been thrown overboard along with his specification, so that if we do not make a good vessel we will have [only] ourselves to blame'.[38]

Upon completion of the *Berenice*'s outward voyage to Bombay, her commander, Captain Grant, sent Napier a copy of his report to the Indian Navy (the navy of the EIC) with the personal assurance that '[y]our noble ship has behaved well, and beat the [English-built partner] *Atalanta* by eighteen days'. To his superior, Grant reported in detail on all aspects of the voyage and affirmed that the ship was 'at this moment in almost as efficient a state as when we left Falmouth, and perfectly capable of undertaking as long a voyage as that she has just now so satisfactorily finished'.[39]

[35] Lambert (1991: 67–87); Leggett (2015: 26–58).

[36] Cain and Hopkins (2002: 278–84). For more nuanced analysis, see Bayly (1989: 155–60) (Scottish 'improving' landlords); Alborn (1998: 12–14, 21–52).

[37] On the Company's long maritime traditions of trading between London and the East, see Miller et al. (1980: esp. 167–9) relating to the decline of the Company's fleet during the early decades of the nineteenth century.

[38] Robert Napier to George Duncan, 15 May 1835, in Napier (1904: 56–62), on 62.

[39] Captain Grant, IN, to Robert Napier, 24 June 1837, in Napier (1904: 64–7) (including official report to the Superintendent of the Indian Navy).

Receiving this kind of favourable testimony from servants of the new-look EIC marked a dramatic elevation in Napier's authority. And it also brought him into personal and social contact with the chief secretary to the EIC Court of Directors, James C. Melvill. In September 1838, for example, Melvill informed Napier that the Court, 'being desirous of extending to you their liberal consideration, have awarded the sum of £700 as an acknowledgment of the sense which they entertain of your conduct' over the *Berenice*'s performance.[40]

That performance, on the testimony of Napier's son James, owed everything to David Elder, who fitted expansion valves that allowed the steam left in the cylinder at the end of the stroke to remain close to the pressure of the boiler and thus to that of the incoming steam at the beginning of the next stroke. Napier's son believed that the consequent 'cushioning' of the steam, avoiding any abrupt changes of pressure, accounted for the superior performance of the Clyde-built ship compared to the London-built consort.[41]

By the early 1830s, a distinctive arrangement had emerged. Napier (with Elder) negotiated the contracts, constructed and fitted the engines, and subcontracted the building of the wooden hulls to Port Glasgow shipbuilder John Wood. In a letter to Wood in 1841, Napier told his shipbuilder that he had 'uniformly in England and Scotland held you and your work up as a pattern of all that was excellent, and I have never yet had it proved to me that I was mistaken'. Soon after Wood's death, Scott Russell (Chapter 7) – himself indebted to Wood – paid tribute in the highest terms to the Port Glasgow shipbuilder: '[h]e was a consummate artist in shipbuilding, and every line was as studied and beautiful as fine art could make it. John Wood was in fact a pattern shipbuilder'. The character of the man and the character of his work had become morally indistinguishable. Of 'a very modest and retiring disposition', Wood had established a reputation as, quite simply, 'the best builder on the Clyde'. At the 'Disruption' (1843) in the Church of Scotland, Wood followed Chalmers at the creation of the Free Kirk, even participating in the construction of a 'floating church' to serve Highland congregations around the remote Loch Sunart on the West Coast.[42]

[40] James C. Melvill to Robert Napier, 7 September 1838 (transcript), NC.

[41] Napier (1866: 99–101); Smith and Scott (2007: 481–8). Napier noted that Elder was as well acquainted in his own way with the practical advantages of the 'cushioning' of steam as Professor Rankine 'with his theoretical deductions on the same subject' (p. 99).

[42] Napier (1904: 51, 94–5); Russell (1861: 145). 'A pattern shipbuilder' in Wood's case meant that he laid down the 'pattern' for the new ship – perhaps the forerunner of a new class – by deciding on the form – 'the model' – of the hull and thus for the frames and planking. Neighbouring shipbuilders might then, by agreement, follow exactly the same model thus established by copying each frame. The first four Cunard ships were constructed in this manner (Chapter 5). With the rise of a separate profession of naval architects, the term was probably no longer extant. On Wood see 'Wood, John (1788–1860)', *ODNB*. See also Ritchie (1985: 168–9) (on floating churches).

Napier's engine-builder, on the other hand, followed a different spiritual path. With a 'keen interest in theological discussions' due to his 'vigorous intellect' and 'early religious training', David Elder duly forsook his Presbyterian roots to become a member of the Church of England where he found a place for his 'strong and broad religious sympathies'.[43] Two years prior to the Disruption, the Revd Norman Macleod had noted the Anglican Church's attractiveness to '[t]he meek and pious souls who love to eat their bread in peace, and who, weary of the turmoil in our Church [of Scotland], flee to the peace of the Church of England, which seems to reflect the unchangeableness of the Church invisible'.[44] Macleod was later to represent the Elders, father and son, in just such a Christ-like light of peace and humility (Chapter 17).

In 1833 Napier acquired some eighteen acres of land at Shandon near Row on the eastern shores of the Gareloch, and initially constructed a cottage mainly for summer use. Eventually this building yielded to West Shandon House, a turreted and quasi-medieval structure completed around 1852. The fabric consisted of stone brought by way of the Forth and Clyde Canal. The Govan men fashioned the woodwork throughout Napier's newest privileged site. A large picture gallery, with organ built by David Elder, a library and a museum formed part of the ground floor. Everything inside and outside the house spoke of trust and confidence in steam navigation. The mansion overlooked the sheltered and scenic sea loch where many of his, and the Clyde's, new steamers undertook their engine trials and adjusted their compasses before embarking on careers of crossing oceans. In the world of this new merchant prince, guests could wander round the impressive art and book collections or admire the craftsmanship of the shipwrights and joiners displayed in the building itself. And they could gaze across the waters that stretched out to form a veritable nursery for steam shipping.[45]

Evangelical friendships also formed an important component in Napier's personal and business networks, none more so than that of the Melvills. As EIC Secretary with considerable powers of patronage in the metropolis, James Melvill had already acknowledged the quality of Napier's *Berenice*. A Christmas Eve letter from Melvill in 1856 showed the durability and depth of the friendship. Were James Watt alive, he told Napier, 'he would designate my friend Robert Napier as the man who above all others living has now given practical effect to the inventions of Watt and has shown to the world the great efficiency of steam navigation – I in my conscience believe that the best vessels afloat are those with which you have had to deal'. Napier instead chose to honour Henry Bell with generous contributions to conspicuous monuments

[43] [Anon.] (1891: 17).
[44] Macleod (1876: vol. 1, 155–6).
[45] Napier (1904: 56–7, 167–78).

at Dunglass (visible to every passing vessel on the River Clyde) and on Helensburgh seafront.[46]

Napier formed a close friendship with Canon Henry Melvill of St Paul's (James' brother and Principal of the East India College at Haileybury from 1844). When Robert's brother Peter became minister of Glasgow's Blackfriar's Church (known as the College Church) around 1844,[47] Robert sent Henry a copy of Peter's sermons, to which the Canon responded in humorous vein that confirmed a shared distaste for extreme evangelicals: '[t]hey [the sermons] are excellent both in matter and style, quite good enough for Episcopalians; I had almost said too good for Presbyterians. Certainly if the hearers of such sermons object to the preacher they ought to be doomed to some ranting raving fellow who will wear out a red cushion in twenty-four hours'.[48]

Nor was the Cambridge-educated Henry Melvill a stranger to maritime concerns. In a sermon delivered before the Corporation of Trinity House in 1840 – the year of Cunard's first transatlantic steamers – the Canon preached on the theme of 'Christianity [as] the guardian of human life'. By acting on Christian principles the Corporation had become pre-eminently the guardian of lives at sea and an illustration of 'the truth that Christianity is a life giving thing' designed to throw 'fresh ardency into the conflict with death'. Most of all, pilots, which the aptly named Trinity House 'authorise ... as guardians of property and life', demanded 'incessant attention' to prevent the admission of those 'whose unworthiness might have been known'. Thus 'the pilot who cannot steer the labouring ship, like the pastor who cannot guide the wandering soul, is risking men's eternity; the one may cut off opportunities of repentance, as the other may fail to impress its necessity; both, therefore, may work an everlasting injury'. Rising to the occasion, Melvill held out a vision of the day of judgement and the end of all things: 'Then shall many a noble ship, freighted with reason, and talent, and glorious and beautiful things, be broken into shreds ... And the only vessels, which shall ride out the storm, shall be those which, having made the Bible their map, and Christ their light, steered boldly for a new world, in place of coasting the old'.[49]

[46] Melvill to Napier, 24 December 1856, DC90/52, NP; Osborne (1995: 247–8); MacLeod (2007: 187–8) (Bell memorials).

[47] Peter had graduated MA from Glasgow College in 1810 and was described in the *Scottish Pulpit* as being 'In private life diffident, and to strangers distant ... the peculiarity of his theology, [is its] catholicity, scriptuality, and practicableness. His manner, though destitute of the winning arts of the orator, is natural and dignified'. See [Anon.] (1876). I thank Anne Scott for supplying this material.

[48] Henry Melvill to Robert Napier, 5 February 1845, in Napier (1904: 163–4).

[49] Melvill (1846: 20–4). The volume includes five such sermons, delivered at two-yearly intervals, between 1838 and 1846.

As a result of his and Elder's work, Napier became 'the great authority on steam navigation'.[50] Napier and Elder's guiding principle was to build confidence into all their steamers. In order to implement this goal, they focused on a threefold process of system building: first, to ensure that the system of engines and boilers was designed and constructed to the highest standards of accuracy and reliability; second, to entrust hull construction only to shipbuilders with a reputation for excellence both in the quality of the work and in the design of the ship; and third, to oversee the practical integration of both engines and hulls to achieve a vessel capable of fulfilling the purposes for which it was intended. The completed vessels subsequently demonstrated their fitness in service, performing with consorts on long coastal and cross-channel steamship lines, and even showing early potential on the new EIC mail service between Suez and Calcutta.

[50] Napier (1904: 138); Smith and Scott (2007: 481–8).

3 'A Swarm of Projectors'

Promises of North Atlantic Steam Navigation

[A]nd we now have a swarm of projectors, much more largely supplied with zeal than knowledge, who, not content with advancing in the march of improvement with that calm deliberation and salutary caution so necessary to ensure a permanently profitable issue for any great undertaking, would rush to their ends without even informing themselves of the means at their disposal, and proceed *per saltum* from a channel trip to the circumnavigation of the globe.

Dionysius Lardner, in the pages of the Whig Edinburgh Review, anonymously indicts the new fashion for launching transatlantic steamship projects[1]

Job's patience is much celebrated but I don't think that he ever undertook ... to establish a steam [ship] company.

Junius Smith reflects in 1836 on the serious difficulties of persuading British and American investors to take up shares in his projected transatlantic steam navigation company[2]

Summary

From the early 1830s projectors of steam navigation began to persuade investors and the wider public of the promise of reliable and frequent links between the Old and the New Worlds. Their proposals claimed that the transatlantic crossing under steam power could now be made as regular as coastal steamship passages such as those between Dundee and London or between Glasgow and Liverpool. Leading the promotional campaigns, New Englander Junius Smith raised his ambitions to a projected fleet of eight large steamships linking New York with London and Liverpool. Increasingly beset by a major rival venture and threatened by financial instability in America, he faced a further challenge from a controversial voice claiming well-grounded scientific authority.

3.1 'Steam is Up and No Mistake': Envisioning Transatlantic Steam

Inspired by the popularity of river and coastal steamboats on both sides of the North Atlantic, visionary enthusiasts for ocean steam navigation are not hard to find in the late 1820s and early 1830s. In November 1832, for example,

[1] [Lardner] (1837: 118).
[2] Pond (1927: 63).

the *American Railroad Journal* carried an anonymous article (written by New England civil engineer Ithiel Town) imbued with an American faith in progress. The author declared that he had recently crossed the Atlantic twice under sail. During the second passage, which took nearly forty days 'much of which was calm weather', he had reflected 'upon the advantages of a Steam-ship'. Fully convinced of its practicality, he asked his readers if 'in this truly enlightened age' we might not anticipate very soon 'a new and most interesting era in the progress of travelling by water'.[3]

Compared to coastal and river steam navigation, this era would 'be of a much higher, more noble, and more astonishing kind in the estimation of the world'. It would involve 'crossing the Atlantic with much greater safety, ease, pleasure, and despatch'. It was an era that would 'divest the present good mode [sailing packet] ... of at least one half of its dangers, its average time for the performance, its privations, sufferings and various other disagreeable circumstances'. And by the saving of time and expense it would at least double the number of passengers that would otherwise cross the Atlantic.[4]

The new age would also 'induce more of the better class of Europeans to visit and to emigrate to our extensive country'. It would 'afford the facility and inducement for more of our countrymen to visit the "Old World", and enjoy, while there, the pleasures of its various agricultural beauties, its improvements, antiquities and classical associations'. These travellers would upon their return 'bring with them not only those improvements in Agriculture, the Arts, Manufactures, and Commerce, but bring also the strongest and most clear conviction of the perfection, as well as the superior justice and equality, of the Constitution and Government of their own country'.[5]

The author anticipated criticism that experiments had already been tried and found wanting in safety and economy. His reply was that 'such ships and such experiments were not *real* ships and experiments'. He instead envisioned a ship of 1,500 to 2,000 tons with up to six independent steam engines, each connected to an iron or copper 'waterwheel' to drive the vessel. Designed principally for passengers rather than freight, the ship's many subdivisions would permit a strength two or three times that of 'what is now deemed a good ship'. He suggested a payload of up to 350 passengers, divided into four classes with corresponding passage money in the range £15–£30. Safety measures included enclosing each boiler with a strong wood-and-iron partition, spare suits of first-class-quality sails adaptable 'to any emergency', and the selection of

[3] Reprinted in Pond (1927: 26–32, esp. 26). Pond points out (p. 24) that the Connecticut-born author, guarding his professional reputation in the event of failure, did not reveal his identity until 1838 after the celebrated Atlantic voyages of the *Sirius* and *Great Western*. An architect and bridge builder, Town had offices in New Haven and New York.

[4] Pond (1927: 26).

[5] Pond (1927: 26–7).

officers from 'men of the best nautical talents and experience in the country'. Furthermore, risks of fire at sea would be reduced by minimizing the quantity of combustible articles on board, protection against piracy would be provided by mounting cannons on deck, and damage from collision with an iceberg or another vessel would be contained by steam-driven pumps.[6]

At this time, fellow New Englander Junius Smith began to project his own scheme for Atlantic steam navigation. Born in 1780 and graduating from Yale College in the class of 1802, at first Smith practised law. His brother-in-law, David Wadsworth, and his firm of merchants, traded from London and it was to the British capital that Smith took passage in 1805. Britain had seized as a war prize a vessel in which Wadsworth's firm held shares, but Smith won substantial damages on behalf of the owners and duly established himself as a city merchant engaged in the transatlantic trade. Following David's death in 1825, Junius worked with David's son Henry under the name of Wadsworth & Smith. The business had particular interests in shipments of manufactured goods to New York from the industrial towns in northern England. In 1827–9 Smith focused his mercantile activities in Liverpool while living across the Mersey in Woodside not far from the Laird family's iron shipbuilding works in Birkenhead.[7]

Smith had been brought up in the Congregational Church in New England. Throughout his correspondence, this puritan inheritance shone through. In 1836, for example, he chastised his nephew for being 'agitated by the violence of temptations' to the extent that 'the God of this world [mammon] has blinded your eyes, and therefore obtained the mastery'. While in London in the 1830s, he worshipped with Revd Melvill's evangelical Church of England congregation (Chapter 2). At his Sydenham residence, Smith would call together the entire household for morning and evening prayers and New Testament readings. In later years, he professed to his nephew that his puritan creed meant that 'I have no idle time, all fully occupied in some way' and that when 'it shall please the Master to call me I hope to be found in some beneficial duties which will extend their beneficial influence to distant ages'. He also wrote of how 'events the most interesting and important are brought about by the Providence of God without our aid'.[8]

In January 1833 Smith landed at Plymouth, England, after a visit to his native land where, he later claimed, his proposals for ocean steam navigation had been declined by several eminent New York merchants 'with the ever memorable characteristic declaration, "Go back to London, and if you form a company

[6] Pond (1927: 28–32).
[7] Pond (1927: 18–22).
[8] Pond (1927: 19 (parents' membership of Congregational Church in Plymouth, CT), 72 (chastising his nephew), 271 (Melvill), 15–16 (household prayer), 253 (puritan values)).

there, and succeed in the enterprise, we will come and join you"'.[9] The winter crossing of the Atlantic took thirty-two days, half of which were spent beating against strong easterlies in the western approaches to the English Channel. 'Any ordinary seagoing steamer would have run it ... in fifteen days with ease', he told his nephew. 'I shall not relinquish this project unless I find it absolutely impracticable'. As exemplar, he looked to an 800-ton steamer (with two 100-hp engines) under construction on the Thames, most probably the 835-ton HMS *Medea* completed at Woolwich Dockyard around 1833. If his own commercial version came into being, he explained, 'it will be experimentally, and if found to answer, another of the class will be added'. Two more steamers with a New York pedigree would complete a line of four.[10]

Smith quickly despatched a detailed letter to the London and Edinburgh Steam Packet Company with the goal of either chartering one or two of its steamers or entering into a partnership. Not one of the many American coastal and river steam packets he had experienced, he asserted, was 'calculated for a sea voyage' despite their size, capacity, comfort and speed. In contrast, modern British coastal passenger steamers possessed just that fitness 'for the high seas' that the American steamers lacked. The London and Edinburgh Steam Packet Company, however, declined his proposal, declaring that 'all their vessels are otherwise appointed'.[11]

Confident that 'if our steamers work well they will run sailing ships off the road' for passengers, specie and fine goods, Smith issued a first prospectus in June 1835 for a 'Union Line of Steam Packets' to be known, more formally, as The London and New York Union Steam Packet Company. 'No part of the world presents so great an opening for the successful employment of Steam Ships as the Line from London to New York', it began. 'The Continent of Europe, the Continent of America and the British Dominions, make this route their common road'. With 50,000 persons landing from Britain in New York in 1834, 'the scope for such an undertaking is sufficiently ample'. Taking Portsmouth as the final port of departure, access from the Continent would be easy and the traditional sailing packet course 'by way of Liverpool will be in some measure superseded by the greater certainty ... of Steam Navigation'.[12]

This prospectus envisioned two British and two US steamers in order to circumvent the navigation laws of the respective nations. These laws effectively prevented goods deemed foreign and colonial being shipped into Britain by non-British ships and goods deemed foreign being shipped into the United States by non-American vessels. Each projected steamship would be 1,000 tons,

[9] Pond (1927: 233).
[10] Pond (1927: 33–5); Banbury (1971: 87) (*Medea*).
[11] Pond (1927: 36–9 (correspondence with the London and Edinburgh Steam Packet Company)).
[12] Pond (1927: 44–9).

cost £35,000 and require engines delivering 300 hp. Four such vessels would make as many passages in a year as eight sailing vessels. The initial capital sought was £100,000. The prospectus set out estimated expenses based on the 'minute calculations of practical gentlemen well acquainted with the trade'. These gentlemen, however, remained unnamed. The expenses per vessel were collated in three categories: annual total wage costs (£1,928); wear and tear (including insurance) estimated at 20 per cent of the capital cost (£7,000); and 4,000 tons of coal for six round voyages at 25 shillings per ton (£5,000), making a total annual cost of almost £14,000.

For each outward passage to New York, the prospectus estimated receipts deriving from 400 tons of goods at £2 per ton (£800), 60 cabin passengers at £30 each (£1,800), 80 second cabin passengers at £20 each (£1,600) and 100 steerage passengers at £10 each (£1,000). Reduced fares for children removed £550 from the receipts. It omitted specie to and from New York but suggested that this freight would compensate for a fall in passenger numbers in winter. The passage time from Portsmouth to New York would average fourteen days but it allowed two months for each round trip, including refitting time in the ports. Receipts for the outward passage would total £4,837. Assuming this figure as a monthly earning for each steamer regardless of outward or inward direction, the monthly gross expenses (adding provisions, port charges and management costs to the crew, wear and tear and fuel expenses) came to £2,120, making a monthly profit of £2,717. The potential investor would have deduced that these estimates appeared to point to a yearly profit from the fleet of around £130,000! And indeed the prospectus 'recommend[ed] the subject as one calculated for a safe, permanent and profitable investment'.[13]

This comfortable paper exercise bore little resemblance to Smith's struggles to launch a joint-stock steamship company. During a hot July, he admitted that 'the patience and labor of forming a company in London is beyond all that you can imagine'. It was the 'worst place in the world to bring out a new thing' even though it was the best place 'when it is done'. One hundred thousand pounds might be just a drop in the ocean to the 'great monied interest of London', but the challenge 'is to overcome the affinity which that drop has for its old birth and to induce it to flow in a new channel. Do that, and it comes in a flood'.[14]

Head of Reid, Irving & Co., 'one of the first [banking and trading] houses in London', Sir John Reid was MP for Dover and a director of the Bank of England. Smith noted that he 'is quite an enthusiast in our favour and if he could possibly have devoted any time to it would have been a director'. He nevertheless gave Smith 'every assurance of support by the weight of his influence by taking shares and by doing all he can ... He says he has no doubt

[13] Pond (1927: 45–9).
[14] Pond (1927: 52).

whatever of the success of the undertaking'.[15] John Irving of the same firm was a leading merchant banker and also an MP with close links to the EIC, especially through his partner Thomas Reid who served for a time as chairman of that powerful institution. Reid, Irving & Co.'s activities included monopolistic schemes to import large quantities of silver to England from Mexico during the Napoleonic Wars and they later played a very significant role in the establishment of the Royal Mail Steam Packet Company (RMSP) with Irving as chairman (Chapters 7–10).[16]

By autumn 1835, Smith told his nephew that he had secured around nine directors. Of these, Isaac Solly was to occupy the chair. '[A]n old stager long known to the public as a government contractor', he was head of the London Dock Company with which Smith was already familiar through his own involvement as agent for the 'Union Line of Sailing Ships' operating between the off-river London Docks and New York. Solly's chairmanship of the Great Birmingham and London Railway Company gave him valuable credibility, not least because railway companies were especially adept at raising large amounts of capital for their joint-stock projects. Smith judged that '[n]o man in England [is] better known, and just the sort of man for a thing of this sort'.[17]

Most of the other directors fell into the category of wealthy gentlemanly capitalists. Allen had recently retired from the EIC, Winn was a Portsmouth banker and Doran was a director of the General Steam Navigation Company (GSNC) who was 'quite familiar with all the details of ship building, &c'. Ruckir, 'a most respectable man and highly esteemed', was a West Indies sugar merchant, Glover was an auditor and partner in London-based American ship-brokers and Hodges was 'a man of ample fortune' worth around half-a-million pounds. Smith, however, asserted that in the choice of directors he had 'shunned all ship builders and engineers so as to stand clear for open competition'. In this decision, he differed starkly from two subsequent rival schemes associated with Brunel and Cunard, both of whose lines had intimate links with their respective builders. And unlike Cunard and his partners, Smith's project lacked powerful patrons within the Government, especially the Admiralty (Chapters 5 and 6). After two meetings of the directors, however, Smith considered that 'a more powerful and respectable direction cannot be found of any company in London'.[18]

Buoyed up by the support he had received in the metropolis but fearful of potential rivals taking advantage of any delay, Smith now made his plan public. It was scarcely coincidence that in October 1835 Brunel and his Bristol

15 Pond (1927: 52–3).
16 Thorne (1986: vol. 4, 285–6). I thank Grayson Ditchfield for this reference.
17 Pond (1927: 54). On the attractiveness of railway companies for investors, see Taylor (2006: 136, 156–7).
18 Pond (1927: 54–5, 59–60). On GSNC, see Palmer (1982: 1–22).

engineering friend Thomas Guppy, engaged in engineering the Great Western Railway (GWR) from London to Bristol, made their initial move to project a line of 'steamers between Bristol and New York. They enlisted Captain Christopher Claxton, RN, 'as a practical nautical man' to investigate the prospects. Satisfied that 'the leading gentlemen connected with the railway and some of the leading men in the city were ready to come forward if a fair case were made out', Claxton wrote, the key projectors undertook a journey 'through all the great steam ports of the empire'. In fact, they 'visited all the principal coasting steam ports, and made passages on most lines' within the British Isles. Accompanying the party was Bristol shipbuilder William Patterson 'in whose abilities ... the utmost confidence could be placed and who was known as a man open to conviction and not prejudiced in favour of either quaint or old-fashioned notions of shipbuilding'.[19]

On 1 January 1836 Claxton submitted his report to 'a committee of gentlemen' formed for the purpose of getting up the steam line to and from New York. It insisted that 'vessels built expressly for their stations, modelled upon scientific principles, and propelled by efficient engines, may be capable of performing long voyages, and may encounter the heaviest gales'. His report also identified 'a vessel of at least 1200 tons' as best suited to 'the purpose of carrying cargo as well as passengers, [to] the most speedy and certain passage, [with] the greatest economy of power, and the best assurance of a profitable return for the capital invested'. Compared with the ten-year average of thirty-six days westward and twenty-four days eastward of sailing packets, Claxton, more cautious than Smith, judged that such a vessel should make the westward passage in less than twenty days and the eastward crossing in about thirteen days.[20]

This looming competition, however, failed to reduce Smith's public and private optimism. His own calculations pointed to 'a profit of 50 per cent per annum, with just half the complement of passengers the ships are entitled to take'. In November 1835, he had reported to Henry that '[s]team is up and no mistake'. 4,300 out of 5,000 shares had been taken up, though probably only partly paid with a £10 deposit. '[T]he plan meets with universal approbation excepting always those whose nautical interests are likely to be materially affected', he wrote. Seeing for himself the attitude of London's entrenched maritime cultures, Smith told his nephew that '[a]ll the old sailing interest of course is against me, because their craft is in danger. They know it is in danger and feel it to be so'.[21]

19 SC (1846: 7); Pond (1927: 97–8); Buchanan (2002: 56–7, 60) (Guppy).
20 Select Committee of the House of Commons (SC) (1846: 7–8).
21 Pond (1927: 52).

He and his friends had meanwhile increased the target capital to £500,000. By early December he reported 'a million sterling subscribed'. The evidence thus suggests that by the close of 1835 Smith had overcome early resistance from London investors and was on track to raise sufficient funds provided New York also rose to the occasion. American enthusiasm, however, for the complementary New York-based line failed to materialize. New Yorkers thought it 'too great in its object to be carried into practice and looked upon it as a vision of a fanciful brain'. Even allowing for the 'little experience you [in America] have had in deep sea navigation compared with the extent to which it is already carried in this country', he still felt disappointment that 'the activity and force of my own countrymen' had not been manifest. 'Few shares are as yet taken by American merchants or those connected with the American trade', he admitted to Henry in November 1835.[22]

The US-based line also faced formidable opposition. Smith warned his nephew that 'those in the trade, ship owners and [those financially] interested in ships, will oppose you'. Henry would thus encounter 'the most determined opposition' if he attempted to do anything in steam, most notably from the Old Line (the celebrated Black Ball Line of transatlantic sailing packets) 'and all connected with them without one single exception'.[23] New York, after all, was the centre of the transatlantic sailing packet system, trusted for the strength of its vessels, the skills of its commanders and the regularity and safety of its services. Fox claims, for example, that despite the year-round frequency of the voyages (twenty ships from New York to Liverpool, twelve to London and sixteen to Le Havre), there were only two disasters in a twenty-year period up to the late 1830s, the worst of which involved the loss of forty-six persons.[24]

Smith, however, refused to bow to the Manhattan elites, insisting that '[t]he public will never submit ... to be forty or fifty days going to New York as many of the packets are, when they can make sure of a passage in fifteen'. He therefore urged his nephew to adopt a fresh strategy, consistent with rising 'democratic' values in the United States, of appealing directly, over the heads of the mercantile and ship-owning houses, to a wide literate public with money to invest as they chose. He instructed Henry to 'prepare the public mind' and to win over the best writers and editors of the papers. 'You must have clever fellows associated with you', he urged. 'You must keep printing, all America reads, and when they have done very likely to be reprinted here and all Europe reads'. But, he cautioned, in maintaining a regular indirect fire, 'be careful that nobody sees you, lest you get shot yourself'.[25]

[22] Pond (1927: 53–60).
[23] Pond (1927: 60, 69).
[24] Fox (2003: 15). He extracts this information from Albion (1938).
[25] Pond (1927: 64–5).

Vested interests were nevertheless not so readily bypassed. Seeking incorporation for the now-named American Steam Navigation Company in spring 1836, Wadsworth & Smith, supported by city merchants and businessmen, presented a memorial to the New York legislature, only to find it returned on the grounds of monopoly. 'I cannot see any monopoly in a steam navigation company any more than in a railroad company', a frustrated Junius wrote. But Smith seemed to accept that the American company was now going nowhere. 'I suppose your steam is all blown off, and enterprise defunct', he suggested to Henry with a rare note of ironic pessimism.[26]

3.2 'One Grand Transatlantic Steam Navigation Company': Junius Smith's Ambitions

Vocal British opposition to Smith's project came far less from sailing ship interests than from a scientific voice raised in sceptical tone against direct steamship links between England and America. In a public lecture delivered in Liverpool towards the end of 1835, the Revd Dionysius Lardner, FRS, warned that a project for direct voyages from Merseyside to New York under steam was 'perfectly chimerical, and they might as well talk of making a voyage from New York or Liverpool to the moon'. The rhetoric, however, concealed his principal agenda. He was instead advocating the revival of a project using Valentia at the south-west tip of his native Ireland as the key departure port fed by a chain of rail and steamer lines to London. Steam vessels of 800 tons and 200 hp would then cross the ocean between Valentia and New York. There is also some evidence that around 1836 the well-established City of Dublin Steam Packet Company (CDSP) (Chapters 4 and 12) was planning a similar project.[27]

Educated at Trinity College Dublin and the recipient of an LLD from Cambridge University, the speaker was in the process of building scientific credibility with wide public audiences. As editor of the *Cabinet Cyclopaedia* series (published between 1830 and 1844 in some 133 volumes, of which he himself wrote seven), as a professor of natural philosophy at University College London, as a popular lecturer on engineering and physical science topics and as an active figure in the British Association for the Advancement of Science (BAAS), Lardner was a significant scientific authority. Indeed, his *The Steam Engine Familiarly Explained and Illustrated* passed through seven editions during his career and was translated into French, German, Italian and Danish.[28]

[26] Pond (1927: 70–1, 76).
[27] *Liverpool Albion*, 14 December 1835 (lecture); Bonsor (1975–80: vol. 1, 47–8); Tyler (1939: 37–8) (CDSP).
[28] 'Lardner, Dionysius (1793–1859)', *ODNB*.

Lardner, however, won himself no friends in Liverpool. His immediate opponent had credibility in most matters of steam navigation. Son and brother of the Merseyside shipbuilders William and John Laird, respectively, Greenock-born Macgregor Laird's authority derived from his personal involvement with a new kind of marine engineering: iron-hulled steamers designed for imperial river projects in Africa and Asia, but also used as experimental ocean-going craft. As a key participant in a famous 1832–4 expedition from Liverpool to the River Niger, Laird co-authored a book-length published account. In passages of high-flown rhetoric, he justified the African venture in terms of a moral duty 'to raise their fellow creatures from their present degraded, denationalised, and demoralised state, nearer to Him in whose image they were created'. Laird proclaimed that we now had 'the power in our own hands, moral, physical and mechanical' to do so. The Bible offered the foundation for the moral power, the adaptation of the Anglo-Saxon race to 'all climates, situations and circumstances' provided the basis for the physical power and 'the immortal Watt' had bequeathed the key to the mechanical power.[29]

Inseparable from moral and physical power, mechanical power was intertwined with missions of Christian regeneration, with bringing light to dark places of cruelty and tyranny, and with 'improvement' and 'progress' through commerce. The Laird-built, 70-foot *Alburkah* (whose name translated as 'blessing'), powered by a 16-hp steam engine, embodied these complementary goals of commerce and moral regeneration as the vessel steamed through the Niger Delta and into the interior of Africa itself. But the Lairds had a prior mission for the iron steamer which 'formed in herself the test of an experiment of the most interesting kind'. That trial was whether or not 'she could withstand the wear and tear of a sea voyage of four thousand miles'. While fitting out in Birkenhead under Laird's supervision, the projectors had been ridiculed by local maritime critics who 'gravely asserted that the working in a sea-way would shake the rivets out of the iron', that 'the heat of a tropical sun would bake alive her unhappy crew as if they were in an oven' and that 'the first tornado she might encounter would hurl its lightnings upon a conductor evidently sent forth to brave its powers'.[30]

On 19 July 1832 the *Alburkah* and two accompanying vessels (one steam and one sail) departed from Liverpool. Reaching Milford Haven, they took on coal and supplies for the deep-sea run to West Africa. In their preparations, they were greatly supported by Captain Edward Chappell (Chapter 7), then agent for His Majesty's Post Office steam packets at the well-sheltered south-west Wales port. Packet stations of this kind, located principally on the

[29] Laird and Oldfield (1837: vol. 1, vi, 397–8). The first part of the work consists of Laird's narrative. See also Headrick (1981: 17).

[30] Headrick (1981: 6–7, 15–16).

Irish Sea and English Channel coasts, had been established between 1821 and 1826 to provide year-round cross-channel mail steamer routes. Some twenty-six Post Office-owned steamers were in use, but losses of almost £155,000 in the period 1832–6 prompted the Government to replace the system from 1837 with Admiralty contracts awarded to private operators.[31]

Departing for the Niger, Laird now seized the opportunity to witness and record the performance of the steamboats. As a result of his observations, Laird devised a set of practices for steam at sea. For vessels powered primarily by steam, he recommended an arrangement of fore-and-aft sails as in a schooner rig 'to steady her in a beam sea'. For a combination of steam and sail, on the other hand, he suggested 'loftier masts with lower yards square' in order to maximize the contribution from wind power, especially with wind abaft the beam. He also noted problems with the paddle boxes 'checking her way materially at every roll the vessel made'. To obviate as much of this retarding action as possible, he therefore advised that the paddlewheels, with floats remaining in place, should be free to rotate.[32]

The performance of the iron-hulled *Alburkah* also confounded the critics on shore. 'In spite of those wise opinions, her rivets are yet firm in their places, as the fact of her not having made a cupful of water sufficiently proves', he reported in his account. As a 'universal conductor' of heat, 'she was always at the same temperature as the water in which she floated; and for the same reason, though lightning might play around her sides, it could never get aboard her'. The *Nautical Magazine*, however, had already noted that no profitable return resulted from the African part of the venture. And as Laird and Oldfield recorded, of the thirty-nine European men who embarked on the two steamers, only nine returned. At least eight local 'Kroomen' also died serving the expedition.[33]

Soon after his return in 1834, an undaunted Laird appeared as an authoritative witness before a major House of Commons Select Committee on steam navigation to India. Questioned about the Niger expedition, he seized every opportunity to claim the superiority of iron over wood for riverboat hulls with respect to health and comfort, freedom from vermin, strength, watertight qualities under all conditions and greater durability. At around 50 per cent of the weight of an equivalent wooden vessel, the iron riverboat also had the inestimable advantage in working shallower waters with the same amount of freight. In fact, he admitted having nothing at all to say in favour of wooden hulls.[34]

[31] Daunton (1985: 154).
[32] Laird and Oldfield (1837: vol. 1, 15–16). The problems and character of sea-going marine engineers from the 1830s are investigated in Hamblin (2011).
[33] [Anon.] (1834: 626–7); Laird and Oldfield (1837: vol. 2, 410–11).
[34] SC (1834: 56–9); Headrick (1981: 23–30, esp. 28–9).

Together with his close ties to the Laird family's iron shipbuilding yard, these experiences gave him the ammunition with which to attack Lardner's 1835 lecture. Signing himself 'Chimera' in an address published in the *Liverpool Albion*, Laird challenged Lardner's size limit. He instead argued for a scaling up of an existing 420-ton, 140-hp cross-channel steamer, the *Leeds*, running from Dublin to Bordeaux for the City of Dublin Company since the late 1820s, to a 1,260-ton, 300-hp Atlantic steamer with a capacity for about 500 tons of coal. The engine power coincided with Smith's projected steamers but with 25 per cent more tonnage. At a consumption of thirty tons per day and a speed of ten knots, Laird reckoned to land his passengers at the other side in less than fifteen days, thus proving 'that it is easier to go from Portsmouth or Liverpool to New York than to the moon'.[35] Laird's reference to Portsmouth clearly suggested his knowledge of the Junius Smith project, and indeed in the course of the next year Laird was appointed Secretary to the Company. The change coincided with Liverpool interests joining the London merchants, the doubling of the planned capital and the raising of the tonnage of each projected steamship from 1,000 to 1,800 tons.[36]

In contrast to Smith's London networks of merchants, bankers and other establishment figures, Laird's Merseyside network had direct experience of steamship construction and operation, and thus added markedly to the credibility of the company. Crucially, Laird was not a lone figure in this regard.[37] Smith reported in May 1836 that they now had in town several gentlemen from Liverpool and Cork 'desirous of uniting with our company and establishing a line of steam ships from Liverpool to New York, as well as London, to touch out and in at Cork'. These gentlemen included Joseph Robinson Pim, James Beale and Paul Twigg, each connected with the St George Steam Packet Company. All of them soon joined Smith's board.[38]

Pim, a Dublin Quaker, and William Laird (Macgregor's father who had established the family's Birkenhead works) had earlier set up a Liverpool and Dublin Steam Navigation Company around 1824. The line challenged Charles Wye Williams' newly founded CDSP, but by 1826 the latter had absorbed the former and assigned seats on the Board to Pim and Laird. Beale, a Cork-based Quaker, was, like Pim, a shareholder in the St George Company running principally to Cork. Beale's family had interests in iron mills and later in iron shipbuilding. These powerful networks of Quaker merchant families not only had

[35] *Liverpool Albion*, 28 December 1835; Harcourt (2006: 27).

[36] Pond (1927: 75, 84–7). It is difficult establish the precise date of Laird's joining the company but his name first appears as Secretary in the new Prospectus issued around January 1837.

[37] Fox (2003: 74) makes this point with respect to Laird alone and also blurs the timing and contexts for Laird's involvement in Smith's project.

[38] Pond (1927: 75–7). Pond prints the 1837 Prospectus (pp. 84–7) that now listed Pim, Beale and Twigg as Directors.

strong interests in Irish Sea shipping, but also in Irish glass-making, railways and banking.[39]

Pim warned Smith 'not to expect the support of merchants in Liverpool connected with the lines of sailing ships to New York'. Smith agreed. At the same time, he took heart from his own conviction that in 'deep sea navigation of steam ships all the world is far behind England ... The government has taken up this thing and [is] extending [it] to the Mediterranean, West Indies, East Indies and in short wherever their navies float'.[40]

The Great Western Steamship Company (GWSS) came into being between January and June 1836 following the Claxton committee's favourable report. Five out of nine directors were also GWR committee members. With a capital of £250,000 readily subscribed, the company had the sternpost of their first vessel erected at Patterson's Bristol yard in July.[41] A month later, however, a second verbal storm broke over Atlantic steam navigation, this time in the very heart of GWR territory when the BAAS met in Bristol. Addressing Section G (mechanical science), Lardner again attempted to show, with quantitative data, that a 1,600-ton steamer would use up all the coal after little over 2,000 miles. Brunel challenged Lardner's data, but the very reporting of the controversy scarcely promoted investor and public confidence in the project.[42]

In the second half of 1836, there were ominous signs of a looming financial crisis in the United States. Smith acknowledged that commercial affairs in America looked 'like a cart in a snow storm, wheels all clogged and no faculty of motion, hard pulling and no corresponding headway'. Publicly unfazed by knowledge that construction of the *Great Western* had already begun, Smith 'proposed to augment the capital to one million in 10,000 shares of £100 each'. The plan was now to have eight 1,600-ton steamships (each with two 225-hp engines) with four running out of London and four out of Liverpool. The first vessel would cost an estimated £60,000. '[I]f successful we shall ultimately combine the great commercial interests of Great Britain in one grand transatlantic steam navigation company', he assured his nephew, 'which will be sufficiently powerful and sufficiently extended to ward off all opposition'. By October 1836 the order for the first steamer's hull had been placed with the Limehouse yard of Curling & Young on the Thames, with the engines to be constructed in Glasgow. Laird would have oversight of the contract at every stage.[43]

[39] Irish (2001: 19–23) (Irish Quaker merchant families and their networks), 107–08 (Beale); Harcourt (2006: 20–7) (Pim, Laird and Williams).

[40] Pond (1927: 65–6, 75–7).

[41] Bonsor (1975–80: vol. 1, 60); Buchanan (2002: 58–9); Tyler (1939: 39); Fox (2003: 70–5).

[42] Morrell and Thackray (1981: 472–4); *The Times*, 27 August 1836.

[43] Pond (1927: 77–8, 79–80).

The new prospectus appeared around January 1837 under the fresh title 'British and American Steam Navigation Company'.[44] By the spring, the American financial crisis had become a panic, sending Wadsworth & Smith into bankruptcy. In June, the Clyde engine-builders for the now-named *British Queen*, Claude Girdwood, also failed. To add to Smith's woes, Lardner published a substantial but anonymous account of his sceptical views in the widely read, reform-minded *Edinburgh Review* in April.

At the Bristol BAAS meeting, Lardner had claimed that one ton of coal per horsepower would propel a vessel up to 1,800 miles. Thus a 400-hp steamer would consume 400 tons in traversing that distance. Joshua Field, of the London engine builders Maudslay Son & Field, who were supplying the machinery for the GWSS vessel, criticized him for using outmoded steamship data. In order to address this shortcoming, however, Lardner had since gained access to log-books from nine Admiralty steamers, one Post Office packet and one private packet for which the horsepower was known. The documents yielded 'the actual quantity of fuel consumed in long intervals of time in which these vessels made numerous voyages'. From the total number of hours steaming at full power and the total consumption of coal in pounds during that time, he calculated the number of pounds of coal consumed per horsepower per hour for each of the eleven steamers. From the log-book data on total distances covered, he also estimated the average speed in nautical miles per hour for each vessel. Finally, he defined what he had named the 'locomotive duty' as the number of miles one ton of coal per horsepower drove a vessel. The coal consumption (converted from pounds to tons) per horsepower per hour drove the vessel the number of nautical miles given by her average speed. Dividing the speed by the coal consumption in tons per hour gave the locomotive duty.[45]

Most of the naval steamers consumed, per horsepower per hour, around ten pounds of coal, while Post Office and private packets showed a higher figure of about twelve pounds. One vessel stood out above the others. HMS *Medea* was an armed frigate designed by Oliver Lang (master shipwright at Woolwich Dockyard) and fitted with a 200-hp Maudslay, Son & Field steam engine. She had been constructed 'with the avowed object of imparting to her every advantage, nautical and mechanical, which the art of navigation by steam can supply'. Her model resulted from 'careful and anxious consideration; her machinery the very best which the most consummate skill and mechanical talent in this country can supply'. Costly and complicated feathering paddlewheels, designed after various experimental trials on Admiralty steamers,

[44] The 'Prospectus of the British and American Steam Navigation Company' survives in the Napier Collection GMT (NC). See also Napier (1904: 114). Bonsor (1975–80: vol. 1, 54) states the capital as £500,000.

[45] [Lardner] (1837: 119–20, 129–32).

yielded 'an increased locomotive duty in a higher ratio than that of four to five' compared to ordinary wheels with fixed paddle-boards.[46]

HMS *Medea* consumed only 8.3 pounds of coal per horsepower per hour. Calculating her locomotive duty at 2,105 nautical miles, Lardner showed that not one of the other ten steamers in his list even approached 2,000 miles. One barely exceeded 1,000 miles. Indeed, Lardner deemed Napier's crack *Dundee* and *Perth*, plying one of the longest routes between Dundee and London at almost ten knots, to consume above ten pounds per horsepower per hour. Lardner therefore took the *Medea* as his 'modulus' or standard. The key problem was now to apply that standard to the controversial question of a 'permanent and profitable line of steam communication across the Atlantic'.[47]

The earlier parts of Lardner's review identified where mechanical power was either wasted or acted to retard a paddle vessel. In contrast to the efficient working of a river or lake steamer in smooth water, he explained, '[a] vessel in the open ocean is at all times liable to the action of waves of greater or less height, and will pitch or roll so that the wheels will be immersed to varying depths ... A waste of power must ensue, and will be greater according to the increased depth of immersion of the wheels'. Ordinary paddlewheels, acting like undershot waterwheels, had fixed paddle-boards, each of whose inner edge was always positioned towards the shaft of the wheel. These wheels had the advantages of strength and simplicity, together with the merit of being generally repairable at sea. Feathering paddlewheels had boards arranged to move independently of the wheel's framing in such a way as to enter and leave the water with minimum expenditure of power. They would thus shed most of the water that simple boards would tend to lift above the surface of the sea or lake. Less strong and more complex than their simple counterparts, these wheels suffered from greater likelihood of damage and, concomitantly, were more difficult or even impossible to repair at sea. Lardner thus pointed to the paradoxical conclusion that 'it is precisely in the long sea voyages that the common wheel is most injurious and the feathering wheel most necessary'. Nevertheless, he attributed the *Medea*'s superior performance in large measure to the use of feathering paddlewheels.[48]

Other internal problems reduced performance. First, the lack of fresh water for the boilers usually meant a resort to salt water. The practice of 'blowing out', whereby the boiler water with an increasing concentration of salts was allowed to escape into the sea, prevented the boiler water reaching saturation point. This process, however, only slowed down the encrustation of salts on the iron surfaces, leading to a reduction in the conduction of heat, and thus to

[46] [Lardner] (1837: 140).
[47] [Lardner] (1837: 120, 133, 136–7).
[48] [Lardner] (1837: 121–2, 133).

waste both of fuel and of the iron. Lardner acknowledged the recent introduction of condensers designed to distil freshwater from seawater, but admitted that the method was still 'upon its trial'. Second, the soot produced from the fuel similarly tended to deposit a hard crust on internal boiler passages and flues, impeding the conduction of heat and again wasting fuel. On long voyages, this process required stopping the engines and putting out the fires at sea. Only when the boiler had cooled could men enter and clean the flues. The whole stoppage was likely to exceed twenty-four hours. The additional coal needed to restore steam to the engines, Lardner claimed, 'would generally be sufficient to propel the vessel a distance of about fifteen miles'.[49]

Lardner also addressed external conditions on the Atlantic, principally the prevailing westerlies accompanied by a heavy sea with long swell, quite different from the short, steep waves of land-locked seas. In his quest for 'some degree of numerical precision', he looked to the '*Liners*', that is, the packet-ships which were '[s]ome of the finest and most efficient sailing vessels which have ever crossed the ocean'. Using data on the performance of the 'Liverpool liners' over the previous three years, Lardner arrived at an average year-round westbound time of thirty-five days and seventeen hours, an average eastbound time of nineteen days and seven hours, and an overall average of twenty-seven and a half days. Consequently, an average westward passage under sail took almost 25 per cent longer than the average of the two passages.[50]

Lardner's aim was now to reason by analogies between the known performances in seas already navigated by the most powerful and efficient steamers (such as *Medea*) and the unknown performances of projected Atlantic steamers under conditions of wind and wave similar to that encountered by the 'liners'. He invited readers to consider the most favourable instance in which the *Medea*'s known performance could be transferred to the case of the Atlantic, for which no practical data existed. He first assumed for the steamer the same 25 per cent saving in time for the eastbound crossing over the average for the round voyage for the 'liners'. With a known locomotive duty of 2,000 miles for overall average performance, the westbound locomotive duty would reduce to 1,500 miles. The Atlantic crossing of 3,200 miles between Liverpool and New York would then require 2.1 tons of coal per horsepower. Allowing 12 per cent extra as spare fuel, the steamer would have to carry 2.4 tons per horsepower, equivalent to some 530 tons for the 220-hp engine. With a total load capacity of 800 tons, of which 220 tons represented engines and boiler space, there was almost nothing left for paying passengers, still less freight!

On the same basis, that is, in the most favourable, but in practice impossible, conditions, Lardner showed that a 1,200-ton steamer would have sufficient fuel

[49] [Lardner] (1837: 122–6).
[50] [Lardner] (1837: 134–7).

only for two-thirds of the passage. He therefore concluded that 'in the present state of steam navigation ... the vessel which performs it, whatever may be her power and tonnage, must be capable of extracting from coals a greater mechanical virtue, in the proportion of three to two, than can be obtained from them by the combined nautical and mechanical skill of Mr Lang ... and Messrs Maudslay and Field'.[51]

Lardner's review riled Smith in the extreme. He tried desperately to persuade Henry that 'Dr Lardner is a perfect Quack ... He knows nothing upon the subject but what he picks up here and there like an old ragman'. Actual experience since then, he claimed, disproved the facts and basis of Lardner's arguments. One of the two new EIC steamers, *Atalanta*, had steamed 2,400 miles without refuelling and thus exceeded Lardner's latest limit of 2,200 miles. The other, Napier's *Berenice*, had run from England to Fayal (Azores) under sail alone in the fast time of eight days, demonstrating that steamships could sail just as well as any sailing ship. 'The article in the *Review* was begun, continued and ended with a view to opposing the British & American Steam Company', he raged. 'It won't do'.[52]

In the same vein, Lardner addressed the Liverpool BAAS meeting of 1837. This time, however, steamship interests closed ranks for a seemingly co-ordinated counter-attack. The first, humorous salvo came from a Mr Rotch in whose experience 'government data were the very worst, and as a practical man, he would say that government steamers were the veriest tubs afloat'. Among private steamers 'of the latest and best construction, up went his maximum [locomotive duty], and the London and Dundee steamers went far beyond it'.[53]

Thomas Guppy fired a complementary salvo. He accepted that the *Medea* was 'a fine ship, sailed and steamed well', but reminded his audience that during the past five years 'science had not stood still, either in practical ship-building or practical engineering'. Like Smith, he held up the published performances of the EIC steamers. But Lardner responded plausibly that calculations for both vessels had not been made for steam alone, but rather under conditions when both sail and steam had been in operation. He also stated that he had not been able to obtain the fuel consumption of the Dundee steamers from Napier. In the absence, therefore, of a 'more favourable modulus', he upheld his choice of the *Medea* 'because she afforded the greatest average – most favourable to the [Atlantic] project'.[54]

This clash of science with commercial projection came just months before the first major practical demonstration of Atlantic steam navigation in the

51 [Lardner] (1837: 138–40.

52 Pond (1927: 94–5).

53 [Anon.] (1837a: 692–5) (reprinting the report from the *Liverpool Standard*).

54 [Anon.] (1837a: 695–9). Logs, and discussions of the performances, of the East India steamers were published in [Anon.] (1837b, 1838).

spring and summer of 1838. The opposing authorities had as yet no certain knowledge of the outcomes of that demonstration, still less of those voyages that might come after. If the projects failed decisively, Lardner's views would emerge to show the triumph of scientific judgement over practical ignorance. If the crossings were deemed successful, his opinions would be relegated to the waste bin of maritime history and his reputation forever tarnished with the brush of reactionary prejudice against the great tide of Victorian engineering progress. It is therefore to the fashioning of those demonstrations that we now turn.

4 'This Noble Vessel'
Realizing the Promises of Transatlantic Steam

> The advantages will be incalculable; no more petty rivalries, or national antipathies; no odious misconstructions and paltry jealousies, but a mutual love and respect growing out of an accurate knowledge of one another's good qualities, and a generous emulation in the onward march of mind, genius, enterprise, and energy, towards the perfectibility of man, and the amelioration of our physical, social, moral, and commercial condition. Such are among the prominent features of the bright and exhilarating vision brought into birth by this most auspicious event, and by which the minds of our fellow citizens have been so excited. They are founded in fact, and have nothing Eutopian [sic] about them, and are as deducible from positive data, as any demonstration in ... [Bacon's] *Novum Organum*, or any solution in ... [Laplace's] *Mecanique Celeste.*
>
> *James Gordon Bennett interprets the historic significance of the Sirius' arrival in New York in 1838*[1]

Summary

'The light of science and revelations of truth, blending their rays, and beaming upon barbarism, will soften down its character, and hasten the advent of more glorious times', Junius Smith wrote to Benjamin Silliman in the aftermath of the first westbound Atlantic crossings by steamships.[2] He aimed his high-flown rhetoric at those, like Lardner, who would use their science to undermine the ambitions and pretensions of transatlantic schemes. In varying degrees and in different ways, however, Smith and his rival projectors faced the daunting and simultaneous challenges of engineering the steamship systems, of raising the capital amid the economic uncertainties of the time and of negotiating the sensitivities of political interests in Britain, in the North American territories and in the United States. Realizing the promises of transatlantic steam thus carried no guarantees and no self-evident historical destinations.

[1] [Pray] (1855: 235).
[2] Smith (1839: 162–7) (letter dated 5 September 1838); Pond (1927: 147).

4.1 'No Longer an Experiment': Rendering the Atlantic Crossing a Fact

Three years before Smith and Brunel launched their respective transatlantic steamship projects, Robert Napier had investigated the challenges of steering boldly by steam 'for a new world, in place of coasting the old' (Chapter 2). Having considered, but rejected, the option of constructing a suitable steamer on speculation to his own account, Napier met with a golden opportunity in the mid-1830s to fulfil his manifesto of transatlantic promise. As secretary of the new British and American Steam Navigation Company, Macgregor Laird, with whom Napier was already well acquainted, negotiated with the engineer for building the machinery of the line's first steamer.

Napier quoted £50 per nominal horsepower (hp) and offered to oversee the hull construction for a fee of £1,000. By October 1836 he put in a tender of £20,000 for the 400-hp engines, including, against his better judgement, Samuel Hall's patent surface condensers designed to eliminate salt accumulation that otherwise had to be removed manually from the boilers every few days (Chapter 3). Smith's Company, however, rejected Napier's tender in favour of a lower price quoted by Clyde engine builders Claude Girdwood and Company. Their subsequent bankruptcy in June 1837 allowed Napier to take over the contract at an increased price.[3]

Elder and Napier issued special instructions to the London builders of the hull to ensure the strength of the engine beds. A high-quality yard, Curling Young and Company, had built East Indiamen before turning to steam in the mid-1830s and building, to Admiralty specifications, the paddle-steamer *Iberia* for mail service to Portugal and Spain for the Peninsular Steam Navigation Company (Chapter 12).[4]

As projector of the 'steam frigate' provisionally named *Royal Victoria*, Smith intended the ship to be the largest and 'the most splendid steam ship ever built'. By late April 1837 he noted that the spectacle of the wooden hull, rising on the north bank of the Thames at Limehouse, 'now excites a good deal of public curiosity. A vast many people visit the yard to see what has not been seen before upon the same scale'. Two months on, he told his nephew that the *Victoria* 'is now in frame and the finest frame I ever saw. The best timber and the best workmanship'. He also gleefully reported that the 'nobility are beginning to cast their eyes upon her'. He had recently met the Marquis of Blandford, eldest son of the Duke of Marlborough and heir to Blenheim, 'one of the most magnificent palaces in Europe'. This Churchill family member,

[3] Napier (1904: 114–15). See also Fox (2003: 71–6).
[4] Banbury (1971: 162–3).

Smith insisted with satisfaction, had been 'much delighted with the ship and called her the St Paul's of naval architecture'.[5]

By the spring of 1838, however, the new steamer had been on the stocks for well over a year. The mounting delays, coupled with the imminent completion of the rival *Great Western*, induced Smith to charter the coastal steamer *Sirius* from the St George Steam Packet Company, three of whose directors (Pim, Beale and Twigg) were also on Smith's board (Chapter 3). Built by Robert Menzies of Leith on Scotland's east coast and fitted with side-lever engines by Glasgow's Thomas Wingate, the 703-ton *Sirius*, designed for a Cork to London service, was similar in size and layout to Napier's celebrated Dundee to London and Liverpool to Glasgow steamers. The ship was, moreover, only slightly smaller than the projected vessels in Napier's 1833 transatlantic plan for Wallace (Chapter 2).[6] Napier himself appears to have subscribed financially to the Atlantic trial of the *Sirius* in part because of delays to the *Victoria*, but also because the voyage would yield important data for use in the refinement of his and Elder's marine steam engines. Commanded by the experienced Lt Richard Roberts, RN, the *Sirius* was a substantial vessel of a type well-suited to handling the Atlantic gales sweeping the south-western approaches between Cork and the English Channel.[7]

Brunel's *Great Western*, however, looked invincible as the steamer prepared to depart from the Thames. Hailed by the *Athenaeum* as 'almost a living thing' whose 'huge whale-like sides' heaved with impatience, the 'paddles instinctively dash into the water, as a war-horse, when he hears the trumpet, paws the ground'.[8] The ship's prestigious London engine builders, Maudslay, Son and Field of Lambeth, however, had altered the planned layout to place the boilers and funnels forward, rather than abaft, the engines and paddlewheels. Writing to Scott Russell in November 1840, the builder explained that the iron ballast added in the after part of the hull 'was not for the purpose of correcting any deficiency in her stability, but for the purpose of correcting her trim which had been greatly deranged by Messrs Maudslay's new plan of putting the boiler before the engines'.[9]

As the steamer departed the river at the end of March 1838, fire broke out in the region of the funnel, filling the engine room and forward boiler room with smoke. During efforts to extinguish the blaze, Brunel suffered a serious fall

[5] Pond (1927: 75–7, 90–3).
[6] Tyler (1939: 46); Bonsor (1975–80: vol. 1, 54–5, 59).
[7] Pond (1927: 132). Fox (2003: 76–80) makes the unwarranted claim that the *Sirius'* crossing was 'just a heedless, dangerous publicity stunt, a desperate gambit by sore losers'. See also Burgess (2016: 36).
[8] *Athenaeum*, 31 March 1838, quoted in Tyler (1939: 48). Griffiths et al. (1999: 20–6, 90–104) give a very thorough account of the genesis and career of the *Great Western*.
[9] RMSP Minutes (26 November 1840); Nicol (2001: vol. 1, 15–16).

when the rungs of the stoke-hole ladder gave way. Confidence in the new vessel dipped. Several angry stokers deserted and fifty prospective passengers cancelled, leaving only seven or eight for the Atlantic crossing.[10] Napier's warning to ocean steamship projectors in 1833, that accidents and failures could rapidly erode trust, had been justified (Chapter 2).

One of the remaining passengers, a citizen of Philadelphia, boarded the Bristol steamer from the tender out on the River Avon on a day of increasing wind and rain. He found 'spars, boards, boxes, barrels, sails, cordage ... stirred well together ... Captain scolding, mates bawling, men growling, and passengers in the midst of all, in the way of everything and everybody ...'". But following departure, confidence rose: '[s]uch stability, such power, such provision against every probable or barely possible contingency, and such order presented itself everywhere on board, as was sufficient to allay all fear'. The optimism dissipated as the steamer encountered the force of the Atlantic to the 'dance of pot-hooks and frying pans [that] was nothing in din to the glorious clatter among the moveables that accompanied it'.[11] The *Sirius*, too, had no easy passage. Evoking images of discomfort and fear, the log's terse entries noted in turn 'heavy head-seas', 'vessel labouring heavily', 'shipped a great deal of water', and 'stopped engine three-quarters of an hour to fasten screws'.[12]

In April 1838 the *Sirius* arrived first after a crossing from Cork lasting some eighteen days (Figure 4.1). The press immediately seized on the event. 'The *Sirius!* The *Sirius!* The *Sirius!* Nothing is talked of in New York but about the *Sirius*. She is the first steam vessel that has arrived here from England, and a glorious boat she is ...', claimed the *New York Weekly Herald*. The *New York Express* spoke of how the arrival was 'universally considered the beginning of a new era in the history of Atlantic navigation' witnessed by thousands of people who gathered to see the *Sirius* anchored off the Castle on a fine clear afternoon. Then, fifteen days out from Bristol, the runner-up approached. '[L]ong before her hull was visible', indeed, the *Great Western*'s smoke had been sighted on the horizon 'ascending in black volumes'. On entering port, the *Great Western* received a twenty-six gun salute from the fort, sailed around the *Sirius*, and steamed up the East River to anchor near Pike Street.[13]

The *Express* also observed how this 'successful experiment of steampackets between New York and England gave life and joy to all'. Indeed, the *Great Western* had brought England nearer to New Yorkers 'than many parts of our

[10] Tyler (1939: 49–50); Bonsor (1975–80: vol. 1, 61); Griffiths et al. (1999: 25–6); Burgess (2016: 36–7). See also Buchanan (2002: 58–9); Cantrell (2002: 18–38) (Henry Maudslay); Ince (2002: 166–84) (Maudslay, Son and Field).

[11] Quoted in Tyler (1939: 50–1).

[12] Pond (1927: 102–03).

[13] Pond (1927: 104–6).

Figure 4.1. The *Sirius*, first westbound steamer into New York (1838)

own country' and had shown that 'steam navigation across the Atlantic is no longer an experiment, but a plain matter of fact. The thing has been done triumphantly'. A passenger aboard the vessel echoed the sentiment at the conclusion of the voyage: '[e]xperiment then ceased; certainty was attained; our voyage was accomplished'. Scientific trial, it seemed, had as good as guaranteed the future commercial success of ocean steam navigation over sail. The celebrated statesman of the New World, Daniel Webster, put the matter more imaginatively. 'It is our fortune to live in a new epoch', he proclaimed after a feast aboard the *Great Western* hosted by Captain Hosken and the company agent. 'We behold two continents approaching one another. The skill of your countrymen, sir, and my countrymen, is annihilating space'.[14]

On the other hand, there were frequent reminders of the fragility and lethal power of the new steam navigation systems. Scalded by steam the day after the *Great Western* had arrived, chief engineer George Pearne died of his injuries a few days later. And from Cincinnati came the distressing news that some 125 persons had been killed by a boiler explosion aboard a river steamboat taking part in a race.[15] Against such uncomfortable reports, congratulatory talk of progress and power could sound like empty rhetoric.

[14] Pond (1927: 106–11); Burgess (2016: esp. 7–9) (Webster's remarks).
[15] Pond (1927: 113–14).

One ambitious newspaper editor, James Gordon Bennett of the *Herald*, took a profoundly personal as well as commercial interest in witnessing the arrival of transatlantic steam. Scottish-born in 1800 and named after a local Presbyterian minister, the Revd James Gordon, he had been brought up as a Catholic near the Moray Firth fishing port of Banff. 'I was educated a Catholic, in the midst of a Protestant community', he later wrote, '... yet both Catholic and Protestant breathed the moral atmosphere of the Scriptures'. Around 1818 he visited Glasgow. On a Saturday he went down to the Broomielaw. 'The Clyde was transparent like a mirror', he wrote, 'and here I first saw a steam-boat, and could hardly believe my own eyes'. On the Sunday he took up an invitation to hear Chalmers. 'What deep, absorbing eloquence fell from the lips of that excellent man!' he exclaimed. On Monday evening he attended the theatre which he found dull, empty and gloomy. 'I far preferred the kirk to the theatre, and Dr Chalmers sank deeper into my mind than any player there', he confessed. As a consequence of this experience, by the age of eighteen he had replaced 'all the ridiculous superstitions of the church of Rome' with a resolve to read nature and the scriptures for himself. Twenty years after he had first seen a steamship, he proclaimed the advent of the *Sirius* and *Great Western* as embodying a watershed between the old epoch of dogmatic authority and the new epoch of human progress towards perfection (see Epigraph).[16]

No mere armchair enthusiast for steam, Bennett took passage on the *Sirius* on 1 May as the vessel returned from New York to Falmouth. On board were twenty-eight first-class and twenty-one second-class passengers, an indication of modest confidence in transatlantic steam. 'All went merrily the first week', recorded one of the passengers, the Chevalier Wickoff. 'Then stormy weather set in ... She was dreadfully knocked about, but was staunch and steadfast in the worst gales'. With coal nearly exhausted, the vessel had to steam at half speed and later to burn 'whatever could be spared'. Dense fog in the western approaches only just cleared in time to reveal, dead ahead, the Isles of Scilly, notorious for their toll of unsuspecting vessels.[17] After a respectable passage of eighteen days, the *Sirius* arrived at Falmouth from where Bennett travelled overland to deliver to London's editors and publishers the first news of the successful voyages across the Atlantic.[18]

While in London, he witnessed the choreographed launch of the *British Queen* on Her Majesty's birthday. 'There was a very numerous assembly of visitors, consisting of fashionables from the west end of the town and persons connected with the shipping and commercial interests of the city', *The Times* enthused. Its correspondent (who had doubtless been well briefed and fêted by

[16] [Pray] (1855: 221, 231–5, 276–7).
[17] Pond (1927: 115).
[18] [Pray] (1855: 236–7); Pond (1927: 115–16); Tyler (1939: 61).

the Company) reported that every manner of craft from larger vessels decorated with flags and streamers down to skiffs and wherries crowded the Thames. The paper named some five peers (including the First Lord of the Admiralty the Earl of Minto), five knights (including former Governor of Canada Sir F. B. Head and City merchant Sir John Reid), the American and Dutch ambassadors and two admirals. There were also present several unnamed MPs. Bennett seized this special opportunity to meet some of the dignitaries present, including radical Whig MP, novelist and poet Sir Edward Bulwer Lytton.[19]

At two o'clock in the afternoon, 'the signal to strike away the last shore was given by the firing of a pistol; the usual ceremony of breaking a bottle of wine on the bows of the vessel immediately followed, and the superb ship slid gradually and majestically out of the dock[yard] into the river, amidst the most enthusiastic and deafening cheers of the assembled thousands'. As the ship turned broadside in the river, the *British Queen*, thirty-five feet longer than any vessel in the Royal Navy, took centre stage: '[s]he actually seemed a floating Colossus, and every other vessel appeared diminutive in her presence', *The Times* continued. No serious accident occurred, it reported with satisfaction. And the only minor calamities arose when the spring showers 'wetted a few of the ladies bonnets, and an adventurous dandy or two was here and there exposed to the pelting of the elements'.[20] At long last, Smith's project appeared in the material form of a large wooden hull afloat on the Thames, but still far from fulfilling the purposes for which it was intended.

4.2 'With the Speed of a Hunter': Lobbying the Whigs for Patronage

Twenty days out of Halifax, the Admiralty mail brig *Tyrian* stood becalmed in light spring airs. Aboard was the Nova Scotian newspaper owner, orator and political leader Joseph Howe and the lawyer and satirist Judge Thomas Chandler Haliburton (see Introduction). Howe described the scene as the *Sirius*, black-hulled with red paddlewheels and white sails, closed 'with the speed of a hunter, while we were moving with the rapidity of an ox-cart loaded with marsh mud'.[21] He and Haliburton accompanied the mails as they were transferred by ship's boat from brig to steamer. Aboard the *Sirius*, they drank a glass of champagne with Captain Roberts and, doubtless, Bennett. Much impressed by the steamer's performance, they discussed the possibilities of Atlantic steam navigation for British North America during the remaining

[19] *The Times*, 25 May 1838; [Pray] (1855: 237).
[20] *The Times*, 25 May 1838.
[21] *Novascotian*, 12 July 1838, quoted in Tyler (1939: 71). Howe published a full account of the voyage as 'The *Nova Scotian* afloat' in his paper between 5 July and 8 August 1838. See Roy (1935: 62).

voyage aboard the brig. Keenly aware of the likelihood of future transatlantic mails passing through New York rather than Halifax, they agreed to 'bestir themselves, and not allow, without a struggle, British mails and British passengers, thus to be taken past their very doors'.[22]

On landing at Falmouth on 20 May, Howe and his *Nova Scotian* companions toured Cornwall and Devon before arriving in Bristol one week later. There they endeavoured to persuade the GWSS to operate to Halifax provided that the British Government offered a mail subsidy. On 29 May Howe reached London where the following day he dined with Halifax merchant and ship-owner Samuel Cunard (Chapter 5).[23] Cunard was at the end of an annual business visit to Britain and already had extensive experience of mail contracts out of Halifax by sailing vessels. He had also held a financial stake in a small wooden paddle-steamer, *Royal William*, which had crossed the Atlantic during the late summer of 1833 in a bid to find a European buyer for the loss-making vessel.[24] Howe and Cunard clearly had a shared interest in promoting a regular transatlantic mail and passenger system between Britain and its North American territories.

Howe further conferred with fellow countrymen William Crane and Henry Bliss. The latter also had strong Halifax family connections but was now Agent for New Brunswick. He had already twice petitioned the British Postmaster General, the Earl of Lichfield, for a weekly Atlantic mail steamship service. The Government's response to the first letter, coming as it did before the *Sirius* and *Great Western* crossing, was that it 'wished to await the result of the steam experiment then in progress'. The second letter arrived soon after the 'steam experiment'. Lichfield now doubted that the small number of voyages undertaken thus far scarcely supplied justification for the large expense proposed for a regular mail service. The Lords of the Admiralty also preferred to await the results of winter trials.[25]

If Viscount Melbourne's Whig Government as a whole appeared hesitant about the Halifax project, one rising member of the administration took a very different view. Francis Thornhill Baring, grandson of the founder of Baring Brothers' celebrated banking house, had been born in Calcutta as the son of an EIC servant. He was, therefore, thoroughly familiar from an early age with the

[22] Chittick (1924: 219); Tyler (1939: 71); Chisholm (1909: vol. 1, 187–8); Pond (1927: 116–18).

[23] Roy (1935: 65–73). Roy's scholarly account relies on his having deciphered Howe's personal diary. Fry (1896: 61) writes that 'Cunard, on hearing of the efforts of Howe and Haliburton, met them in London with his own proposals'. Fry had personal knowledge of Atlantic steamships and their owners. He had also witnessed the 'launches' of the *Great Western* and *Great Britain* (p. vii).

Grant (1967: 78–88), provides a very detailed but unattributed version of Howe's meeting with Cunard. Langley (2006: 67–9), largely repeats previous undocumented accounts.

[24] Bonsor (1975–80: vol. 1, 51–2) (*Royal William*). This steamer should not be confused with the CDSP's later *Royal William* (1838).

[25] Tyler (1939: 76).

politics of empire. Although a lifelong and committed Whig, he had supported George Canning's liberal Tory ministry in the mid-1820s over the premier's foreign policy that particularly welcomed the independence from Spanish rule of South American territories as an opportunity for British mercantile influence (Chapter 11). Now Baring formed part of an inner circle of Whig leaders in the 1830s. Serving as Financial Secretary to the Treasury between 1835 and 1839, he was in a strong position to exert considerable power even before he was elevated to the post of Chancellor of the Exchequer in August 1839.[26] And that power manifested itself in the form of state support for private mail steamship lines serving the empire. 'Mr Bliss's letter is theatrical and exaggerated', he wrote on Henry Bliss' missive, 'but the importance of the case itself is very great. ... I think we should take immediate steps to enable ... boats to start [monthly] for Halifax in Spring ...'.[27]

In late August 1838 Howe and Crane lobbied the affable Lord Glenelg, Colonial Secretary and evangelical Presbyterian, for the introduction of an Atlantic mail system to Halifax. Their petition carried Howe's authority as eyewitness to the mid-Atlantic encounter between sailing brig and steamer. It drew attention 'to the inhabitants of those Provinces [Nova Scotia and New Brunswick], to their more intimate connection with the mother country and to their peace and security as dependencies of the Crown'. What it deemed 'the successful voyages' had 'solved the problem of the practicability of steam navigation across the Atlantic'. And it highlighted the 'fearful destruction of life and property [and] the serious interruption of correspondence, consequent upon the loss of so many of the ten-gun brigs' engaged in the monthly Admiralty mail service between Halifax and Falmouth. This last point alone, the letter argued, 'ought at once to determine the Government ... to replace them by a very superior description of vessels'.[28]

The appeal to empire formed the core of the letter. Its message to Government has to be seen against an immediate context of political unrest in British North America, anxieties within Government about the effectiveness of its political system for control of the colonies and specific attempts to strengthen the overland mail services between Halifax and Quebec, Montreal and Toronto:

If Great Britain is to maintain her footing upon the North American Continent – if she is to hold the command of the extensive sea coast from Maine to Labrador, skirting millions of square miles of fertile lands, intersected by navigable rivers, indented by the best harbours in the world, containing now a million and a half of people and capable of supporting many millions, of whose aid in war and consumption in peace she is

[26] 'Baring, Francis Thornhill (1796–1866)', *ODNB*.
[27] Tyler (1939: 72–7) (quoting Baring on p. 76).
[28] Chisholm (1909: vol. 1, 188–9).

secure – she must, at any hazard of even increased expenditure for a time, establish such a line of rapid communication by steam.

Such a question was thus not 'a mere pecuniary' one, for upon its satisfactory solution depended 'the speedy transmission of public despatches, commercial correspondence and general information, through channels exclusively British, and second to none in security and expedition'. It was, in short, a measure 'which cannot fail to strengthen and increase the prosperity of the empire'.[29]

The GWSS was also engaged in a quest for state patronage based on its recent track record. With sixty-six passengers aboard, the *Great Western* completed the first return passage to England in fourteen days. The *Great Western* had departed on a second voyage to New York with fifty-seven passengers when Claxton gifted a formal 'record' of the celebrated voyage to the First Lord of the Admiralty, the Earl of Minto, to whom it was dedicated. In an accompanying letter dated 4 July 1838, Claxton expressed the Company's indebtedness to the Admiralty for 'the kindest assistance'. In the Company's Annual Report (1838), the directors had indeed expressed their obligation to the Board of Admiralty for placing plans, drawings and calculations relating to naval steam vessels at the disposal of the Company, and specifically thanked chief surveyor Sir William Symonds 'for important suggestions' and master shipwright at Woolwich Dockyard, Oliver Lang, 'for continual calculations of the most valuable character'.[30]

Omitting any direct reference to monetary matters, Claxton now stated rather obliquely that the 'only return which would be adequate[,] they [GWSS] have made ... [through] conduct[ing] their undertaking [in such a way] that it might be useful to the two greatest maritime nations of the world, beneficial to science, and honourable to their country'.[31] These three symbolic 'returns' from the success of the enterprise provided the foundation for the Company's urgent campaign for an Admiralty mail contract.

The campaign took the form of a memorial from a committee of Company directors dated September 1838 to the Lords Commissioners of Her Majesty's Treasury. It highlighted both the 'great cost and risk' involved in establishing the first line of steam communication between Britain and the United States, and the success of the *Great Western* in departing on schedule for the six consecutive passages so far completed across the Atlantic (averaging under fifteen days out and thirteen days five hours home). On each passage, the steamship had carried Government despatches 'of great importance' without remuneration

[29] Chisholm (1909: vol. 1, 189–91); Roy (1935: 75).

[30] SC (1846), 8 (transcript of Claxton's correspondence and extract from his report); Buchanan (2002: 58–9) (quoting from the GWSS's Annual Report, 1 March 1838).

[31] SC (1846), 8.

and, at a rate of two pence each, had delivered 'a very large number of letters' with 'great punctuality within periods never before attained'.[32]

That rate, however, was 'by no means adequate to the extraordinary service performed by the *Great Western*'. The 'extremely heavy expense' indeed 'exceeded by several times the cost of maintaining sailing vessels of the first class'. In short, the *Great Western*, despite passenger numbers both ways averaging over one hundred during summer passages and paying some £40 per person westbound, was operating at a loss. The memorialists therefore appealed to 'the custom of Her Majesty's Government to make large money allowances to parties conveying Her Majesty's mails to Lisbon, Cadiz, Gibraltar, and other places on the Continent ... at a maximum distance from England of one-third that comprised in the voyage to New York, and therefore requiring for their maintenance less than half the outlay'. This claim alluded specifically to the award of an Admiralty mail contract in the previous year to the new Peninsular Steam Navigation Company engaged in trade with the Iberian Peninsula (Chapter 12). Now the GWSS sought to use the credibility built up over the summer of 1838 to argue for a similar contract in recognition of the 'great risk in opening a source of communication of incalculable national advantage'.[33]

The memorialists also made clear that the *Great Western* was only the first of 'other large and expensive ships they are about to build', all of which stood to suffer from 'the insufficiency of the earnings'. That insufficiency threatened to arrest steam communication across the Atlantic, with an unspecified consequence of 'great general detriment'. On the other hand, the Company was 'prepared to bring every effort of science and enterprise to bear upon producing the greatest possible facilities to the public service, provided just and reasonable remuneration be granted them'. It therefore sought recompense for the carrying of the despatches and for the carriage of the Royal Mails. Moreover, the memorialists expressed the willingness of the proprietors to promote, subject 'to their own security and fair profits', the service of mails and despatches 'by other lines of steam communication to be established' between Britain and New York, or between Britain and American colonies and the West Indies.[34]

The Government moved fast. By late-autumn 1838, it was advertising for the conveyance of mails by steamship between England and Halifax with an onward link to New York. The conditions required a sufficient number of ships for a monthly service, steamers of nothing less than 300-hp, a naval officer to accompany each vessel with powers to oversee the mails and interpret the

[32] SC (1846) 9.
[33] SC (1846) 9, 30 (passenger numbers for each outward and homeward passage). Burgess (2016: 40) claims that the *Great Western* 'sailed on ... to the profit of her owners'. The evidence suggests otherwise.
[34] SC (1846), 9.

contract, penalties for stopping gratuitously and a start date of April 1839. The contract appeared to be of one year's duration.[35]

The only bidders were the Great Western and St George companies. Having returned to Halifax in mid-summer, Cunard was as yet oblivious to the advertisement while apparently attempting, unsuccessfully, to interest mercantile investors first from Halifax and then from Boston in a transatlantic steamship project. The GWSS responded with a projected service from Bristol to Halifax, linked by feeder vessels to New York. Under its terms, the service would be supported by a seven-year mail contract worth £45,000 per annum, while the projected steamships would be 350-hp and either 1,000 tons if they were iron hulled or 1,200 tons if wooden hulled. But the complete system would take up to two years to construct. The St George's Company already had two wooden-hulled 300-hp steamers and so could propose a Cork to Halifax service starting as required – again for £45,000. A Cork–Halifax–New York service would need £65,000. The Company attempted to set its own additional conditions such as pilotage and port charge exemptions, and the right to depart from Halifax without waiting for return mails. With the *British Queen* still not ready, Smith himself did not submit a bid.[36]

4.3 'A Scandalous Thing': The Fracturing of Trust in Atlantic Steamships

Little more than a month after the launch, the *British Queen* was on the way under tow to Glasgow. Smith had earlier hoped to have the steamer in service by September, but by then he accepted that '[t]he enormous work of putting the engines on board ... is likely to delay her departure until November or December'.[37] Apart from the difficulty of taking over an engine contract two-thirds complete, Napier and Elder faced the challenge of placing the two massive engines in the hull of the world's largest ship, freshly arrived in a raw state from the Thames. As such, the arrangement differed markedly from their well-established practice of integrating Elder's engine-building skills with the hull-building craftsmanship of John Wood (Chapter 2). As a result of his own additional preparatory work on the ship's hull in Glasgow, Elder would assert that 'whatever else of the vessel would break up, if she happened to get wrecked – this part of his [the engine beds] ... would stick together'. But the delays – occasioned by work to strengthen these very engine beds and by the installation of Halls' new patent surface condensers – threatened Napier's

[35] Quoted in Tyler (1939: 77). SC (1846), 10, notes that the contract was for one year only.

[36] Tyler (1939); Hyde (1975: 2–5); SC (1846), 8 (Claxton's answer on 13 December to the advertisement of 7 November 1838).

[37] Pond (1927: 125–6, 129–31).

cherished engineering reputation. Fierce controversy also erupted in the pages of the press over the decision to engine the London-built ship in Scotland.[38]

Although the *Sirius* made a second round voyage that summer, the charter was not renewed. Smith's directors felt that the £3,500 loss on the *Sirius*' first voyage had been justified, 'quite satisfied that they gained their object'. Even though those directors who also sat on the St George board had made 'a good thing at our cost', Smith nevertheless felt that the result of the *Sirius*' voyage had vindicated his own advocacy of large steamers. Directors, who had previously begun 'to quake about the size of the *British Queen*', had now fallen into line such that the second steamer, *President*, would be at least as big in tonnage. Offering a still-positive perspective to his nephew in New York, he remained wedded to his vision of four ships out of Liverpool and four out of London.[39]

A new rival, meanwhile, was in the offing. The CDSP, with Charles Wye Williams (Chapter 12) at the helm, put its year-old Irish Sea steamer *Royal William* on the Atlantic as an advance guard for a planned service due to start as soon as they took delivery of a large purpose-built steamer. The *Royal William* had been constructed by Merseyside builders William and Thomas Wilson as the first sea-going steamer fitted with watertight bulkheads. The 617-ton wooden paddle-steamer, smaller than *Sirius*, made the most of the summer of 1838 for the first of three round voyages from the Mersey to New York, although the voyage times increased with the onset of winter to over twenty-one days westward.[40]

Never one to fall for promotional optimism, the anonymous 'Mercator' told readers of the *Nautical Magazine* that 'these trips ... seem to have been made gratuitously, just to show that the thing could be done'. He stressed that the owners had kept back 'the important fact, that they were not attended with "profitable results"'. And while disclaiming any wish to disparage 'these two very fine steamers' in their original roles as channel steamers, they were 'not the sort of ships fit to make the Atlantic passages with profitable results' due to the fact 'that their consumption of fuel, in proportion to their bulk, is outrageously out of all reason'.[41]

By autumn 1838, however, CDSP had set up the Transatlantic Steam Ship Company to operate the first of three large projected steamers. The venture was closely associated with Sir John Tobin, a very wealthy and controversial Merseyside merchant whose long career had included rising from deckhand to

[38] Napier (1866: 100–1) (on Elder); Napier (1904: 116–17); 'Mercator' (1839a: 43–4) (criticism of sending hulls north to Glasgow or Liverpool); Fox (2003: 80–1) (correspondence in the *Mechanics Magazine*).

[39] Pond (1927: 127–8); Tyler (1939: 67).

[40] Bonsor (1975–80: vol. 1, 55, 60–1, 67–8).

[41] 'Mercator' (1839a: 40).

shipmaster, involvement in the Atlantic slave trade, a fortune made in trading West African commodities (especially palm oil), investment in the LM&R, Irish Sea steamship ownership and involvement with William Laird in a large Wallasey property deal. He had also been elected as a Tory Mayor of Liverpool amid serious accusations of bribery. With his command of capital, Tobin was initially the principal owner of the 1,150-ton *Liverpool* constructed on Merseyside. Before the *Liverpool*'s first voyage, a *Railway Magazine* correspondent reported that the Transatlantic Company had purchased the steamer from Tobin for £44,000 at a time 'when the world went mad about transatlantic steam navigation' (Chapter 13).[42]

Ostentatiously boasting not one but two funnels and 'warm and cold baths', the new ship had an inauspicious beginning. At the BAAS in Liverpool (1837), Oliver Lang had spoken of witnessing at first hand 'the great Bristol steamer [*Great Western*], and the one building here for the same purpose [*Liverpool*]'. The former he judged 'well designed' and he 'liked her best', but 'the one here not so well, she was too full behind'. In the summer of 1838 Smith privately concluded that '[t]he ships *Royal William* and *Liverpool* from Liverpool, the latter Sir John Tobin's ship, will come to nothing good. Too small, 2,000 to 2,500 tons small enough'.[43]

These prophesies gained credibility. One observer of the new systems, Liverpool cotton broker George Holt (the elder), noted in his diary in late September 1838 that the *Liverpool* had been 'obliged to put into Cork, owing to a miscalculation of the expenditure of her fuel & the wind continuing so violent as to prevent making more than 4 knots per hour'. 'Mercator's' verdict was similarly severe. This vessel's machinery 'consume[s] coals at such a rate, as would render it utterly impossible she can ever make a passage across the Atlantic, unless she be so loaded with "the locomotive power", and consequently so destructively expensive, as to entitle her to be set down as a failure'.[44]

The unfortunate *Liverpool* took a total of thirty-four days to reach New York on that maiden voyage, almost the same as the average of the sailing packets contending with head winds (Chapter 3). The steamer went on to complete seven round voyages. The final Atlantic crossing to Liverpool still took twenty-seven days, including a diversionary call for coal at the Azores. The Transatlantic Company closed down the service after little more than one year and entered into a merger with Peninsular Steam to form P&O (Chapter 12). In contrast, the *Great Western* continued to provide a reliable service, though

[42] 'Tobin, Sir John (1761–1851)', *ODNB*; *RM* (1843), 6–7 ('Fair Play').

[43] [Anon.] (1837: 698) (Lang); Pond (1927: 130); Tyler (1939: 63, 65–6); Bonsor (1975–80: vol. 1, 69) (cold and hot water baths). 'Too full behind' meant that the ship's hind quarters from midships to the stern lacked fine lines, just as 'too full forward' meant bluff bows and big shoulders.

[44] George Holt Diary (GHD) (26 September 1838); 'Mercator' (1839a: 40). Note the Lardneresque term 'locomotive power' (Chapter 3).

at a loss. After completing five round voyages in 1838 averaging about sixteen days westbound and thirteen days eastbound, the ship completed six round voyages in 1839 with a similar average performance. Even 'Mercator' declared the ship 'a pattern for all seagoing steamers to improve upon, perhaps, but certainly to look up to, as the first of sea-going steam ships of the present day'.[45]

Ever optimistic about his own projects and keen to counter Lardner's critiques, Smith launched his own bid to confer scientific authority on steam over sail in the wake of the *Sirius*. He chose to submit his analysis in the form of letters, published early in 1839, to Silliman's *American Journal of Science*. Founded in 1818 with Yale-based Professor Silliman as editor, the journal had long included articles and reports pertaining to steam navigation, including discussions of various types of steam engines, the safety of steamboat boilers and the use of iron for hull construction.[46]

Smith's letters aimed to show that 'steam power is both safer and more philosophical than the power of wind in navigation'. His argument rested on a distinction between the actions of wind and steam. Considering the masts of a sailing vessel as vertical levers, he claimed that when wind is applied to their sails the direct tendency 'is to upset, instead of to propel the ship'. Especially dangerous was the case of a square-rigged sailing vessel 'taken aback' by a sudden change of wind from a direction abaft to one forward of the beam, such that 'she is hurried stern first into the depths of the ocean'. In a steamship, however, 'the power is applied to a combination of short levers, acting horizontally upon the body of the ship always propelling the ship forward, and never losing power by a collateral [secondary] motion'.[47]

He then argued that 'recent experiments' had shown the practicality of navigating the Atlantic by steamships and had thus 'confounded the theoretically wise'. But these supposedly 'scientific champions' (such as Lardner) had taken up the new position of accepting the *practicality* while denying the *profitability* of transatlantic steam navigation. Smith advanced the proposition that '[w]hatever article of produce or manufacture can be exported or imported in a sailing ship, at a remunerating freight, can be exported or imported in a steam ship at a greater or equal profit, independently of passengers'. Admitting that the proposition was one that 'the public mind is scarcely prepared to credit', he launched into a comparative analysis of the 'working power of steam and sailing ships'. Weighing projected annual freight earnings against expenses, he concluded that a single large steamer doing the work of eight small sailing vessels in the New Orleans to Liverpool cotton trade would generate, on account of lower operating costs, over £3,000 *more* profit per annum than the sail-only

[45] 'Mercator' (1839a: 43–4); Bonsor (1975–80: vol. 1, 61, 69–70).
[46] For example, [Anon.] (1821); [Anon.] (1822); [Silliman] (1831).
[47] Smith (1839: 160–2) (first letter dated 30 July 1838); Pond (1927: 139–40).

fleet. Smith's quantitative estimates took as his standard the ubiquitous 400-ton sailing vessel rather than the big sailing packets of 800–1,000 tons coming on to the Atlantic in the late 1830s.[48]

Beneath this seeming confidence, Smith suddenly lost faith in those upon whom the project crucially depended. First, his trust in Napier as an engine builder broke down. 'From the extraordinary and most unjustifiable delay', Smith wrote to his nephew in January 1839, 'I think no confidence can be placed in him'. Napier had 'promised me [in] the middle of September that she should be ready in two months', he continued. 'The truth undoubtedly is that he has been working for Russia and our government and neglecting our work. It was not our intention to deceive the public nor to be deceived ourselves, but so it is'. 'A scandalous thing!' Junius exclaimed a month later on receiving a supporting verdict from a member of the Laird family with respect to Napier's seeming deception. 'Scotch again, no matter who suffers or how much, so their own ends are served'.[49]

Smith's indictment, however, resonated with a second and more destructive breakdown of trust. At first he appeared to feel that Laird was attempting to usurp his own role as founder of the Company and even as the 'legitimate father' of transatlantic steam navigation. A pamphlet, published in New York in the wake of the *Sirius*' achievement, promoted the Company but, in Smith's reading, it gave Laird a role in pioneering Atlantic steam. Smith also condemned Laird as 'no man of commercial business'. By the close of 1838, his relations with Laird slipped further. '[C]onnected as he is with his brother [John] no doubt great efforts will be made to build iron ships', Smith wrote prior to rehearsing an argument in favour of wooden over iron hulls. In long vessels of up to 222 feet, he claimed, wooden hulls possessed an elasticity (aided by the oakum caulking) that ensured the vessel sustained heavy seas. Ships constructed of iron plates riveted together, however, had no elasticity and thus 'the rivet holes must rent, the ship bursts asunder and nothing can save her'. By June 1839 he denounced Laird to his nephew as 'a faithless, designing, smooth, cunning fellow, not nice about serving his own interests at any expense ... I have not the slightest confidence in him, although recently he has been remarkably civil'.[50]

Such fractures temporarily faded when, in the summer of 1839, the long-expected *British Queen* arrived back in London from the Clyde. *The Times* offered a concise appreciation, plausibly supplied by the owners, of the steamer's splendid accommodation, including an Elizabethan-style dining room

[48] Smith (1839: 162–7); Pond (1927: 141–7).

[49] Pond (1927: 167–8, 171). Napier had been building steam engines for two Admiralty vessels, *Vesuvius* and *Stromboli*, delivered in 1839. See Napier (1904: 70).

[50] Pond (1927: 119–38, 160–3) (wood versus iron), 177.

Figure 4.2. Junius Smith's 'noble ship' *British Queen* (1839)

sixty feet long and thirty feet wide. The staircase, with a double flight of stairs, displayed rich carvings in English oak while the drawing room was decorated in white with gold mouldings and arabesque hangings. Together, these spaces offered a vista nearly one hundred feet in length.[51]

Arrival on the Mersey took place a few weeks later. Halifax's *Acadian Recorder* reported the sensational event. As cannon boomed along the river, 'the populace rushed from every district of the town to the river's side'. The 'Behemoth stranger', steaming at a rapid rate, completed 'with consummate accuracy, the difficult task of tacking in a river'. The ship was so large that 'her colossal proportions appeared to diminish' the breadth of the estuary as the steamship let go the anchor abreast the port. So great was the reception, indeed, that the *British Queen*'s 'equipment and model have been the theme of unmixed and enthusiastic approbation, and it may be confidently asserted that a more splendid specimen of the progress of European arts and manufacturing never provoked competition or elicited applause' (Figure 4.2).[52]

By the end of July the new ship had docked in New York, sending 'the whole city in[to] a state of the wildest excitement'. With 147 passengers – including Samuel Cunard, ship-owner and merchant of Halifax – and about 100 crew, 800 tons of goods and 600 tons of coal aboard, 'property to the amount of $750,000

[51] *The Times*, 15 July 1839.
[52] *Acadian Recorder*, 3 August 1839.

(connected with her) was afloat; the richest ship and cargo that ever braved the ocean wave'. And lying stern on to the *Great Western* at the wharf, the world's largest vessel made Brunel's steamer look like a pigmy in comparison.[53]

In contrast to these rhetorical expressions of wonderment, there were persistent reports that the big steamers – *British Queen* and *Great Western* – lacked the necessary design strength to withstand the hostile seas of the Atlantic. 'Mercator' was the most articulate contemporary critic, especially of the *British Queen*. Had a vessel of similar length, he argued, been constructed as a sailing vessel, it would never at any time 'be subjected to the trial a steamer [would] ... because ... she never would be forced against a head sea ... of such magnitude as the steamer *must* be opposed to'. The steam vessel 'must be dependent, altogether, on her own *unbending* strength, fore and aft, for holding together', he insisted. '[And] I know many people to doubt it being in the power of man, to form any thing of sufficient strength to bear this'. In particular, he called into question the resistance to bending provided by fir for planking the ship above the turn of the bilge. And he vigorously challenged the practice of using 'treenails', requiring large-diameter bored holes in the timber, to secure the strakes to the frames, instead of metal bolts or spikes with much smaller diameter holes which would therefore cause less weakening of the timbers.[54]

With the *British Queen* in service in 1839–40, 'Mercator' returned to the subject with a heightened sense of urgency in a letter addressed to the RMSP directors who were projecting a fleet of fourteen large transatlantic steamships (Chapter 7). Now, he implied, there was confirmation of the consequences of head seas on large Atlantic steam vessels with a full and heavy forebody, 'the effect of which must be to twist and loosen the ship, destroy all connection, and render the fabric crazy'. Indeed, 'Mercator' concluded that the 'wonder is, that the ships should have made even one voyage without complaining, not that they should be shaken to pieces the first time they really have bad weather to contend with'. Lt Roberts, then the *British Queen*'s master, defended 'this noble vessel' against 'Mercator', claiming that 'notwithstanding the extraordinary severe weather of our last voyage home, she strained and worked as little as any ship I have ever sailed or steamed in, during 20 years' experience at sea'.[55]

The 'fact' that crossing the Atlantic by steam had been demonstrated in 1838 prompted the new Admiralty Comptroller of Steam Machinery and Packet Service to send out circulars soliciting tenders for a mail service by steamer between England and British North America. As a former Arctic explorer, the devout evangelical Sir Edward Parry had close friends within mercantile

[53] *Acadian Recorder*, 3 August 1839.
[54] 'Mercator' (1839b: 187–90).
[55] 'Mercator' (1840a: 114–18) (letter to RMSP); 191–2 (Captain Richard Roberts' reply).

communities on both sides of the Atlantic. One such friend was George Burns (Chapter 1). Heavily involved with David MacIver in running his own network of coastal steamers, Burns did not immediately respond to the circular. Back in 1816 while stationed in Halifax, Parry had also become acquainted with Cunard, the Nova Scotian merchant already well versed in the political arguments for prioritizing a service between Britain and its North American possessions.[56] All of these men were only too well aware of the moral and financial uncertainties of embarking upon such an enterprise.

[56] Hodder (1890: 191–209); Grant (1967: 41) (Parry).

5 'Giving Rich Promise of Serious Intentions'

Mr Cunard's Line of Steamers

> It is Contracted, Agreed, and Ended between Samuel Cunard, Merchant in Halifax, Nova Scotia, and Robert Napier, Engineer in Glasgow, in manner and to the effect following: That is to say, the said Robert Napier Binds and obliges himself and his heirs executors and successors to Build and construct with the best materials, for the said Samuel Cunard, ... Three good and sufficient steam-ships, each not less than Two hundred feet long keel ... with cabins finished in a neat and comfortable manner for the accommodation of from sixty to seventy passengers ... each of which vessels shall be fitted and finished with two steam-engines ... all to be delivered on the Clyde
>
> *Contract with Robert Napier for the first Cunard steamers (18 March 1839)*[1]

Summary

Samuel Cunard was neither a lone genius nor a gambler-capitalist. When this colonial gentleman arrived in England in 1839 to promote a line of steamers for a mail service between Britain and Nova Scotia, he quickly created a network of trusted friends. Held together by shared values, it worked to plan, build and manage the new system. In contrast to Smith's venture, Cunard and his associates ensured the closest ties with shipbuilders (especially John Wood) and marine engine-builders (David Elder and Robert Napier). The mutual trust also brought George Burns and David MacIver into the Company as managers, located respectively in Glasgow and Liverpool. As a result, Cunard's line of steamers soon constituted the first, and so far only, such system on the North Atlantic. Yet, despite such confidence, when Dickens embarked as a passenger on the *Britannia* in 1842, he gave voice to the multiple anxieties of many a transatlantic traveller.

5.1 'Not Born With a Silver Spoon in His Mouth': The Values of Samuel Cunard

'A Mr Cunard from Halifax at present in London wants some steam vessels, and has, through a friend of his [William Kidston] in Glasgow, applied to me regarding them, and for a reference to some person in London', Robert Napier told James Melvill, EIC secretary, in late February 1839. 'I have mentioned

[1] Napier (1904: 251–2).

to him that I did the *Berenice* for you. If he calls on you it will do me an act of great kindness that you deal as leniently and favourably with my character to him as you can with propriety do'. Melvill was already acquainted with Cunard who, from his Halifax base, had been an agent for the EIC since the early 1820s, with distribution rights for China tea throughout British North America. Advertisements for the high-quality tea frequently appeared in Halifax's *Acadian Recorder*. Having crossed to Falmouth on an Admiralty sailing packet in January 1839 with the goal of tendering for the Atlantic mail contract, Cunard duly received from Melvill the recommendation to 'put himself in Napier's hands'.[2]

With Quaker origins and a strong loyalty to Britain, an earlier generation of the Cunard family had migrated from the United States to Nova Scotia in the wake of the American War of Independence. Samuel's parents strengthened their commitment to British values by attending the Episcopal (English) Church in Halifax. On the other hand, Samuel's wife Susan, devoutly religious, had strong Scottish Presbyterian roots. Her father, William Duffus, hailed, like James Gordon Bennett, from the fishing port of Banff on Scotland's harsh north-east coast and married the daughter of a Church of Scotland minister, James Murdoch. The Revd Murdoch's forebears in the late seventeenth-century had supplied William of Orange's troops at the relief of the siege of Derry during the celebrated Irish campaigns that culminated in the Protestant victory of William over the ousted Catholic King James II at the Boyne near the Irish port town of Drogheda in 1690. Yet again, the loyal British pedigree could not have been better demonstrated.[3]

Aged thirty-two, Susan passed away in 1828. Her Presbyterian values of independence, hard work and dislike of waste and ostentation, however, remained with the widowed Halifax merchant. Following his wife's death, Samuel continued to live according to her precepts, sending his younger brothers to a private school run by a Presbyterian minister.[4] He and his partners would also embody those values in their steamships.

Cunard's surviving correspondence bears out a powerful commitment to independence, prudence and providence. '[Y]ou are at liberty to draw for one hundred pounds if you want it and with your large & increasing family I dare say you will need it – and if you do not exactly want it you will not waste it – I wish it were in my power to be more liberal to you but we must only do the

[2] Napier to Melvill 28 February 1839 (transcript), NC; *Acadian Recorder*, 8 March 1828; Napier (1904: 123–4). See also Langley (2006: 28–31).

[3] Duffus genealogy, MG1 vol. 3013 no. 14, Nova Scotia Archives (NSA) (citing *The Maritime Merchant*, 7 March 1929, 85–6). See also Grant (1967: esp. 17, 33) (Cunard family and religion); Fox (2003: 39–55) (who draws on the Cunard family researches of Phyllis Ruth Blakeley to claim that Cunard's mother, Margaret Murphy, had Irish Catholic roots).

[4] *Acadian Recorder*, 9 February 1828; Fox (2003: 45, 49).

best we can and be thankful', he wrote to his daughter Jane in late 1849. A few months later he told her that he was glad 'you keep clear of debt', it being 'an additional mark of your good sense' and 'a good thing to have a little money in store to meet contingencies which may occur'. He also assured her that he would 'take care that your prudence and good sense shall not be forgotten'.[5] Similarly, he agreed with Charles MacIver in 1858 on the financial hallmark of their mail steamship company in contrast to many of its late rivals: 'We have a large Fleet of Ships free from any debt and if additional funds should be required they can be had'.[6] Prudence was a conspicuous hallmark of the Presbyterian, whether in Scotland or Nova Scotia, and whether in connection with family, business or the safe navigation of ships.

Cunard's deeply held cultural values permeated all areas of family and business life. In 1863, for example, he wrote to Jane concerning a possible career in the army or navy for her son George. Although the 'merchant does not always succeed', her father explained, yet with 'patient industry he generally does – there is one thing certain: that no one succeeds without application and close attention to the business he is intended for ...'. Especially for anyone 'not born with a silver spoon in his mouth', however, promotion in government service meant 'nothing but honor'.[7] Cunard's values thus elevated the mercantile life to a noble status that bestowed independence from poverty and subservience. Indeed, the words with which he frequently characterized his goal both for family and ships was 'comfort' – in opposition to poverty on the one hand and extravagance on the other.[8]

To the end of his life, Cunard maintained this strong belief in his own 'providential' role in relation to the well-being of family members as well as ships. It was a belief sustained by a broad faith in divine Providence, without reference to particular doctrines that often divided Christian denominations. '[W]e cannot look into futurity', he told Jane towards the end of his life, 'we must leave our fate to a higher power'. In his final days he made clear to his family that he had reflected deeply upon religious questions and had come to identify with the evangelical faith of his Scottish partners. Welcoming his son Edward's suggestion that a clergyman – 'who he saw in Paris some years since & who my father thinks very highly of' – visit him, Samuel stressed that he would not confess to any religious doctrines with which he did not agree or which

[5] Cunard to Jane Francklyn, 29 December 1849, 24 May and 24 June 1850, Cunard Papers (CP), MG1 vol. 248, nos 45, 48, 49, NSA.

[6] Cunard to Charles MacIver, 1 January 1858, CP, Sidney Jones Library, University of Liverpool (LUL).

[7] Samuel Cunard to Jane Francklyn, 23 December 1863 and 10 November 1864, CP, MG1 vol. 248, nos 50, 55, NSA.

[8] Samuel Cunard to Jane Francklyn, 24 March and 10 November 1864, CP, MG1 vol. 248, nos 51, 55, NSA (uses of 'comfort' and 'comfortable').

he did not feel to be true. In Edward's words, his father 'could say nothing to him [the clergyman], that he did not feel and admit & believe'. To his father's old ship-owning partner, George Burns, Edward was explicit about Samuel's evangelical credentials: '[h]e expresses his firm belief in the mediation of our Saviour, and feels that he can only be saved through Him'.[9]

Fox takes Cunard's remarks to his son as a deathbed confession of his lack of religious faith, rather than as a desire of someone unwilling to subscribe to mere religious platitudes. This misreading is inconsistent with Edward's tone in general and with his assurance to George Burns in particular. Samuel's remarks simply reflect his earnest desire to subscribe to those doctrines he believed to be true and not to pretend to subscribe to those he felt to be doubtful. Fox's interpretation leads him to represent Cunard as believing only in himself with a 'hard, driving, ruthless, tireless engine at the core of his being' that rendered him 'a quite modern personality, focused intensely and narrowly on the ongoing prosperity of his enterprises'. For Fox, Cunard measured 'his success by profits and numbers that he could see and weigh and count' and 'trusted nothing but his immediate family and his own unquenchable ambitions'. Talk of this 'stealthy' individual cutting 'the deal of his life' and stealing 'the game away from its earlier players' further reinforces the image of gambler-capitalist, motivated by an ambition to defeat the sailing packet lines. As such, it is an inaccurate portrayal, removing him from his political and cultural networks.[10]

A few days before its report of Susan's death, the *Acadian Recorder* had carried an announcement by the Nova Scotian Commissioners of Light Houses, Samuel Cunard and Thomas Maynard, of the completion of a new light – coloured red to distinguish it from any other on the coast – at the entrance to Halifax Harbour. The concerns of the merchant to minimize waste of capital and commerce complemented perfectly the concerns of the Christian gentleman to save human life and of the Christian evangelist to save human souls. The duties of commissioners of lighthouses carried heavy moral, as well as commercial, freight. Their responsibilities thus found a ready echo in the sermons of Henry Melvill, ecclesiastical brother to the EIC's secretary (Chapter 2).[11]

Cunard had learnt the business of ship-broking in Boston and around 1815 had established a contract packet service with sailing vessels linking Halifax with Newfoundland, Boston and Bermuda. His commercial interests ranged widely over whaling, land, ironworks, canal projects, lumber, fishing and coal

[9] Cunard to Jane Francklyn, 16 April 1864 and Edward Cunard to Jane Francklyn, CP, MG1 vol. 248, nos 53, 61, NSA; Edward Cunard to George Burns, 28 April 1865, in Hodder (1890: 427).

[10] Fox (2003: 55, 84).

[11] *Acadian Recorder*, 4 January 1828. See Grant (1967: 71–2) (Cunard's policy of using a distinctive colour scheme for individual lighthouses to serve as daymarks); Langley (2006: 91–3). Thanks to Simon Schaffer for the useful trope 'freight'.

mines, giving him power and influence in diverse fields. His capital by 1830 was estimated at some £200,000 with a considerable increase by the time he approached Napier nine years later, despite the financial storm that had so affected Junius Smith (Chapter 3).[12] During his 1838 visit to London he carried a letter of introduction from the Governor General of Nova Scotia, Sir Colin Campbell, to the Colonial Secretary, Lord Glenelg (Chapter 4), in Lord Melbourne's Whig Administration. It summarized his credentials as 'one of the firmest supporters of the measures of the Government' and pronounced that 'his being one of the principal Bankers and Merchants and Agent of the General Mining Association, & also Commissioner of Light Houses, gives him a great deal of influence in this community'.[13]

As one of the most powerful and respected of Nova Scotian citizens and merchants, Cunard possessed both the practical experience and the symbolic capital to realize Howe's and Crane's vision of enhanced imperial order through mail steamers in the face of recent instabilities in British North America. Returning to London early in 1839, he now worked to secure Government patronage for the mail steamship project. Reciprocally, however, aristocratic Whigs were quick to bestow patronage upon this colonial merchant who combined impeccable loyalist credentials, gentlemanly trustworthiness and long mercantile experience. Fanny Kemble's representation of Cunard as a 'shy, silent, rather rustic gentleman from the far-away province of New Brunswick [sic]' suggests a lack of metropolitan sophistication with no automatic entry into aristocratic circles. Yet such deficiencies were advantageous. Here was a self-made British North American who could be moulded into fully gentrified form to serve and strengthen the causes of empire.[14]

Winning Admiralty and Treasury patronage was critical to the realization of the as-yet visionary transatlantic project. Francis Baring was now Chancellor of the Exchequer with effective control of the Treasury. Parry, a naval hero who in the mid-1810s had been stationed at Halifax to enforce a blockade against the United States, occupied the new Government post concerned with the steam packet services. Equally significant for Cunard's project was the Secretary to the Admiralty, Charles Wood. As Cunard told a Parliamentary Committee a few years later, both Baring and Wood had 'spent a great many hours at different times in going through the calculations and routes with me'.[15] On 11 February 1839 Cunard submitted to the Lords of the Admiralty a written proposal for a twice monthly mail service 'from a point in England' to Halifax with 'steam boats' of at least 300 horsepower. He proposed that 'Branch boats'

[12] Hyde (1975: 1–5, 78–9); MacMechan (1931: 151–4) (capital accumulation).

[13] Tyler (1939: 80) (from Colonial Office records). See also Fox (2003: 88).

[14] Kemble (1878: vol. 1, 286); Hyde (1975: 5–6).

[15] Parry (1858: 35–7); SC (1849: 133); Fox (2003: 88).

of not less than 150 horsepower would convey the mails onward to Boston and also from Pictou to Quebec when the St Lawrence was free of ice. His bid was for £55,000 per annum for ten years. He further committed himself to making 'such alterations and improvements [to the vessels] as their Lordships may direct' during that period. The Admiralty deemed the sum 'not unreasonable for the proposed service, if the Lords of the Treasury are prepared to entertain a proposition of this kind'.[16] Whig patronage was gathering pace.

5.2 'For I Have Not the Science Myself': The Glasgow Network

In February 1839 Cunard contacted his Scottish agent, Kidston, who had spent much of his life in Halifax as a merchant before returning to Glasgow to develop his shipping and ship-owning interests. Kidston was also a friend of Howe and the latter probably visited him after taking passage from Liverpool to Glasgow aboard the Napier-engined Burns and MacIver steamer *Commodore* in the summer of 1838.[17] Kidston told Robert Napier that Cunard wanted 'one or two steamboats of 300 H.P. and about 800 tons' and explained that the Halifax gentleman had heard that 'Messrs Wood and Napier are highly respectable Builders and likely to be enabled to fulfil any engagement they may enter into'. In particular, Cunard had been told that the Dundee steamer *London* (1837) 'is a fine vessel'. He required his own vessel 'to be of the very best description and to pass a thorough inspection and examination of the Admiralty ... a plain and comfortable boat, but not the least unnecessary expense for show'.[18] Cunard's distinct moral and spiritual values meant that he quickly established trusted friendships and business relationships with the Glasgow Presbyterians.

Napier's reply to Kidston listed virtually every ship-owner that he and Wood had supplied, in several cases adding the names of the steamers and the manager. The geographical spread ranged from the North Sea coasts of Scotland and England to the Irish Sea and North Channel, but also included the Red Sea and Indian Ocean:

- EIC (*Berenice*): James Melvill (Secretary)
- Dundee Shipping Company (*Dundee*, *Perth* and *London*): George Duncan (Chairman)
- Inverness Shipping Company (*Duchess of Sutherland*): Thomas Davidson (Manager)
- Aberdeen and Leith Shipping Company (*Sovereign* and *Duke of Richmond*): Robert Mitchell (Manager)

[16] Tyler (1939: 80–1).

[17] Roy (1935: 63–4, 83). This section and the next draw on Smith and Scott (2007: 471–96).

[18] Cunard to Kidston and Company, 25 February 1839, CP, LUL. Printed in Napier (1904: 124–5). See Hyde (1975: 5–6); Harvey and Telford (2002: 10).

- Isle of Man Steam Packet Company (three steamers)
- Londonderry Steam Packet Company (three steamers)
- Belfast Company, Glasgow (three steamers)
- City of Glasgow Steam Packet Company (*John Wood*, *Vulcan*, *City of Glasgow* and one building)
- Thomas Assheton Smith (*Menai*, *Glowworm*, *Fire King*).

'To any of these parties', he told Kidston, 'you are at full liberty to apply in order to ascertain the manner I fulfilled my contract for these [23] vessels'. His closing sentence expressed his willingness, if Kidston's 'friend' was 'really in want of vessels', to meet Cunard face-to-face in London where Napier had 'no doubt but that we would in a very short time understand one another'. He also affirmed that, given 'the great accommodation I have for doing work, I could at present undertake to build and finish in twelve months two or more steam-vessels, were I favoured with the order soon'. He stated as to cost that 'good vessels, warranted to stand any inspection and give entire satisfaction both as to the vessel and machinery, cannot be done for less than from £40 to £42 per ton, – this for the vessel ready for sea, with cabins, sails, rigging, anchors, cables, &c'.[19]

By travelling to Glasgow instead, Cunard demonstrated that he was 'really in want of vessels'. This early March meeting took place in Napier's home, Lancefield House, close to the works (Chapter 2). Cunard now committed himself to ordering three vessels on the understanding that Napier would lower his initial offer of £32,000 (£40 per ton) to £30,000 per vessel. In a letter to G. & J. Burns a year or so later, Napier gave an account of the result of the meeting: 'seeing that he [Cunard] was prepared at once to give me an order if my terms pleased him, I at once said at the rate of £40 per ton'. To this offer, Cunard responded that 'he considered it fair and reasonable, but as he had three vessels all of one size (and that similar to what I had in hands for the City of Glasgow Steam-Packet Company [the 205-foot, 700-ton *Admiral* delivered in 1840 for David MacIver's Glasgow to Liverpool express service]), he said if I took £30,000 for each of the vessels, he would give the order before he left me. This I agreed to ...'. Cunard then returned to London, leaving Napier to write to the London merchant house Reid, Irving & Company (financially involved with Junius Smith as well as with RMSP) seeking validation for his initial judgement that Cunard 'seems from the little I have seen of him to be a straight-forward business man'.[20]

[19] Robert Napier to Kidston and Sons, 28 February 1839, in Napier (1904: 125–9). The identity of the 'Belfast Company' is unclear but may be a reference to G. & J. Burns and their associates. See Duckworth and Langmuir (1977: 2–5, 86).

[20] Napier to G. & J. Burns, 28 January 1841, in Napier (1904: 129–31); Napier to Reid, Irving &Co., 12 March 1839 (transcript), NC.

Prior to Cunard's second visit to Glasgow in mid-March 1839 to collect copies of plans and specifications drawn up in the interval, Napier reconsidered the project. He concluded that the proposed ships were too small. While Cunard resisted on the grounds of increased capital cost, Napier's dread was not cost but 'failure', the loss of confidence generated by a failure of the steamers 'to gain the different objects in view'. This fear echoed his 1833 letter to Wallace (Chapter 2). It was now his view that 'if these small vessels did not succeed they would do him more injury in character than any money [that] he could gain would benefit him'. Napier therefore put forward an agreement acceptable to Cunard whereby the tonnage would rise to 960 and horsepower to 375. 'I am of opinion Mr Cunard has got a good contract [dated 18 March 1839], and that he will make a good thing of it', Napier told Melvill. 'From the frank off-hand manner in which he contracted with me, I have given him the vessels cheap, and I am certain they will be good and very strong ships'. Under the agreement, the ship-owner would pay an additional £2,000 'for the alteration of the vessel and work connected therewith' while the engine-builder would 'then make him a present of all my part of the work for the enlarged size of vessels'. Cunard accepted the recommendation, feeling 'confident the Boats would be much improved', and adding with Nova Scotian pride that he wanted 'to shew the Americans what can be done in Glasgow and that neither British [English] nor London Boats can beat them'.[21]

Early in their relationship, Cunard drew Napier's attention to patents for surface condensers which promised to prolong boiler life by ensuring a supply of fresh rather than saltwater. Two inventors, Samuel Goddard and Samuel Hall, had been lobbying the ship-owner. The latter insisted that his patent for surface condensers 'is very superior to Mr Goddard's and that boilers made under his [Hall's] plan will last eight or ten years & that consequently engineers are not fond of recommending them as it interferes very much with their trade and profits'. A perplexed Cunard turned to Napier, asking 'am I not right in saying that you are to give me everything upon the best & most improved plan'?[22]

Particularly after his difficulties with Hall's condensers on the *British Queen*, Napier now replied with contempt towards patents.

> I was quite prepared for your being beset with all the schemers of every description in the country. [I] think it right to state that I cannot and will not admit of anything being done or introduced into these engines but what I am satisfied with is sound and good. Every solid and known improvement that I am acquainted with shall be adopted by me, but no patent plans.

[21] See Napier (1904: 131–3); Napier to James C. Melvill, 19 March 1839 (transcript), A. C. Kirk Collection (ACKC); Cunard to Napier (1839), CP, MG100 vol. 129 no. 20a, NSA (copy).

[22] Samuel Cunard to Robert Napier, 21 and 25 March 1839, NC.

Cunard felt a sense of relief. He would continue to refer everyone marketing patent schemes to Napier on the grounds that 'I am entirely ignorant on the subject but from your science you will be enabled to appreciate the value of their improvements'. He subsequently told a Select Committee of the House of Commons in 1846 that 'any credit that there may be in fixing upon the vessels of proper size and proper power is entirely due to Mr Napier, for I have not the science myself'.[23] Consistent with his earlier views on the importance of securing and preserving public trust in his work, Napier rigorously excluded experimentation from commercial passenger steamers, while remaining committed to experimental trials with other craft, most notably steam yachts (Chapter 2).[24]

Cunard also relayed to Napier the news that he was receiving rival offers from Liverpool and London builders intent on destabilizing the Clyde's reputation. 'You have no idea of the prejudice of some of our English builders', he wrote, 'and when I have replied that I have contracted in Scotland they invariably say "You will neither have substantial work nor completed in time"'. He admitted, however, that the Admiralty 'agree with me in opinion [that] the boats will be as good as if built in this country' and that he had 'assured them that you will keep to time'. He also reported that both the 'Admiralty and Treasury are highly pleased with the size of the boats' and that he had 'given credit where it is due to you and Mr [John] Wood'.[25]

Napier's reply stressed the moral value of honest work rather than words. 'I am sorry that some of the English tradesmen should indulge in speaking ill of their competitors in Scotland', he told Cunard. 'I shall not follow their example, having hitherto made it my practice to let deeds, and not words, prove who is right or wrong'. Rather, he admitted to saying no more than to 'court comparison of my work with any other in the kingdom, only let it be done by honest and competent men'. Napier's style echoed that of his late fellow-Presbyterian James Watt who had urged his English partner Matthew Boulton to 'let us be content with *doing*' rather than with showmanship. Napier further offered the guarantee, important for Cunard's state patrons, that his engines '*will be made similar in construction to those I am at present making to the Admiralty*'.[26]

The ship-owner's increasing confidence in his contractors had several foundations. First, it rested on Melvill's word as a trustworthy gentleman at the heart of the old British establishment. Second, Cunard had also seen Napier – and Napier's works – for himself during his two visits to Glasgow. And third,

[23] SC (1846: 23); Napier to Cunard, 27 March 1839 (transcript), ACKC (printed in part in Napier (1904: 136–7); Cunard to Napier, c. 1 April 1839, NC; Bonsor (1975–80: vol. 1, 56) (*British Queen*).

[24] See Marsden and Smith (2005: 112–15) (Collins Line use of experiment in contrast to Cunard).

[25] Cunard to Napier, 21 March 1839, NC; Hyde (1975: 6).

[26] Napier to Cunard, 27 March 1839, in Napier (1904: 136–7); Marsden and Smith (2005: 60) (Watt).

he now knew of Napier's already-high reputation as a marine engine-builder and of Wood's standing as first-class shipbuilder. In reply to the criticisms of English builders, for example, he could simply point to the model of Napier and Wood's recent steamers *City of Glasgow* (1835), *London* (1837) and *Commodore* (1838). The formal contract, indeed, cited each of these vessels by name. The Cunard steamers were to be 'equal in quality of hull and machinery to the steamer *Commodore* or the steamer *London* ... and equal to the *City of Glasgow* steamer in the finishing of the cabins'.[27]

From the outset, then, Cunard placed his faith in Wood and asked Napier to tell his builder that 'if he does not build them all [in his own yard] I shall still look to him to see that they are well and faithfully built'. As ship-owner, Cunard knew he was purchasing neither a commodity nor a consumer item. And his strategy of developing close relations with his contractors contrasted strikingly with Junius Smith's professed policy of distancing his project from specific shipbuilders. Cunard's belief in the skilful exertions of Wood and Napier remained undiminished over the coming months. When he considered a simultaneous order for smaller ships to feed the main transatlantic service on the North American coast, he told Napier that he was 'not disposed to go beyond you and Mr Wood if you can accommodate me'. And a short while later he expressed his contentment with the arrangements for building the Atlantic steamers: 'Let who will build the vessels. Mr Wood and you must look to their being well done. I rest my faith in you two and know that I shall not be disappointed'.[28]

Napier, however, remained concerned that the model for the three Atlantic steamers – only a little larger than the coastal *Sirius* or *Commodore* – would not prove 'fully fitted for the trade'. Cunard resisted again, ostensibly because the Admiralty and Treasury seemed well satisfied. He complained to Melvill that his builder 'was always proposing larger boats'. Melvill, however, insisted that 'to ensure success the adoption of Napier's views was imperative, as he was the great authority on steam navigation, and knew much more about the subject than the Admiralty'. Ruefully, Cunard had to admit to Napier that their EIC friend 'takes a lively interest in your welfare'.[29] By 4 May 1839 Cunard had signed an initial seven-year mail contract worth £55,000 per annum for a service from Liverpool to Halifax and Boston.[30]

He soon confessed to Melvill, however, that his principal problem was raising capital, especially given the increases in size and power. As he probably never intended to provide all the capital himself, his challenge was how best

[27] Cunard to Napier, c. 1 April 1839, NC.
[28] Cunard to Napier, 21 March and 1 April 1839; 4 April 1839, NC.
[29] Samuel Cunard to Robert Napier, 2 April 1839, NC; Napier (1904: 137–8).
[30] Hyde (1975: 8).

to raise that capital in a period of economic instability. He chose not to follow an EIC model of a corporate, joint-stock system that other steamship lines had adopted. The RMSP, P&O and PSNC secured Royal Charters that protected proprietors against unlimited liability, whereas the GWSS and Smith's Company did not. London investors especially may have felt, with good reason, that ocean steamship ventures without a Royal Charter and without a mail contract tended to be over-promoted with rhetorical promises that were difficult to fulfil.[31]

Cunard's other option was to form a partnership principally of family and business friends built on mutual confidence. The wider public, and especially speculators, would be kept well outside.[32] But here, too, there were no guarantees of winning private supporters. Thus, even after he had offered his Glasgow friends, the Kidstons, the management agency in Glasgow, he could not persuade them to invest on the grounds that 'they had no experience in steamers'. They did, however, recommend the Burnses, whose experience of steamers was unrivalled.[33] How that connection was formed is revealing.

Napier adopted a strategy to unlock the capital of Glasgow's wealthiest merchants, many of whom had little or no knowledge of steamships themselves. Ahead of the rest, James Donaldson's wealth had earlier backed David MacIver's City of Glasgow Company (Chapter 1). According to George Burns' later account, Napier led Cunard personally to talk to Donaldson who told Cunard that he 'never did anything without consulting a little friend'. That 'little friend' was none other than Burns who, through his friendship with Parry, had already received and declined an invitation to submit a bid himself for the mail contract (Chapter 4). Donaldson duly 'came trotting down from his office' to Burns' office to tell him that Cunard and Napier were waiting for him, and had 'proposed that we should do something to get up a concern for carrying the North American Mails'. As a result, Donaldson introduced Cunard to Burns and left the two men alone to talk over the proposal.[34] Building confidence into the project thus preceded the raising of capital and the launch of the first ship.

Although Burns 'entertained the proposal cordially', MacIver – who joined Cunard and Burns for dinner that day – 'went dead against the proposal'. Early difficulties centred on the Admiralty's penalties over voyage delays. Cunard then invited the Glasgow pair to join him and Napier for breakfast the next morning at Lancefield House. Agreement to proceed rested on the confidence placed in the power and reliability of Elder's marine engines to meet the Admiralty's schedule. The same day Burns began his campaign to

[31] Alborn (1998: 13–52) (EIC structure). Fox (2003: 86–92) discusses Cunard's own financial problems in Nova Scotia.

[32] Taylor (2006: 24–5) (on partnerships).

[33] George Kidston to James Napier, 25 February 1895, ACKC.

[34] See Napier (1904: 138–40); Hodder (1890: 196).

persuade Glasgow investors and spoke to William Connal, head of a large firm of Glasgow tea and sugar merchants, who told him: 'I know nothing whatever about steam navigation, but if *you* think well of it, I'll join you'.[35]

Burns now informed Cunard that he and MacIver 'could hardly take up such a large concern ... without inviting a few friends to join us' and that Cunard should feel free to 'make any arrangements he thought best with his own friends'.[36] The fact that Cunard readily agreed to a month's delay in order to know the outcome of Burns' invitations suggests that his other options, if any, were less favourable. Ten days after the signing of the first mail contract with the Admiralty, Cunard accepted as partners Burns and MacIver, who together would take a half of the value both of mail contract and steamers and who could admit other investors into their share. As a result, nineteen Glasgow merchants had by late July 1839 joined Burns and MacIver to form 'the Glasgow Proprietory in the British and North American Royal Mail Steam Packets, established for the purpose of carrying mails, passengers, specie and merchandise between Britain and certain North American ports'.[37]

The major shareholdings soon assumed the following profile: Cunard himself held £55,000, Donaldson £16,000, Connal £11,500, the Burnses £10,500, the MacIvers £8,000 and Napier £6,000. Apart from Cunard, virtually all the capital came through the Glasgow connections of Burns, MacIver and Napier. A later addition that year brought in another seven Glasgow shareholders and four from Manchester. The capital of the newly formed British and North American Royal Mail Steam Packet Company now stood at £270,000.[38] Its public culture belonged overwhelmingly to the second city of Empire, Glasgow, and to the second port of Empire, Liverpool. But it remained for some forty years an otherwise very private concern. As the *Liverpool Albion* explained in 1852, '[i]ts proprietors are limited in number, and generally to large capitalists, who arrange their interests in private meetings, the results of which are not made public'.[39]

Napier firmly believed that confidence in a line of steamers depended on 'honest and competent men' not only in steamship design and construction, but also in management. 'I told Mr Rodger [Glasgow banker]', he explained to Cunard in early April 1839, 'if he could get any of these two Companies [Burns or MacIver] to take it up, the vessels would be well and honestly managed, and ... make money'. He added that were this to happen, he himself would take

[35] Hodder (1890: 196–7); Napier (1904: 140–3); Hyde (1975: 9–10); *Glasgow Herald*, 7 August 1893 (Connal).

[36] Hodder (1890: 197).

[37] Hodder (1890); Hyde (1975: 11).

[38] Napier (1904: 141–2) (list of thirty-two original subscribers). Hyde (1975: 10–13) unscrambles what he admits was a confusing and complex process of building up the early capital.

[39] *Liverpool Albion*, 2 February 1852; Maginnis (1892: 30–1).

a small interest. Once these arrangements had been agreed around mid-April 1839, Cunard told Napier that he had 'left with Mr Burns & Mr McIver [sic] to do as they and you may think right as to enlarging the boats'. Burns took the Glasgow agency, MacIver the Liverpool agency and Cunard – although centred in London – established the Halifax and Boston branches of the Company. Such was the growth in confidence that the original order increased to four ships, ensuring greater future reliability in the system should one vessel require major repairs.[40]

Under these arrangements, the size of the vessels now rose to around 1,120 tons and over 400-hp. They approximated in size to the *Great Western*. Scott Russell, who had worked in Greenock close to the yards where the Cunard ships were under construction and who knew intimately all the Clyde builders (Chapter 7), later observed: '[t]he moment Mr Cunard saw that she [*Great Western*] had succeeded, he rushed in and built four *Great Westerns*. These he built all alike, all copies of one another, with the machinery all identical ...'. Cunard, however, had already publicly denied that his vessels had been built 'with reference to the size of the *Great Western*'.[41]

With Wood's *Acadia* as the 'pattern card', the other hulls were subcontracted to small, family-owned Greenock, Port Glasgow and Dumbarton shipbuilders (Robert Duncan, Robert Steele and Wood's brother Charles) known for the quality work of their shipwrights. Strength was everywhere apparent in each hull with massive frames, planks and fastenings. Napier told Burns and MacIver that 'she is filled up solid in the bows between the timbers with strong beams & knees & water tight bulkhead to prevent accidents should the vessel strike ice. She is also belted and doubled with hard wood planks from the bilges & the binds well fastened with copper ... A great many extra iron straps & trusses. Strong diagonals of wood. Extra strong stringers going round the vessel within for strengthening the sides & fitting up an extra deck for carrying soldiers or goods ... the whole when completed will make the vessels without doubt by far the strongest and best steamers ever fitted out for any station'.[42]

The steamers' formal specification included engine bearers of African oak 'twenty inches deep & made up to about thirty inches deep under the Engines; by fifteen in[che]s broad, in long lengths ... & bolted thro' every frame under the Engines, with 1½ in [diameter] copper bolts ... Bottom planks one strake on each side of the keel 8 ins thick and 14 ins broad'. The thickness of other planks varied from four inches upward depending on position. The ships were specified, as in the original contract, to be 'at least equal in quality & quantity

[40] Napier to Cunard, 2 April 1839 (transcript), 4, 8, 30 April 1839, NC.

[41] Russell (1863: 16–17); SC (1846: 23).

[42] Napier to Burns and MacIver, 12 February 1840, ACKC. See also Napier (1904: 143–4).

to the "Commodore" steam ship ... but in proportionate strength & quantity to the extra size of the vessel[s]'.[43]

Named by Isabella Napier (Robert's daughter), the *Britannia*, first of the quartet, was launched from the Greenock yard of Robert Duncan in February 1840. It was an event, *The Times* prophesied, that 'will probably turn out a very marked feature in the annals of this country, inasmuch as it is the commencement of a mode of communication between the old and the new world which brings them within one-half of the former passage, and thus strengthens the relation in which they stand to each other as parent and child'. The paper also observed that Napier's name gave 'a sufficient guarantee that, when completed, nothing better will be found in the kingdom. ... from the keel to the topmast, everything will be found substantial, and adapted to the course the vessel is to take'.[44]

The *Britannia* departed from Liverpool with Cunard himself aboard on American Independence Day 1840. The westward voyage to Boston via Halifax took just over fourteen days. The *Acadian Recorder* reported the ship's arrival at Halifax with fifty-three passengers including the Bishop of Nova Scotia and noted approvingly that the steamer 'is built long and narrow' with 'a felicitous combination of grandeur, elegance, speed and durability in her construction and *material*'. Somewhat optimistically, it also claimed that the saloon dining table was large enough for 400 people (Figure 5.1).[45]

In a letter published in the *Acadian Recorder*, Junius Smith invoked politics, religion and geography to denounce the project as deluded. 'The idea of running little steamers to Halifax is perfect folly', he wrote disparagingly

> So long as we can run directly to New York and take letters for nothing I should be glad to know who upon earth would be such a fool as to send his letters to the United States by the way of Halifax and pay the postage? ... It is ten times more easy, and less expensive to reach any part of Canada by the way of New York

Moreover, '[n]ature, God's own works, place an impassable barrier to this preposterous plan – frozen regions, eternal fogs, crags, rocks, and an inhospitable coast'.[46] A sharp reply to this impassioned critique pointed out, implicitly indicting Smith's navigational authority, that *every* steamer bound for New York would normally pass not far off Halifax and Boston en route. And after less than a year, the same Halifax paper reported that the *Britannia* had arrived with some 11,000 letters.[47]

[43] 'Specifications of Steamer Britannia', NC; Fox (2003: xiii–iv, 93–4).
[44] *The Times*, 8 February 1840.
[45] *Acadian Recorder*, 18 July 1840. See also Fox (2003: xiv, 93) (lack of ornament in the Cunard steamers).
[46] *Acadian Recorder*, 11 July 1840.
[47] *Acadian Recorder*, 15 May 1841.

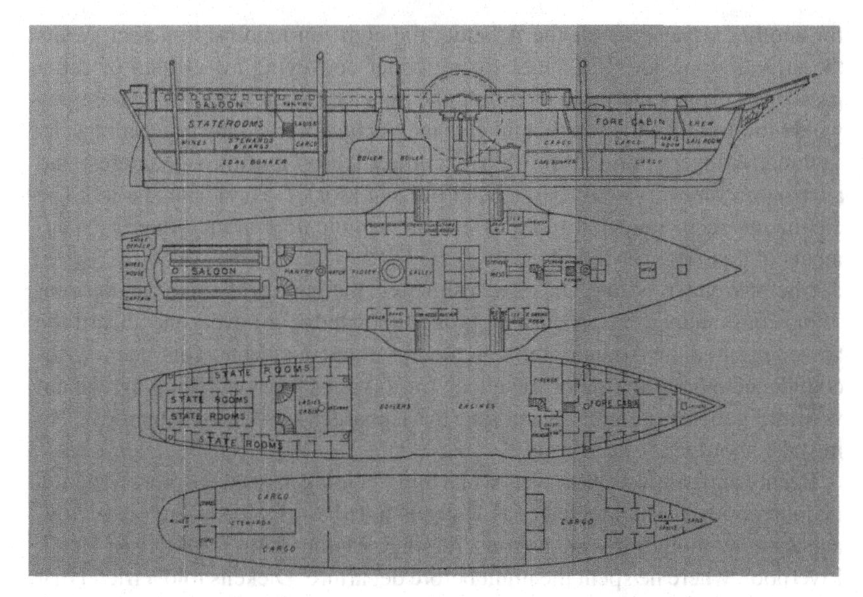

Figure 5.1. Cunard's *Britannia* (1840): General arrangement plans

The citizens of Boston, flattered but in no way subservient to Halifax marked the *Britannia*'s arrival with 'a magnificent public banquet, at which their enthusiasm found vent in speeches of a most complimentary nature'. They apparently extended some 1,873 dinner invitations to Samuel Cunard. The homeward passage from Halifax took only ten days.[48] By early January 1841 the last of the Cunard quartet, *Columbia*, left Liverpool on the maiden voyage and thus finally implemented the promise of a year-round, regular mail service by steam vessel to North America. Except for the winter months from November to February when sailings were monthly, the Cunard ships departed from Liverpool on the 4th and 19th of each month to maintain a fortnightly cycle for the rest of each year. Consistent with Burns' strict Presbyterian credentials, sailings which fell on Sundays were postponed by one day, thus avoiding disturbing the Sabbath rest with the noise and excitement of embarkation (Chapter 1).[49]

5.3 'He Believes She is a Very Strong Ship?' Embarkation Anxieties

Dimly visible through the gathering fog of an early January day in 1842, the Cunard steamer *Britannia* lies at its moorings in the tidal Mersey. Almost ready

[48] Hodder (1890: 202–05); Fox (2003: xiv–xvii).
[49] Bonsor (1975–80: vol. 1, 74).

for another voyage across the Atlantic, the eighteen-month-old vessel awaits the arrival alongside of a small steam tender conveying a full load of cabin passengers from Liverpool, including the novelist Charles Dickens whose pen would immortalize the westbound passage. As he observed the proceedings, he saw many fingers point at the ship, accompanied by murmurs of interest and admiration on every side: 'How beautiful she looks'! 'How trim she is'! The comments seem intended more as efforts at mutual assurance than as statements of genuine conviction.

One passenger, whom Dickens designates 'the lazy gentleman' on account of his cool manner, has, according to the 'knowledge' of everyone aboard the tender, crossed the Atlantic 'thirteen times without accident'. This 'fact' raises confidence among the other travellers. So too does his nonchalant agreement with the general opinion concerning the Cunarder's attractive qualities: 'No mistake about *that*', he affirms.[50]

Beauty and trimness, however, is one thing. 'Going across' the North Atlantic is quite another. The language, as Dickens noted wryly, made the ocean voyage seem as though it were a river crossing. Writing from the Adelphi Hotel, Liverpool, where he spent the night before departure, Dickens told a friend that these 'Cunard packets are not very large, you know, but the quantity of sleeping berths makes them much smaller, so that the saloon is not nearly as large as one of the Ramsgate boats'.[51] Prospective passengers, Dickens among them, had recently inspected the *Britannia* in the non-tidal safety of Liverpool's Coburg Dock where the vessel was loading cargo and fresh provisions, including a cow, pale suckling pigs, calves' heads in scores, beef, veal and pork and 'poultry out of all proportion'. Stunned at first by the 'utterly impracticable, thoroughly hopeless, and profoundly preposterous box' that was his stateroom and by the hearse-like qualities of the saloon, he and his companions had gradually persuaded themselves – against the evidence of their senses – that, somehow, these threatening spaces would fulfil all the unlikely promises embodied in an artist's depictions hung in the agent's offices in the City of London.[52]

Unlike such routine dock visits, however, the tender's departure from shore marks a point of no return. A sombre but stoical mood prevails. One passenger then shocks polite conventions by inquiring 'with a timid interest how long it is since the poor *President* went down'. Everyone knows that the largest steamship in the world, the *President*, left New York for Liverpool in March 1841 and disappeared (Chapter 6). For his presumption, the passenger is 'frowned upon by the rest, and morally trampled upon'. Sometime later, the same passenger, well-wrapped up, asks 'with a faint smile that he believes She [sic] is

[50] Dickens (1842: 7).
[51] Storey et al. (1974: vol. 3, 10).
[52] Dickens (1842: 5–7).

a very strong ship [?]'. The casual gentleman looks 'first in his questioner's eye and then very hard in the wind's' eye before answering 'unexpectedly and ominously' that '[s]he [would] need [to] be'. As a result, the same gentleman, previously trusted, 'instantly falls very low in the popular estimation, and the passengers, with looks of defiance, whisper to each other that he ... clearly doesn't know anything at all about it'.[53]

'She', meanwhile, continues to offer some reassurance. The 'huge red funnel is smoking bravely, giving rich promise of serious intentions'. The officers, 'smartly dressed, are at the gangway'. Elsewhere, however, chaos reigns. Passengers and their baggage 'are to be met with by the dozen in every nook and corner', settling themselves into the wrong cabin, turning out again and 'bent upon opening locked doors' or 'forcing a passage into all kinds of out-of-the-way places'. The casual gentleman, meanwhile, 'lounges up and down the hurricane deck, coolly puffing a cigar'. This 'unconcerned demeanour' restores some of his reputation as a man whose judgement – based on hard-won experience – can be trusted. And so 'every time he looks up at the masts, or down at the decks, or over the side', nervous fellow-passengers 'look there too, as wondering whether he sees anything wrong anywhere, and hoping that, in case he should, he will have the goodness to mention it'.[54]

Finally, the captain's boat arrives alongside. Captain John Hewitt (born c. 1812 in Liverpool) has good credentials. He has commanded the CDSP's *Royal William*. He has also been chief officer on the problematic *Liverpool* in 1840 (Chapter 4). He is now the perfect complement to the *Britannia*: 'A well-made, tight-built, dapper little fellow; with a ruddy face, which is a letter of invitation to shake him by both hands at once; and with a clear, blue honest eye, that it does one good to see one's sparkling image in'.

After a two-hour delay to await the last mails, the captain appears on one of the paddle-boxes with his loud hailer, the officers take up their stations, and all hands, including the engineers below, stand by for orders (Figure 5.2). Morale is restored: even 'the flagging hopes of the passengers revive'. Coinciding with the first of three cheers, 'the vessel throbs like a giant that has just received the breath of life; the two great [paddle] wheels turn fiercely round for the first time; and the noble ship, with wind and tide astern, breaks proudly through the lashed and foaming water'. Deeply laden with coal, mail and some eighty-six passengers, the *Britannia* pushes steadily northward out of the estuary, turns westward for the Irish Sea and, in due time, sets a course for the open ocean.[55]

[53] Dickens (1842: 7).
[54] Dickens (1842).
[55] Storey et al. (1974: 7n); Dickens (1842: 7).

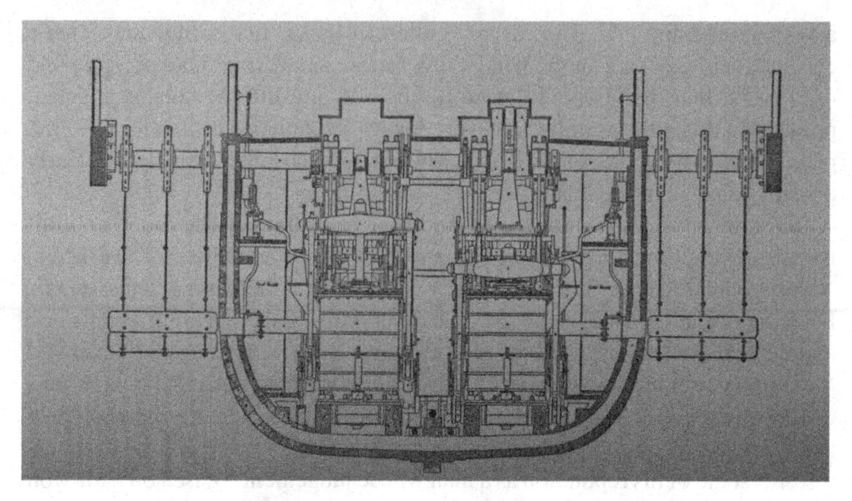

Figure 5.2. First Cunarder's pair of side-lever engines and hull cross-section

Throughout the 1840s, Elder's engines drove the Cunard mail steamers across the Atlantic. 'The machinery of these vessels', Napier's son wrote of the connection between engine-builder and shipping line, '... produced the regularity and gave that feeling of confidence which was so marked a feature of its success'. Elder, while following his earlier side-lever designs, had here bestowed even greater care 'in simplifying and making as accessible as possible every thing in the least degree likely to go wrong or to require attention, either at sea or in a foreign port'. Again guided by minimizing waste, his designs of 'an economical marine boiler are embodied in those flue-boilers' of the Cunard steamers with 'high and roomy furnaces', and water spaces 'large and accessible that every part can be repaired in the vessel'.[56] From the first four mail steamers, Elder's designs increased in size over some sixteen years 'to the culminating grandeur and finish of the great engines of the Persia'.[57]

[56] Napier (1866: 103–05).
[57] [Anon.] (1866: 103).

6 'Proprietor of the Atlantic Ocean'

Politics and Patronage on the Seas

Mr Cunard was not backed by any powerful influence in that House [of Commons]. He was a native of one our North American Colonies; and it would have been much to be regretted, if, in deference to interests in this country [Britain], any unfairness or injury had been done to him, after executing this service with a regularity which had excited the admiration of every one, and showing in all communications had with him, a spirit of fairness and liberality highly honourable to him.

Former Tory Chancellor of the Exchequer Henry Goulburn represents Samuel Cunard as a politically independent colonial gentleman (1846)[1]

Summary

Posted missing without trace, the unexplained disappearance of Junius Smith's *President* early in 1841 occasioned much speculation as to inadequacies in the steamer's design, raised questions as to the credibility of a transatlantic steamship system and brought to an abrupt end Smith's ambitions. In the years that followed, the Great Western Company struggled both with the protracted construction and problematic operation of its iron-hulled screw steamer *Great Britain*. Without a mail contract on its New York run, the GWSS fought hard to resist the Cunard Company's determination to extend its original territory to that of its Bristolian rival. Cultivating the patronage of successive Governments, Cunard and his partners maintained and enhanced a reputation for the reliability of its system and accordingly secured the public trust denied to its ailing rival.

6.1 'Destined to be the Glory and the Wonder': The Disappearance of the *President*

'The *President* will be *ready* for launch on the 23[r]d [November 1839]', Smith told his nephew of the promised birth of the second large steamship for his company. 'She is a prime ship and will do credit to any company'. On 9 December *The Times* reported optimistically that, although two previous attempts to launch the *President* had failed due to a shortfall in the anticipated

[1] *Hansard's Parliamentary Debates* (1846: 1426).

Figure 6.1. The *President*: largest merchant ship in the world in 1840–1

rise of the tide in the Thames, it was 'confidently expected that this day will see her fairly confided to the element of which she is de[s]tined to be the glory and the wonder'. While being towed from London to Liverpool for the fitting of its engines, however, the inadequately ballasted hull was reported to have rolled the masts out of it while off the Cornish coast.[2] But neither the delayed birth, nor the cranky behaviour as an empty, engine-less shell under tow, pointed at the time to a flawed design.

Other contemporary witnesses continued to admire the qualities of this, the world's largest steamer. Early in April 1840 merchant George Holt was taken on board the vessel, lying in Liverpool's Trafalgar Dock to receive its engines from Fawcett, Preston. Alongside lay another Atlantic steamer of promise, the Mersey-built *United States* for the Transatlantic Company. 'Some persons admire the form of the latter ship the most', Holt noted in his diary, 'but my taste leans to the *President*. I think her form better calculated for safety in a storm than the *U*[*nited*] *S*[*tates*] (Figure 6.1)'.[3]

In service from August 1840, the *President* failed to deliver on such great expectations. Even Smith rapidly retreated from his previous endorsement. 'She is his [Laird's] darling', he wrote later that year, 'and unfortunately proved to have more bottom than head'.[4] Already far outpaced by Cunard's

[2] Pond (1927: 195–6); *The Times*, 9 December 1839; Tyler (1939: 105); 'Mercator' (1840c: 792).
[3] George Holt Diary (GHD) (7 March and 6 April 1840).
[4] Pond (1927: 212).

much smaller *Acadia* on the ships' maiden westbound voyages, the *President* covered only 300 miles in six days on the second homeward passage in early November. After putting back to New York for fuel, the ship eventually reached Liverpool and was laid up until February 1841 when, with modifications to the paddle-floats, *President* began the fifth round voyage.[5]

By this time contributors to the *Nautical Magazine* (NM) had expressed concern about the ship's design. In his series of articles questioning the form of such large Atlantic steamers, 'Mercator' referred to a 'great steamer built almost as deep as she is wide' and thus with 'as much tendency to swim on her broad side as on her bottom'.[6] 'Nauticus' reported that 'her performances have been anything but satisfactory; and after having made two trips out and home, she was found so miserably deficient in power as to compel her owners to withdraw her altogether for the present'. While affirming that the construction of the engines 'do the greatest credit to those eminent engineers, Messrs Fawcett and Co., of Liverpool', he placed the blame on 'their power being so utterly disproportionate to her tonnage, so that when contending against a heavy head wind and sea ... they are found next to useless'. At the same time, he observed, Cunard's mail steamers 'have made their outward and homeward passages with admirable quickness and regularity', adding that they 'were all built and fitted by the Clyde ship builders and engineers, and I must say do great credit to them'.[7]

Departing for Liverpool on 11 March 1841 with the experienced Captain Roberts in command, the *President* went missing with all 136 crew and passengers.[8] By mid-April, the *Acadian Recorder* voiced a concern already current among experienced commentators: '[t]his steam ship, which seems destined every trip she makes, to painfully keep alive public anxiety respecting her safety, has not yet made her appearance'.[9] Holt, once so optimistic, recorded laconically the same month: '[s]till no tidings of the *President* steam ship which sailed from America in March'.[10] But the ship's first master, dismissed after making an especially slow voyage, told *The Times* that he would 'stake my reputation on the stability of the ship'.[11]

In early May, Smith decisively shifted the blame to Laird, whose 'pride' the ship had been and who had earlier sought to 'gratify a malignant revengeful disposition' against the Company after finding no public recognition of his role

[5] Bonsor (1975–80: vol. 1, 57–9).
[6] 'Mercator' (1840a: 429–39). Tyler (1939: 109–10) plausibly argues, on the basis of her dimensions, that the reference is to the *President*.
[7] 'Nauticus' (1841: 258–60).
[8] Bonsor (1975–80: vol. 1, 57–8); Pond (1927: 210–11n). See also Fox (2003: 99–100).
[9] *Acadian Recorder*, 17 April 1841.
[10] GHD (12 April 1841); *The Times*, 17 April 1841.
[11] *The Times*, 14 April 1841. Quoted in Tyler (1939: 108).

in transatlantic steam. As a consequence of Laird's sinful thoughts, Smith suggested with more than a hint of the old Calvinism's vengeful deity, 'it seems as if Providence visited his *motives*'. Although by this time Smith feared 'there is but slight ground for hope', rumours of sightings continued.[12] Even as late as June, the *Acadian Recorder* published further reports. But already the Line had ceased payments to families of the missing crew members and had opened the *British Queen* to the public in an attempt to raise funds for the relief of distress and deprivation.[13]

The ship's disappearance sealed the fate of the British and American Steam Navigation Company which was forced to lay up and sell its two-year-old *British Queen*. Indeed, *British Queen*'s last westbound passage from London to New York in March–April 1841 had taken almost a month to complete, the port paddle-floats having been stripped one by one during heavy weather. Bought for £60,000 in August 1841 by the Government of the recently independent Belgium, the *British Queen* began an Antwerp to New York service via Southampton in the following spring. The first westbound passage took twenty-four days from Southampton due to a severe gale during which the steamer, according to *The Times*, 'shipped a tremendous sea which made a complete breach over her, carrying with it part of the paddle boxes and several coal bunkers that were on deck'.[14] Only two more round voyages followed – at an estimated loss of £8,000 per voyage. Sold by public auction in 1844, the once-proud steamer, little over five years after the maiden voyage, was broken up. *The Times* had reported, meanwhile, that the British and American Company had lost between £80,000 and £100,000 since its establishment, a far cry from Smith's original predictions of annual profits of £130,000 (Chapter 3).[15]

In late 1840, the *NM* carried a contribution from the *President*'s builders to a discussion of practices relating to steamship construction. Curling and Young's observations look highly ironic in the light of the steamer's posthumous reputation. 'In some vessels, materials of inferior description and perishable quality are employed', they contended; 'in others, the connexion of the parts is ineffective, the fastenings injudicious or scanty, or the workmanship negligent'. Such defects in the timbers, however, 'are concealed by planking; defects of planking by cabin fittings; and defects of all by painting and decoration, which reduce all to one level of appearance, and by this appearance the public is deceived, and its safety compromised'. Their conclusion highlighted a gulf between public and professional judgement. 'Of the splendour of a saloon and the comfort of a cabin every passenger is a judge', they declared. But '[o]f the

[12] Pond (1927: 213–17n) (sightings).
[13] *Acadian Recorder*, 5 and 19 June 1841.
[14] *The Times*, 16 June 1842. Quoted in Bonsor (1975–80: vol. 1, 58, 176–7).
[15] *The Times*, 7 September 1841. The same report also stated that the Company had received £70,000 from the underwriters for the loss of the *President*.

strength of a vessel, the state of repair, or the general adaptation for a voyage, few are able, under any circumstances, to form a competent opinion, and fewer still have an opportunity for becoming acquainted with those facts, on which alone a just judgment could be formed'.[16]

Consensus subsequently coalesced around the opinion that the hulls of these vessels were too weak for the North Atlantic, especially during westbound crossings against head seas (Chapters 3 and 4). From the vantage point of 1866, for example, James Napier wrote in *Engineering* that 'the fate of the *President* will by those who saw her deformed state, and the means taken to hide it before her departure on her last voyage, be ascribed to a want of anxious caution and forethought on the part of her constructors'. While Napier's intention was to contrast such practices with the caution and forethought 'thoroughly engrained in Mr [David] Elder's character', engineers from the Napier firm would probably, given their role in supplying the engines for the *British Queen*, have witnessed the weaknesses hidden deep inside the *President* as it lay in Liverpool prior to its last voyage. If such knowledge were unavailable to the wider public, it certainly provided the talk of the maritime communities.[17]

Napier was not alone in reporting flaws in the steamer. Engineer Vernon Smith, pondering the future of Atlantic steam navigation in 1857, recalled that the *President* 'had not the reputation of being a strong vessel and ... was supposed to have an element of weakness in her unusual length', the ship's last Commander, Captain Roberts, having 'more than once expressed his doubts on this subject'. Moreover, the *Queen*, seven feet longer, 'had needed repairs, and was said to be even broken-backed'. Thus 'a feeling of insecurity and danger ... had been expressed about both of them, before the final catastrophe of the President'.[18]

6.2 'The Government Line': State Patronage and Cunard Control of the North Atlantic

In August 1846 a House of Commons Select Committee delivered its verdict on the Halifax and Boston mails. Its primary task had been to 'inquire into the circumstances connected with the granting of the present Contract for the conveyance of Mails from England to Halifax and Boston; and also into the circumstances connected with the granting of any new, or the extension of the existing contract, for the same purpose'. Its conclusions were threefold. First, that 'the arrangement has been concluded on terms advantageous to the public

[16] [Anon.] (1840: 852).

[17] Napier (1866: 101).

[18] Smith (1857: 11–12). Smith was probably alluding to 'Mercator's' and 'Nauticus'' warnings that pre-dated the *President*'s loss.

service, and have been most efficiently performed by Mr Cunard'. Second, the Committee declined to express an opinion 'whether a more advantageous one might not have been entered into, had the tender been thrown open to public competition'. And third, the Committee regretted that the arrangements had involved 'consequences injurious to the Great Western Steam-ship Company; and considering the meritorious character of the services rendered ... and its priority of establishment on the New York line, will be glad if, on any future extension of the Royal Mail Service, it receives the favourable consideration of the Government'.[19]

These seemingly conciliatory conclusions masked controversy over the successive awards of North Atlantic mail contracts to one line. The prime mover in the challenge was the GWSS. Its MD, Captain Claxton, delivered the largest volume of witness evidence. Facing a Select Committee on which Bristol interests were well represented, Cunard found himself relentlessly challenged on a number of counts. Two questions in particular seemed to unsettle him. Had he seen the initial advertisement inviting tenders for the North Atlantic mail contract? And how did the sequence of ever-larger contracts that his company had procured justify its claims not only to the Halifax and Boston mails, but now also to New York in direct competition with the GWSS. Cunard's replies, copied down verbatim, often appeared defensive. Moreover, while GWSS begged Prime Minister Peel 'not [to] crush their enterprize by establishing a fresh monopoly with a company ... backed by your Lordships' patronage', Cunard brusquely referred to his company as 'The Government line'. It was small wonder that the sympathetic Fanny Kemble later called him 'a sort of proprietor of the Atlantic Ocean between England and the United States'.[20]

After a difficult first winter the Cunard Company's financial losses had mounted. 'Up to that time trade was very bad and there was a good deal of competition; we had competing with us the British Queen and the Great Western, and we found we were unable to go on', Cunard later told the SC. '[And] but for that additional [contract] sum being granted, we should have given up the contract'.[21] Writing to Cunard in mid-April 1841, MacIver had countered his Halifax partner's pessimism with a more nuanced assessment. '[T]he day must have been cold, I think, or the subject has had a chilling effect over your spirits', he observed. 'It is incumbent on us as shrewd merchants to have our eyes ever open to the dark side of our doings ... there is so much of show, and of the imaginary, in the money matters of steamers'.[22]

19 SC (1846), 3; *Hansard's Parliamentary Debates* (1846), esp. 481–2, 1417–31.

20 SC (1846), 14, 27; Kemble (1878: vol. 1, 287).

21 SC (1846), 23. Fox (2003: 102) (from Admiralty Papers 12/387: 76 (13 May 1841)) notes a loss of some £15,000 for the first nine months of operation. Compared to RMSP and PSNC, this was a minor amount.

22 MacIver to Cunard, 16 April 1841, CP, MG1 vol. 248 no. 43, NSA. See also Fox (2003: 102–04).

Couched in language that reflected Calvinist distaste for showmanship, MacIver was here urging his partners not to be led astray by illusions. With respect to the over-optimistic kind, he cited the rival GWSS who 'thought on their first year's results that all was bright for the future ... [whereas] the base of their error and ruin dates from that'. Three days later, MacIver received from Cunard a letter enclosing the prospectus of an iron steamship company to New York, a project that he forthrightly rejected as another instance of illusion: 'Quackery – not worth noticing – would only do harm to publish it – would excite more attention than it deserves'.[23]

With respect to the over-pessimistic kind of illusion, MacIver turned to their own case. He spoke of 'wise people' like themselves, though not easily led astray by false promises, nevertheless finding 'shaddow [sic] for substance the amount of their doings ...'. He instead offered Cunard a more substantial, less 'shadowy' appraisal of the Company's current position: 'our own doings, up till now, I rate the experimental; and not sound evidence of our true position or prospects'. Thus 'we have paid, like all beginners in new trades of magnitude, thro' costly experience, and are now arrived at that point when we must turn this experience to profit'.

The partners' solution lay with persuading the Government, in its own interests, to increase the payments. MacIver first rehearsed the arguments privately with Cunard. 'Mr Baring must know as well as you or I', he wrote, 'that if we or any other private Coy cannot carry on a Trade to some profit, stop the Trade must, whatever embarrassment may be caused to the Government or whatever the consequences to the Steamboat owners'. He insisted, however, that 'the Government have had good earnest from us of our ability to do the Contract-work well' and that 'it is only by their help we can continue to do so'. The amount of such financial assistance, he contended, should be only 'to the extent of what [the] legitimate sources of the Trade will not supply' in order to permit a moderate return on the capital. Experience had shown that the current contract was inadequate. To do what the Government wished to have done would require an extra £60,000 per annum to support seven, rather than four, transatlantic steamers.[24]

By May 1841 Cunard, MacIver and Burns were in London formally renegotiating the mail contract. Burns communicated to his wife the news that he had seen 'a STRICTLY confidential note and report from [Parry] ... in our favour, and I hope by God's blessing we shall succeed'. The partners continued to benefit from patronage flowing from high places, often Whig evangelical in character. By September the Admiralty increased the previous contract of

[23] MacIver to Cunard, 19 April 1841, CP, MG1 vol. 248 no. 27, NSA.
[24] MacIver to Cunard, 16 April 1841, CP, MG1 vol. 248 no. 43, NSA.

£60,000 to £81,000 per annum on condition that a fifth mail steamer was built. The Line then ordered the 1,400-ton *Hibernia* (1843).[25]

In his private analysis for Cunard, MacIver emphasized that 'the resources of the Trade' consisted of revenue only from passage money, since 'we cannot carry much cargo' and that 'only thro' the summer months'. Cunard put the quantity of freight at fifty or sixty tons per steamer at a fixed rate of £7 per ton. Passage money eastbound (£25) did 'not pay for more than the eating and drinking of the Passengers and Servants' and as such should, 'by arrangement with the [Great] Western Coy' be 'worked up to £30'. MacIver also argued for a fixed rate eastbound of forty pounds and nineteen shillings, regardless of whether passengers embarked at Boston or Halifax since Canadians could obtain no cheaper steamer route than that from Halifax, while New Yorkers could opt for a direct steamer passage if the Boston fare were higher.[26]

When questioned by the SC (1846) on the state of the GWSS, Cunard replied that 'by their own showing, they are in a very bad way, and have been for years. ... I think they never had any profits'. Having made a passage in the *Great Western* himself, he acknowledged that 'she is a good ship, entitled to every credit; she is not quite as fast as ours, but she is a most excellent ship'. Unlike Cunard's line of steamers that 'must go at [year-round] fixed periods', however, the GWSS 'may choose their time', giving them 'some advantage in that respect'. And for the 'last four or five years they have never had a winter voyage'.[27]

One of the principal problems facing the Bristol Company from the beginning was their reliance on one ship in contrast to Cunard's four-steamer system. The GWSS's problems, however, were greatly compounded by the protracted construction of their second vessel, the iron-hulled *Great Britain*. In retrospect, Scott Russell recalled that the GWSS consulted him in the late 1830s 'as to the forms and proportions which I should recommend for the new ship'. His advice was that 'they should make her somewhat larger than the *Great Western*, but in most respects a copy of her'. He drew for the owners 'a finer water-line, so as to give her bow, which was too full, a fine hollow entrance, for speed'. He also recommended a slight increase of breadth 'to give a little more power of standing up under canvas'. Instead, the Company chose not 'an enlarged or improved *Great Western*, of the sort the Cunard Line afterwards adopted', but opted for a hull design, 'entirely revolutionized, and turned into an imitation of Sir W. Symond's new and empirical form of ship'. Indeed, he argued, almost every feature of the vessel's design could be categorized as 'experimental'. The

[25] George to Jane Burns, 15 May 1841, in Hodder (1890: 204–05); Bonsor (1975–80: vol. 1, 75). On mail contracts post-1840 see Daunton (1985: 158–61); Tyler (1939: 97–8); Hyde (1975: 35).

[26] MacIver to Cunard, 16 April 1841, CP, MG1 vol. 248 no. 43, NSA; SC (1846), 24.

[27] MacIver to Cunard (1846: 23–4).

result was, Russell concluded, that the *Great Britain* was unique: an experiment never to be emulated.[28]

Construction of the hull in Bristol required a purpose-built dock costing over £53,000 and designed to eliminate the strains imposed on a long vessel by conventional slipway launches. From an original £76,000, the estimated costs of the ship and engines rose alarmingly to well over £117,000 as specifications changed and building time ran to almost six years.[29]

A major delay arose from the consequences of the decision to change from paddlewheels to screw propeller. The steamship company's engineer, Francis Humphrys, had visited James Nasmyth's Manchester works to discuss machine tools, 'of unusual size and power', for the task of making the paddle-engines. Unable to find a forge hammer powerful enough to shape the thirty-inch-diameter wrought-iron intermediate paddle-shaft, Humphrys asked Nasmyth for advice in late 1839. Nasmyth quickly came up with a plan to construct a steam hammer in which an inverted steam cylinder would use its piston to raise a heavy block of iron (the 'hammer') to a suitable height. Releasing the steam pressure would then allow the block to descend under gravity to administer the first of many blows on the shaft, kneading it 'as if it were a lump of clay'. Humphrys, Brunel and Guppy welcomed the proposal. But the shaft was never forged. Having witnessed Francis Pettit Smith's screw propeller at work, Brunel persuaded the directors to replace the well-advanced paddle-engines with a new arrangement that Humphrys was now tasked to design. But Humphries died soon after, carried off by 'a brain fever'.[30]

As early as 1841, 'Nauticus' in the *Nautical Magazine* questioned the decision to deploy a screw propeller rather than paddlewheels. '[I]t does appear to me – a disinterested person in the affair, – a rather hazardous thing on the part of the Company, to adopt an invention, the success of which they are not as yet fully satisfied', he observed. 'One voyage, however, from Bristol to New York, will be quite sufficient to decide this question'.[31]

In contracting for the steamer, the GWSS had 'determined upon building a ship that would double the number of passengers and double the quantity of freight'. But competition from Cunard's line to Halifax and Boston very much decreased the *Great Western*'s returns on the New York route with the result that 'loss was staring us in the face'. In the aftermath of Cunard's enhanced mail contract in 1841, GWSS pleaded in vain with the Chancellor of the Exchequer (Henry Goulburn) in Peel's new Conservative administration for some form of grant as recognition for the benefits the service had 'conferred upon the

[28] Russell (1863: 17–18).
[29] Corlett (1980: 93–6).
[30] Nasmyth (2010: 238–43).
[31] 'Nauticus' (1841: 258–60).

mercantile, financial, nautical, and mechanical interests of the empire'. Bereft of state support, proprietors pressed the reluctant directors early in 1844 to sell the *Great Western* to P&O. The sale, however, collapsed.[32]

Unlike the relatively obscure birth of the early Cunard mail steamers, spectacle accompanied the *Great Britain*'s construction. The naming and floating out was the occasion for the presence of Prince Albert and aristocracy. Vast crowds occupied all possible vantage points. Upon completion and trials in 1845, the ship spent five months on the Thames, visited by the Queen and Prince Albert, the Lord Mayor of the City of London and the Board of Admiralty before being opened to the public. In Liverpool, some 2,500 visitors per day went aboard in July ahead of *Great Britain*'s maiden voyage. Yet spectacle and novelty were no guarantee of passenger bookings. The public interest and publicity attracted no more than sixty passengers (out of a capacity of over 350) and about 360 tons of cargo. Steaming westward against summer gales the big steamer averaged a respectable 9.4 knots and arrived in New York in just under fifteen days.[33]

Although 21,000 visitors boarded the *Great Britain* in New York, many of the crew deserted and only fifty-three passengers joined for the eastbound passage. Problems with a regular steam supply and injuries to crew members when the main topmast broke during a summer squall raised questions about the novel ship's reliability. But with the passenger list exceeding one hundred, the second Atlantic crossing began well in September 1845. The *Bristol Mirror* saw the numbers as 'proof that the prejudice that once existed against her is fast disappearing'.[34] During ten days of head seas, however, the foremast carried away. Over two weeks out, in the captain's words the steamer 'was among the shoals of Nantucket on a thick dirty night', supposedly set some sixty-five miles north of the intended course by causes unknown. With 200 miles still to cover, the steamer's bunkers (filled with 1,050 tons at Liverpool) were well-nigh exhausted. Worse still, the six-bladed screw had been rendered practically useless by the loss of three extension arms on one side. During the return passage, the screw failed again, leaving the ship to demonstrate the excellent performance under sail alone. After twenty days, tugs took charge of the ship off the Mersey Bar. Loyal passengers, however, affirmed that the *Great Britain* 'on this occasion strengthened our confidence and won the admiration of all on board ... [We] express the superiority of the *Great Britain* in a heavy gale ... [and] in every respect' (Figure 6.2).[35]

[32] SC (1846), 10, 31.
[33] SC (1846), 102–14; Bonsor (1975–80: vol. 1, 62–3). See also Burgess (2016: 1–2, 45–52). His sweeping claim (p. 2) that *at sea* the vessel was a 'miserable failure' was not a widespread contemporary opinion.
[34] *Bristol Mirror*, October 1845. Quoted in Corlett (1975: 117).
[35] Corlett (1975: 114–19).

Figure 6.2. *The Railway Magazine*'s representation of the *Great Britain* (1845)

Laid up in Liverpool for the winter, the vessel underwent a major refit of machinery and rigging. But for the third round voyage, only twenty-eight passengers boarded. On the twenty-day westbound crossing, mechanical failure resulted in six days under sail alone. For the return passage, the list rose to forty-two and, despite headwinds, took under fourteen days. The fourth crossing in the summer of 1846 reflected a gradual restoration of confidence with 110 passengers westbound. Encountering thick fog off Newfoundland, however, the steamer scraped a reef but escaped with minimal damage. Navigational errors continued to dog the iron steamer.[36]

By virtue of the ship's size and novelty, the *Great Britain*'s performance attracted relentless attention. In March 1844, for example, Henry Booth told the Liverpool Polytechnic Society that the 'Leviathan of three thousand tons ... may, indeed, accomplish *cargo* as well as speed, but it will be accomplished at enormous cost'. He argued that such multi-purpose steamships, increasing annually in size and expense, generated inadequate returns on capital. He therefore called for the introduction of fast steamers with the precise aim of conveying the mercantile mail bags from Liverpool to New York in one week.[37]

In the autumn of 1845 rumours of a new contract for Cunard steamers sailing to New York began to alarm the GWSS. Bristol MP Philip Miles called on Secretary to the Admiralty Lowry-Corry who assured him that 'there was no intention of entering into a fresh contract with Mr Cunard'. Soon the press reported the contrary. In Parliament, Miles pressed the Tory Chancellor for an answer and received the vague reply that the original contract specified that any

[36] Corlett (1975: 121–5).
[37] Booth (1844–6: 27–8); Marsden and Smith (2005: 100–01).

additional voyages 'were to be performed by Mr Cunard at a particular rate'. The GWSS continued to press the Chancellor in an interview at which it offered to do the work for the direct New York line 'for half the amount which he paid the other Company'. Goulburn simply repeated his point regarding Cunard's right. The GWSS then met with Prime Minister Peel in early June when he revealed his ignorance of the case: 'Sir Robert Peel gave us to understand that his impression was that the Halifax boats always went to New York'.[38]

When the Whigs returned to power later that month, GWSS directors appealed to Prime Minister Lord John Russell. They highlighted the damaging 'patronage of the Government' in 'confirming the principle of monopoly on one of the most important lines of passenger traffic between this and any other country'. In fact, the contract with Cunard had already been formally signed on 1 July. The GWSS's only remaining weapon was to obtain Parliament's consent for an SC to investigate mail steam-packet contracts. The present system of tenders, Henry Berkeley, MP, insisted to the Committee, 'was a most fictitious and illusory system ... Tenders should be demanded to encourage a fair spirit of competition, and contracts given without favour to those who would do the work on the most reasonable terms'. As implemented, 'there was something bad and rotten in that system' that had been carried out in a 'hole-and-corner manner'.[39]

For the first time, however, the two British lines faced potential competition from state-aided projects in America. In October 1845 the United States Postmaster General invited tenders from American steamship projectors to carry the mails between New York and European ports. The successful bid involved the Ocean Steam Navigation Company whose owners (including sailing packet owner Mortimer Livingston) aimed to raise a nominal capital of $1,000,000 for four transatlantic steamers.[40] At the SC, Cunard agreed with Berkeley's suggestion that the new line of packets would 'have the effect of driving the Great Western and Great Britain off the line' to New York. However, the contract awarded to his own line was likely to do just that.[41]

Under further SC questioning, Cunard revealed the manner in which he and his partners had persuaded the Government to grant them a new contract extending the line to New York. Crucial to their strategy was the clause in the original mail contract which already 'provided for an increased number of sailings'. Equally vital to the Cunard Line's case was the prospect of the American project which threatened to 'deprive the [British] Government of half the postage, and deprive me of half the passengers'. While Goulburn and

[38] SC (1846), 11 (quoting from the *Herald*, 13 May 1846), 13–14.
[39] Hansard (1846: 1417–23); SC (1846), 15–16.
[40] Bonsor (1975–80: vol. 1, 186).
[41] SC (1846), 23.

Figure 6.3. Robert Napier's fitting-out dock at Lancefield on the north side of the Clyde

Corry accepted the importance of the case, Goulburn baulked at the figure of around £85,000 that the Company wanted. Cunard countered with the assertion that '[y]ou are bound to give me so much, it is in the contract'. The Chancellor denied Cunard's claim and sought the opinion of the Crown legal advisers who pronounced that if the Government called upon Cunard 'you are bound to give him so much, but if he calls upon you, you are not bound to give it'. In the end, Cunard accepted an additional £60,000. The partnership now stood to receive £145,000 per annum from the British Government. With the start date for the New York service set for 1 January 1848, Cunard went immediately to Glasgow to settle the shipbuilding order. Certain of the Government contract, the Company had already paid Napier £50,000 apiece for the machinery of three new steamers (Figure 6.3).[42]

Cunard's performance at the SC in 1846 scarcely impressed his GWSS interrogators. His answers were often indirect, even evasive and occasionally contradictory. He seemed untroubled by accusations of 'favouritism' and 'exclusive rights'. When pressed on matters of competition and open tenders, he insisted that had the Government made a contract with any other party it would 'virtually be bringing a line into competition with me. I call my line the Government line'. But the SC was not the only place of trial. Cunard and his

[42] SC (1846), 23–5. One of the three vessels was to have been constructed by Napier as an iron-hulled steamer but the Cunard Company changed its mind.

partners knew that their survival as a Company above all depended on their performance within a very different arena of trial: the Atlantic itself.

6.3 'With Undeviating Regularity': Promoting a Reputation for Reliability

Less than two months after the SC concluded its proceedings, the *Great Britain* departed from the Mersey for its fifth round voyage. With a record passenger list of 180, confidence was high. Benefiting from a following wind and tidal stream, the ship's course required only significant alteration when clear of the south-west tip, known as the Calf, of the Isle of Man. The ship could then pass safely through the wide North Channel between Ireland and Scotland. When in line, two lighthouses placed high up on the Calf marked the position of the dangerous Chicken Rock offshore. Their position, however, was prone to low cloud which obscured their beams. Whether through poor visibility, compass error, alcoholic haze or a combination of all or some of these on an early September evening, the master and officers failed to see either the Isle of Man's distinctive mountainous profile or the Calf lights. Instead, the ship continued on a straight course westward. When the officers eventually sighted a single light to starboard, the master ordered a course alteration to the north. Very soon afterwards, the *Great Britain* ran hard ashore.[43]

As the ship's predicament became known on land, so those in command received confirmation of their gross navigational error: the steamer had passed to the south of the new St John's Point Lighthouse on Ireland's County Down coast more than thirty miles beyond the Calf, only to turn north and plough into the shore of Dundrum Bay. The press quickly gleaned conflicting accounts of the likely causes. One passenger, a 'gentleman of Montreal', told the *Liverpool Journal* that the 'Rev Mr Tucker of Bermuda' had asked a ship's officer why he was examining the ship's compass. 'Something the matter with it came the reply'. Accordingly, 'an opinion prevailed [aboard] that the compass was the cause of the disaster'. But then Captain Hosken 'solemnly declared that there was nothing wrong with the compass'.[44]

The master subsequently blamed the stranding on the absence of the St John's Point Lighthouse from his current chart, while the *NM* and *Mechanics Magazine* (*MM*) blamed the captain's incompetence. In the verdict of a contributor to the former journal, 'the Great Britain seems to have darted from her port, anxious to run her course, like a high-mettled racer, and yet from some unexplained reason, to have been allowed to rush headlong on her own

[43] Corlett (1975: 125–6). A wave-washed lighthouse was built on the Chicken Rock in the 1870s to overcome the problems of low cloud over the Calf.

[44] *The Times*, 28 September 1846 (reprinting a *Liverpool Journal* report).

destruction as if she were in the hands of some landsman who had heard of the sea, but knew nothing of the duties of the seaman'.[45]

Eleven months of pounding by the Irish Sea on an exposed coast was not lost on the proponents of iron vessels. But the combined action of saltwater and winter conditions ruined the engines and interior. Also wrecked were the finances of the GWSS. With the *Great Britain* insured for only £17,000, the owners had to sell the *Great Western* to RMSP. The GWSS's grand project had ended.[46]

Reports that the *Acadia* was taking many of the *Great Britain*'s former passengers from Liverpool to the United States highlighted the reputation of the Cunard Company for reliable public performance, based upon well established practices and attention to every detail.[47] As early as 1842, *The Times*, in its 'Money market and City intelligence' column, praised the line's track record. 'Of all the instances of steam navigation on a large scale, none has been more regular in performance, or more perfect in execution, than that termed "Cunard's line"', it noted. Statistical tables showed that few voyages exceeded the allotted time, while many others had concluded more than twenty-four hours ahead of schedule. The same report also cited an article in *The European* which stated that the Cunard Company's ships had altogether now crossed the Atlantic eighty times 'with undeviating regularity'.[48]

After the death of David MacIver in 1845, day-to-day management of the fleet in Liverpool passed to his brother Charles. For him, as for his late brother and George Burns, the key to 'success' in the face of the trials posed by the Atlantic lay with the 'personal fitness' of the men who directed the affairs of both ships and line. Personal fitness aboard ship centred upon the master's trustworthiness. 'The trust of so many lives under the Captain's charge is a great trust', D. & C. MacIver stated in their seven-page 'Captain's Memoranda' (1848 edition), a detailed set of shipboard practices issued to each master. 'We rely on your keeping every person attached to the Ship, both officers and people, up to the high standard of discipline and efficiency which we expect in the service', the managers began. While acknowledging that '[y]our own practical knowledge may be your best guide', they listed no less than seventy-two 'memoranda' which all masters were expected to follow in the practices of navigating and managing a mail steamer. These guidelines included instructions on best practices in watch-keeping, steering, regular soundings, fire prevention and pilotage.

[45] [Anon.] (1846: 616); Corlett (1975: 126–7) (with extracts from the *NM* and *MM*).

[46] Corlett (1975: 139).

[47] *The Times*, 28 September 1846.

[48] *The Times*, 23 August 1842. An earlier report in *The Times*, 25 April 1842, reported complaints against the line concerning recent delays in the homeward mails from Halifax.

The 'Memoranda' conveyed a strong moral economy reflecting the Presbyterian values of all the partners. The evils of waste had to be fought. With respect to feeding the crew, for example, they advised the master to 'Give all plenty – waste nothing – a little interest expressed by you to the Head Steward in his being economical, and to work up the broken meats for the Crew, Firemen, Coal-trimmers, &c., will remind him of his duty; be very distinct in checking with a firm hand any appearance of waste'.[49]

The partners' opposition to 'wanton and extravagant waste' effectively defined 'personal fitness' aboard ship. Affronted by accounts of over-indulgence aboard the *Cambria* at Christmas in 1847, MacIver issued a circular letter to the officers of the mess. 'No man in this concern has had it in his power to say in truth that he has been otherwise than well treated', he asserted, 'but wherever I find a set of men rating themselves only by what they can stow away in their bellies, I have prima facie evidence that they are not the men for the British & North American Royal Mail Service, and the sooner they change to where they can better gratify their appetites the better'. Worst of all had been the indulgences of Christmas Day – a day scarcely the occasion for any festivities in Calvinist Scotland – when the ship's doctor and purser had joined officers in the consumption of mock turtle and venison, thereby 'virtually commit[ing] the same act, as handing things not their own over the ships' side to their friends under shade of night'. He further informed his wayward officers that they would now be charged for the food consumed 'at passenger rates'.[50]

MacIver also issued a set of practices to be followed with regard to practical and moral economy in the sailing (navigating officers'), engineers' (engine room) and stewards' departments. Day or night the officer of the watch 'should count the revolutions of the engines every two hours, and put them on the Log Slate, and afterwards enter in the Log Book'. Engineers must satisfy themselves 'that the principle and practice of weighing coals consumed in the furnace be pretty correct' for 'without this you will never know what you have consumed or have left on board, *a very essential thing to know on long passages*'. Upon arrival in Boston or New York, stewards and purser were to take 'careful account of what ... [stock of provisions] is on board' and the account was 'to be brought home to us'. On the one hand, this accounting 'enables the Steward department to know correctly what is required at Boston and New York for the homeward trip; guessing leads to unnecessary expense, everything higher in America than England'. On the other hand, it enabled the managers 'to see how much a given number of people consume in a certain time'. Even management of the crew's time was subject to strict economy. In the event of

[49] D. & C. MacIver, 'Captain's Memoranda' (1848), CP, LUL.
[50] MacIver to Officers of the Mess, December 1847, CP, LUL.

the Carpenter and Joiner being unemployed, 'fancy and unnecessary work' was to be avoided on account of it 'being little better than a waste of materials, men's time, and loss to the Company'.[51]

Public confidence depended, not on instant images of 'the largest' or 'the fastest' but on promises fulfilled and performance maintained. The strict discipline would assure passengers that should a serious accident occur it would not be the result of Company incompetence. Moreover, if the vessel were imperilled, the training of the officers and men would mean that everything possible would be done to prevent loss of life.

Over its first decade the Line was by no means immune from accidents at sea. In most cases, the Company not only avoided damaging criticism, but also achieved publicity that actually boosted the reputation of the Line. In July 1843, for example, the *Columbia* with eighty-five passengers aboard went ashore in dense fog near Cape Sable at the southern tip of Nova Scotia. Against the total loss of the steamer, the reported rescue of all the passengers, crew and mails turned the shipwreck to the Line's advantage. The *Liverpool Albion*'s account, reprinted in *The Times*, told of the crew and passengers heroically attempting to refloat the stricken vessel at each high tide. Then passengers, luggage, crew, mails, all the moveable objects and even some of the coal were taken ashore. Although the press report made cursory reference to 'one [crew]man missing', almost all subsequent readings assumed no loss of life.[52]

The Line's seemingly blameless record was not without blemish. The worst accident occurred in June 1849 when the *Europa* was in collision with the emigrant sailing ship *Charles Bartlett*. The latter vessel sank with the loss of 135 lives. Not everyone aboard the *Europa* remained uncritical. 'The ships known as the "*Cunard Line*" are exceedingly well managed and well fitted', shipowner and merchant R. B. Forbes of Boston wrote in the *NM*, 'but it would be ... a cruel libel on the intelligence of the proprietors and the superintendents to say that they are perfect'. His concerns stemmed from the potential for more accidents as rival lines vied for passengers by offering greater regularity, increased speed and more luxury. He warned Cunard against complacency and urged the line to set an even greater example: 'let the means of averting damage and of saving life in case of accident be the first study of the projectors and managers of these ships'. Thus 'one or two hundred [pounds] saved in carving, and gilding, and painting, and spent in apparatus for preserving and restoring life, would be much applauded, and give additional popularity to this yet unrivalled line'. Forbes could indeed speak with experience and authority for

[51] 'Captain's Memoranda' (1848: 2–5).

[52] *The Times*, 25 July 1843; Bonsor (1975–80: vol. 1) (data from fleet lists of Atlantic lines 1843–58).

he had personally risked his life in the sea to save one or more of the stricken vessel's complement.[53]

The much-vaunted rival lines barely materialized in the first decade of Atlantic steam navigation. Claxton had told the SC in 1846 that the Ocean Steam Navigation Company 'have not started; they have not got their money; according to the last accounts I had, the scheme had fallen to the ground'.[54] The obituary, however, was somewhat premature. The United States-backed line raised only enough for two ships ($600,000), including substantial investment from North German states. The subsidy correspondingly fell from $400,000 per annum for a fortnightly service between New York and Bremen to half that value. The 1,640-ton *Washington* (1847) and 1,734-ton *Herman* (1848) were large New York-built sailing packets with steam engines by the nearby Novelty Iron Works. From the beginning the engines of these ships required numerous modifications and both vessels fell short of expectations. In January 1848 the *Washington* arrived at New York after a twenty-seven-day passage in severe weather, while in September the *Herman* ran aground on the Isle of Wight.[55]

French projectors fared even less well. In 1847 the *Compagnie Générale des Paquebots Transatlantique* received from the French Government a subsidy of around £16,000, a similar amount on loan as working capital, and four 1,500-ton wooden paddle frigates or transports built about five years earlier in the Navy Dockyards at Brest and Cherbourg. The service between Cherbourg (later Le Havre) and New York began in the same year. Voyage times westbound typically exceeded twenty days, with frequent sail-only runs and diversions (five out of nine passages) for coal. On one eastbound passage, an American sailing packet departed two days later and arrived at Le Havre on the same day as the French steamer *New York*. Reports of trouble and incompetence multiplied. One of the steamers, for example, inflicted $3,000 damage on a barque before hitting a pier in the East River, allegedly due to a lack of understanding between helmsman and pilot. The *New York Daily Tribune* in 1848 asserted the repeated shortcomings meant that no one in England would be likely to 'patronise this line in any shape'. With reported losses of £80,000 by then, the Company was suspended and the ships returned to the French Navy.[56]

Soon after this short-lived state-sponsored venture, the celebrated Black Ball Line of sailing packets put into service the 1,857-ton wooden paddle steamer *United States*, the first American full-powered steamer to arrive in Liverpool. The ship subsequently altered its route explicitly to replace the French service. After a promising voyage to Le Havre and back in early summer, serious

[53] Forbes (1849: 436–9); Bonsor (1975–80: vol. 1, 140–2).
[54] SC (1846), 21.
[55] Bonsor (1975–80: vol. 1, 186–9).
[56] *New York Daily Tribune*, 7 January 1848. Quoted in Bonsor (1975–80: vol. 1, 190–2).

mechanical problems developed on the second westbound passage. The third and last such passage took twenty-seven days in the face of a succession of westerly gales and consequent fuel shortages. By March 1849 the steamer had been sold. The Black Ball Line took no further part in transatlantic steam and remained resolutely wedded to sail until its demise in 1878, by which time it had lost to the steamer lines all but heavy freight such as rails shipped from Europe to feed the voracious appetite for railroad construction in the New World.[57]

Kemble's imaginative description of Samuel Cunard as 'a sort of proprietor of the Atlantic Ocean' was a telling phrase applicable to the partnership of Cunard, Burns and MacIver. So too was Cunard's public depiction of the enterprise as 'the Government line'. In its first decade, the managers of this expanding system of steamers had successfully kept control of Government patronage, avoiding the hazards of a potentially chaotic sea of free competition on the one hand and steering barely clear of a public perception of favouritism on the other. They had selected officers of personal fitness and moulded them to their strict, distinctly Presbyterian practices of moral economy. And through their careful cultivation of the company's public reputation in the press, they had even represented shipwreck as a triumph for the line's practices and principles.

[57] Bonsor (1975–80: vol. 1, 193–4).

Part II

Westward for Panama

7 'Mail-Coaches of the Ocean'

The West India Company Project

> Never at any period of colonial history was it more imperiously necessary, or more imperiously the duty of every one interested in, or connected with the Colonies, and especially the supreme Government, to bind these Colonies together by the closest ties; and the individual who attempts to disunite them in any way, and in any thing, is an enemy to their interest, and an enemy to his country.
>
> *James McQueen appeals to Francis Baring (Secretary to the Treasury) for Government support for West Indian mail steamship communication (1838)*[1]

Summary

Intensively involved in the construction of an unprecedentedly large fleet of fourteen ocean steamers to carry mail, passengers and high-value freight between Britain and the West Indies and Spanish Main, Scottish projector James McQueen was the geographical system-builder *par excellence*. The metropolitan-centred Royal Mail Steam Packet Company (RMSP), however, assumed values very different from those of the private partnership of Messrs Cunard, Burns and MacIver. The RMSP had ambitions not only to serve national and imperial masters through very close links with the Admiralty but also, through its Royal Charter and its Court of Directors based in London, to aspire to the status of the Honourable EIC. In the process of constructing its grand, but often disordered and troubled, system of ocean steam navigation, RMSP took seriously the authority of a young naval architect and natural philosopher, John Scott Russell.

7.1 'The Work to be Performed': McQueen's Global Plan

Lardner's disparaging remarks in 1837 concerning 'a swarm of projectors' promoting ocean steam navigation (Chapter 3) very likely included James McQueen as a prime target. This fifty-nine-year-old lowland Scot was indeed generously possessed of both zeal and knowledge. Eighteenth-century Glasgow merchants had built their wealth on trade in sugar and tobacco with the Americas and especially with the Caribbean slave islands. Arriving in the West Indies in his teens, McQueen duly rose to become manager of a Grenada

[1] McQueen to Baring, 22 August 1838, in McQueen (1838b: 25–6).

plantation, travelled widely in the region and promoted himself as an authority on African geography. Back in Glasgow around 1820, he was both editor and part-owner of the *Glasgow Courier* and a regular contributor to the 'Commercial Reports' section of Blackwood's *Edinburgh Magazine*. These publications often reflected Tory political perspectives, but appealed more generally to mercantile communities eager to view their commercial interests within wider cultural and colonial contexts. 'There are many markets in the world yet to be opened, and which can be opened to our commerce', he claimed in one of his reports. 'Masters of the ocean, we can gain access into every country, and to every land'.[2]

In the summer of 1837, McQueen presented a preliminary plan for a global imperial network of mail shipping lines to the Post Office and especially to the Admiralty 'for examination and report'. No measure, he told Baring a year later, 'ever received deeper or more serious attention than the subject did from the heads of that department, with whom I had the honour and the great satisfaction to meet about it'. He declared the public's special indebtedness to Sir Charles Adam (First Naval Lord of the Admiralty), Sir Edward Parry and Charles Wood (Secretary to the Admiralty) 'for the readiness with which they all of them entered upon the subject, the great labour they took in the matter, and the very great anxiety which every one evinced, to give the commercial world every facility and accommodation, by bringing into operation, as early as they could, a plan as perfect and as extensive as possible'. Thus '[e]very point, every line, and every combination, and every distance, and every calculation, was most carefully, most minutely, and, I need scarcely add, with masterly hands examined'. As a result, the plan was recommended to the Treasury.[3]

In McQueen's original version of July 1837, the plan provided for the conveyance of mails by steam between Britain and the West Indies, both 'British and foreign'. As a result of the discussions with the Admiralty heads, by September McQueen had rendered the plan more precise and more ambitious. The service would operate 'twice every month, to the whole of the Eastern Coasts of America, from the St Lawrence to the Rio de la Plata, including all the West Indian Islands, and all of the Gulf of Mexico'. Presented to the public early the following year as *A General Plan for a Mail Communication by Steam between Great Britain and the Eastern and Western Parts of the World*, it made the island of Fayal in the Azores (known as the Western Islands) the point 'from which steamers were to diverge to Halifax, to Barbadoes [sic], and to Brazil ... while steamers from Halifax, by New York to Havannah [sic], were to connect the West Indian chain with North America'. The *General Plan* further showed

[2] [McQueen] (1819–20: 594); Nicol (2001: vol. 1, 35).
[3] McQueen (1838b: 10–12); 'Adam, Sir Charles (1780–1853)', *ODNB*.

how the West Indian service could 'be made to extend across the Isthmus of Central America to Sydney, New South Wales, and Canton-China'.[4]

'No narrow or parsimonious views on the part of this great country ought to throw aside the plan', McQueen told Baring in a letter of February 1838 that he included as a preface to the published *General Plan*. The undesirable alternative was to 'leave it to be taken and split into divisions by parties, perhaps foreigners, who will then not only command the channels of British intelligence, but be enabled to demand what price they please for carrying a large and important portion of the commercial correspondence of this country'. Failure to secure possession of 'all the channels of communication with the Western Archipelago' and thus to command 'the principal political influence therein' would hand that influence to the United States which was 'now earnestly looking about, and proceeding to acquire and to extend the same in that quarter of the world'. He also pointed out that the Government would possess great advantages in the event of hostilities 'by having the command and direction of such a mighty and extensive steam power and communication, which would enable them to forward, to any point within its vast range, despatches, troops, and warlike stores'. Of high power and speed, relatively low tonnage, broad of beam and light of draught, these projected steamers carrying mail and passengers but little cargo would therefore be 'the mail-coaches of the ocean'.[5]

McQueen also acknowledged his debt to Robert Napier in estimating the costs of the steam vessels to be employed (Chapter 2). 'I have been chiefly guided by, and adhere to, the statement made by that able and practical engineer Mr Napier of Glasgow, in his evidence to the Post-office Commissioners in 1836, that steamboats of 240-horse power and 620 tons burthen, could be furnished at from 24,000l. [pounds] to 25,000l', he told Baring in his February 1838 letter.[6]

In the *General Plan*, he insisted that it was above all 'the power of steam, together with the late great improvements in machinery, [which] can and ought, in a special manner, to secure unto her, her commerce, her power, and her people'. As he indicated in a later footnote, 'engines now made on the EXPANSIVE system, require fully one-third fewer coals, by which so much expense will be saved'.[7] The expansive principle, associated with James Watt, involved cutting off the supply of steam to the cylinder after the piston had moved only a proportion of its distance of travel within the cylinder. The fixed quantity of steam thus supplied would continue to work expansively, driving the piston the remainder of its distance, and effecting a considerable economy in the fuel required to generate the steam in the boiler.[8]

[4] McQueen (1838b: 3–4); McQueen to Baring, 14 February 1838, in McQueen (1838a: iii–iv).
[5] McQueen (1838a: viii–x); McQueen (1838b: 7); Smith (2012: 30).
[6] McQueen (1838a: vi, 114); Nicol (2001: vol. 1, 36–7).
[7] McQueen (1838a: 1–2, 45n).
[8] See Hills (1989: 64–5).

Table 7.1. *McQueen's projected operating costs for sail and steam (1838)*

I	Sailing packets (each)
First cost £9,500 @ 5% interest	£475
Repairs, wear and tear @7½%	£710
Wages (say)	£1,270
Provisions (say)	£730
Insurance @10%	£950
Total	**£4,135**
II	**Steam boats (each)**
First cost £24,000 @ 5% interest	£1,200
Wear and tear [5%]	£1,200
Insurance [5%]	£1,200
Officers, engineers, crew (40 men)	£2,595
Total	**£6,195**

The performance of Napier's two Dundee to London steamers, *Dundee* and *Perth*, as outlined in the Commissioners Report (Chapter 2) provided the standard for McQueen's scheme. Indeed, the credibility of his plan rested heavily on their reputation. In an appendix to his *General Plan*, McQueen calculated the annual running costs for both sailing packets and steamers as shown in Table 7.1.[9]

At first sight, the steam vessels appeared to be only 50 per cent more expensive than the sailing packets. Indeed, the difference could be easily justified in terms of the greater annual distance covered by steamers. But McQueen had not included in this table the fuel consumption which, using his own estimates, was some twenty-five tons of coal per day. At twenty shillings per ton, the annual bill (assuming 300 days of actual working) totalled some £7,500 – more than all the other costs combined. McQueen's most up-to-date estimates of coal consumption derived from 'intelligence lately received' that the Napier-built East India steamer *Berenice* (230-hp) (Chapter 2) had made the 12,000-mile passage from Falmouth to Bombay in sixty-three days at sea – an average of 194 geographical miles per day, but with a maximum day's run of 256 miles. Coal consumption, he reported, averaged fifteen tons per day compared to the Thames-built *Atalanta*'s seventeen tons. He did not, however, address the extent to which these performances had been sail assisted. Taking 200 miles as the average daily performance, he accepted twenty-five tons per day as the average coal consumption for steamers of 240-hp.[10] These figures

[9] McQueen (1838a: 113–15) (including a detailed breakdown of wages for officers and men).
[10] McQueen (1838a: 13, 34–6).

enabled him to derive both the operating costs and scheduling of his projected mail steamship system.

'So soon as the machinery is completed', McQueen explained to Baring in August 1838, 'by which the whole can be put into operation, that island [Jamaica] will become the heart-spring of the whole operations'. His term 'machinery' was less a reference to the commissioning of individual steamers or even to the total projected fleet of six steamers and nine sailing schooners for the West Indies plan. It was above all this system-builder's way of conceiving the whole geographical mail steamship system. Jamaica was to become 'the centre of all the mail communications, all government movements, and all commercial and money operations for the whole Gulf of Mexico, and all the western coasts of South America through the Isthmus of Panama'.[11]

He also reminded Baring of his *General Plan*'s projection of a new, vastly advantageous system over the existing one that lacked direct mail communication with a long list of territories. 'By the present system ...', he explained, 'every thing is imperfect, irregular and uncertain'. By the new plan, '[t]he work, every where, will be efficiently done, and every thing will be regular and certain'. Indeed, communication would then be 'more safe, and regular, and frequent ... with every quarter of the Western World; an object of great importance to all, but more especially to the British Government'.[12]

At the core of McQueen's projected system was the imperative to maximize useful work and productive labour while minimizing expenditure of time. Waste of time, labour and work was to be avoided. McQueen elevated these distinctly Presbyterian values as the guiding moral and political economy of his system at every level and in every component. 'A judicious and proper combination [of services] and regularity in all movements can, with the same machinery, and with little additional expense, perform, in some instances double, and in many instances nearly double work', he wrote early in his *General Plan*. In contrast, the present system not only used small sailing vessels that were 'generally unfit', but also deployed low-powered steamers 'wholly unfit for the service in any way'. Even the best of them, *Carron*, was 'still too small to perform her work in proper or reasonable time'. The new steamers, moreover, had to be designed to be not only roomy and airy for the 'convenience and health of passengers and crew', but also able to be battened down 'in the speediest and securest manner in the event of a hurricane'.[13]

The key to an economy that provided regularity and safety, as well as a saving of time, labour and fuel in the face of a difficult physical environment was power. 'The work to be performed, in every quarter, must not only be

[11] McQueen (1838b: 13–14).
[12] McQueen (1838b: 4–6).
[13] McQueen (1838a: 6–8).

well done, but done within a limited time, in order to render it beneficial and effective', he insisted. 'Powerful boats, that can overcome the distance and the natural obstacles that present themselves, can alone do this. Small-power boats can never accomplish the work'. For the system to fulfil the goals of both regularity and frequency, 'a commanding power is obviously and indispensably necessary.... steamers of 250-horse power each, will be found to be the best and most economical class of vessels to employ'.[14]

McQueen estimated the annual consumption of six West Indies steamers at 30,000 tons of the 'very best quality coals'. On his assumption of twenty-five tons per day at full power and under steam alone, each vessel would consume their annual allocation of 5,000 tons in the 200 working days at sea. Thus he explained that '[t]he time that the steamers are considered to be engaged in actual work is calculated to include the time passed in getting up steam in each voyage, and also to cover all temporary stoppages'. He again assumed an average distance run of 200 miles per day from steamers capable of ten miles per hour in good weather and light breezes. Sails alone, he claimed, would 'impel a vessel at the rate of 2½ miles per hour on a voyage'. Thus, he inferred that sail assistance would save a quarter of coal consumption in a steamer capable of ten miles per hour, reducing the annual coal consumption for the West Indian portion by 25 per cent (equivalent to 7,500 tons saved over the six vessels).[15]

With or without sail assistance, coal had to be supplied to a region deemed deficient in quality fuel. McQueen stressed the importance of having 'always in readiness, and at well-selected stations, a sufficient quantity of coals to supply each boat'. He believed that merchant sailing vessels 'bound to all quarters, so soon as they perceived that they were sure of a market, would take a proportion of coals as ballast'. Other sailing vessels, instead of waiting in idleness at ports such as Liverpool or Glasgow until they obtained a sufficient outward cargo, would be glad to fill a higher proportion of their capacity beyond simply ballast. In this way they would supply coal at moderate rates, benefit from being able to depart 'at short and regularly stated periods' as well as earn a fair profit for themselves, and thus 'become of the most essential service to the commercial interests of the country'.[16]

Early in 1839 McQueen secured the support of the West India Committee, an association of London merchants and slave plantation owners formed in the eighteenth-century. Several City of London merchants on the Committee presented McQueen's detailed plans to the Government. They agreed to the launch of a mail steamship company, conditional upon the award of a mail contract and the raising of capital in the City. A leading figure was Scottish-born

[14] McQueen (1838a: 4, 9).
[15] McQueen (1838a: 34–5).
[16] McQueen (1838a: 4–5).

John Irving, MP, of the influential merchant banking firm of Reid, Irving and Company with strong interests in Central American silver. Robert Napier had consulted the firm as to Samuel Cunard's financial credibility (Chapter 5) and Sir John Reid was also a keen supporter of Junius Smith's Atlantic project (Chapter 3). Out of sympathy with most reforms, Irving's parliamentary politics included opposition to Catholic relief, support for the suspension of habeas corpus and support for Government legislation against radicalism.[17] On 26 September, Queen Victoria granted the Royal Charter to the new 'Royal Mail Steam-Packet Company'. It named Irving as the first Chairman along with ten directors, each of whom were deemed 'to become subscribers of not less than £2,000 each towards the capital or joint stock' of £1,500,000.

The first directors included Thomas Baring, brother of Francis and the head of Barings merchant bank which had large financial interests in capital projects in the Americas. The RMSP's first deputy-chairman, and later second Chairman, Scottish-born Andrew Colvile was married to a sister of George Eden (Baron, later Earl of Auckland) who had been Governor General of India (1836–42) where he presided over the first disastrous invasion of Afghanistan. Auckland had also served as First Lord of the Admiralty under Whig administrations in 1834–5. Another director was George Brown, Chairman of the equally young Pacific Steam Navigation Company (PSNC) projected by William Wheelwright (Chapter 11). Also appointed was banker and merchant Patrick Maxwell Stewart, brother-in-law of the Duke of Somerset. Stewart sat as Liberal MP for Lancaster (1831–7) and the Clydeside County of Renfrew (1841–6). His mercantile interests had included Tobago in the West Indies. Both the Company's first auditors, Thomas Masterman and Abraham Robarts, were members of eminent City banking families.[18] These gentlemanly, often aristocratic Whig or Conservative, connections between members of the Court of Directors and Government, especially the Admiralty, gave the RMSP Company a distinctive establishment character reinforced by the frequent appointment of former Royal Navy officers to command the mail steamers (Chapter 8).

At their first meeting, held at Reid, Irving's City of London offices in Tokenhouse Yard in July 1839, the directors appointed McQueen 'General Superintendent of Affairs'. It was agreed that he be paid £5,000 for his services in forming the plan of operations of the Company and negotiating the mail contract. For his future services he would receive £1,000 per annum.[19] In late March 1840 the Company signed the contract for mail steamship services between Britain and the West Indies. But it provided for a rather different system

[17] Thorne (1986: vol. 4, 285–6). I thank Grayson Ditchfield for this reference.

[18] Nicol (2001: vol. 1, 40); Bushell (1939: 9–10); 'Baring, Thomas (1799–1873)', *ODNB*; 'Eden, George, Earl of Auckland (1784–1849)', *ODNB*. Some later writers, notably Bushell, refer to 'Colville' rather than to 'Colvile' as in contemporary RMSP documents.

[19] Nicol (2001: vol. 1, 40); Bushel (1939: 7–9).

from that proposed in McQueen's *General Plan* published two years earlier. Steam packets would depart twice monthly and call at Barbados, Grenada, Santa Cruz, St Thomas, Nicola Mole, Santiago de Cuba, Port Royal, Savannah la Mar and Havana outwards and a similar pattern homewards. There would be no calls at Fayal. The Halifax and New York transatlantic services were not pursued. The Brazil line remained to be implemented. The RMSP's steam lines would nevertheless consist of some 684,000 miles per annum, for which the Government would pay £240,000.[20] And it became increasingly apparent that, rather than opt for 620-ton, 250-hp steamers, the main line vessels would be around twice that tonnage, although spending much of their time making calls at a large number of Caribbean ports. The fulfilment of this changed requirement, enshrined in the negotiated mail contract, demanded the construction of fourteen ocean-going steamers.

7.2 'To Construct a Perfect Ship': John Scott Russell and RMSP's Mail Steamers

In autumn 1839 RMSP's newly constituted Building Committee met to consider the design of the steamers. Privileged Admiralty sites played an integral role in the process. John Fincham, master shipwright at Chatham Dockyard in Kent and a former instructor at the School of Naval Architecture, submitted the initial drawings. The Committee then passed these plans for comment to Oliver Lang, master shipwright at the Navy's Woolwich Dockyard on the south side of the Thames. The RMSP directors communicated with the Chief Clerk, John Edye, in the Admiralty Surveyor's Office at Somerset House in central London as to the most suitable tonnage and dimensions. Taken together, the Admiralty's involvement at this early stage suggested that the steamers would not only owe much to warship design, but were also intended to play naval roles if required.[21]

Directors also communicated with experienced mercantile shipbuilders elsewhere, notably Charles Wood on the Clyde and Thomas Wilson on the Mersey. They further consulted with the prestigious Lambeth marine engine-builders Maudslay, Sons and Field who were among the select firms approved by the Admiralty. Maudslay's had also supplied the *Great Western*'s machinery (Chapter 4). The Committee now invited Lang 'to furnish a perfect drawing of the vessel he recommends, which is understood to be of the tonnage of 1,350 tons' along with 'a specification for building the vessel as may be sent round to shipbuilders for the purpose of making their tenders'.[22]

[20] Bushel (1939: 9–10).

[21] RMSP Minutes (25, 26 September and 3 October 1839); Smith (2012: 33); Lambert (1991: 72–6) (Lang, Fincham and Edye).

[22] RMSP Minutes (3, 9, 11 October 1839).

'Honest Mr Lang', as Sir Charles Adam scathingly referred to him a couple of years later, was far from universally admired within naval circles. His staunch Toryism that aligned him with many Lords of the Admiralty and his defence of the traditional power of master shipwrights set him against the Whig-appointed Surveyor of the Navy, Captain Sir William Symonds (Chapters 2 and 6), who endeavoured to reduce the roles of the master shipwrights in the dockyards. Lang, who coveted the post of Surveyor himself, actively supported the return of the Tories under Peel in 1841.[23] While the self-styled Court of Directors contained both Tories and Whigs, McQueen's central place on the Building Committee probably strengthened its relationship with Lang and encouraged the construction of steamers that embodied McQueen's vision of 'mail coaches of the ocean'.

Given the small-scale character of Britain's principal shipbuilding firms, and with increasing pressure from the Admiralty for an early start to the mail service, the RMSP adopted in early October a strategy of inviting tenders from a wide geographical spread of shipbuilding yards and engine works on the Thames, Solent, Avon, Mersey and Clyde. While copies of Lang's specifications went out to these sites, the firms could only inspect Lang's 'perfect drawings' by travelling to RMSP's Moorgate Street office in the City.[24] Furthermore, design changes to the drawings were in the offing.

A month later, Charles Wood (Clyde shipbuilder) communicated to the directors a paper 'recommending some alterations which he thought should be adopted to strengthen and improve the vessels'. He had identified problems with Lang's 'perfect drawings'. The precise nature of the criticisms is not known but, according to Russell's later view of the Wood brothers' craftsmanship, Charles always insisted that stern lines be almost as fine as the bow lines. Responding, the directors ordered the Building Committee to meet with Lang and Wood on the subject before they authorized placing the contracts. As a consequence, Lang submitted to the directors at the end of November 1839 fresh specifications for steamers of just under 1,300 tons. Only in mid-March 1840, however, did they finally invite fresh tenders. Successful bidders would then be asked to take shares in part payment for their work. The Admiralty, meanwhile, extended the start date for the mail service to 1 December 1841.[25]

The directors at the same time instructed their assistant superintendent, Lt Kendall, to inform Lang that he would be paid 50 guineas 'for each set of [duplicate] drawings that may be required'. It seems likely that Lang declined to undertake such a menial task. Ten days later the directors ordered Kendall himself to prepare drawings, specifications and 'all that is necessary for entering

[23] Lambert (1991: 78) (Adam's view of Lang), 67–87 (Symonds era).
[24] RMSP Minutes (11 October 1839; 13 , 16 March 1840).
[25] RMSP Minutes (5, 12, 26 November 1839); Russell (1861: 141–3).

into contracts for vessels to be built on the combined plans of Lang and Russell as adjusted by Mr Kendall'.[26] For the first time, the name of John Scott Russell appeared, quite unexpectedly, in the RMSP minutes. He was soon to play a major role in co-ordinating construction of the large RMSP fleet.

Born in 1808 at Parkhead close to Glasgow, Russell was the son of Agnes Scott (who died shortly after) and her husband David Russell. He was baptized John Russell by the Revd Burns of the Barony Church (Chapter 1). Formerly a parish schoolmaster, his father became a minister of the United Presbyterian Kirk, serving in turn congregations in Kirkcaldy, Hawick and Errol during John's childhood and early youth. The United Kirk had split from the established Church of Scotland in democratic protest against aristocratic patronage exerted over the appointment of clergymen to parishes. Intending to follow in his father's footsteps as a minister, John attended St Andrews University for a short time before matriculating at Glasgow University in 1821.

While at Glasgow, Russell read David Brewster's recent edition of the late John Robison's four-volume *Mechanical Philosophy*. Significantly, Robison had occupied the natural philosophy chair at Edinburgh University, had been one of Watt's closest friends and had himself a longstanding interest in shipbuilding. Russell also studied works by Thomas Young and the continental mathematicians and astronomers Laplace and Euler. For his college education in natural philosophy, Russell was indebted to Professor William Meikleham for whom the study of natural philosophy 'extends our power over nature by unfolding the principles of the most useful arts', 'gratifies the mind by the certainty of its conclusions ...' and 'above all leads us to view the Creator as the Great First Cause, and as maintaining the energies of nature'.[27]

Gaining his Master of Arts, the seventeen-year-old graduate taught mathematics and natural philosophy at Edinburgh's South Academy, the Leith Mechanics Institute and (after the death of the incumbent natural philosophy professor John Leslie in 1832) at the University of Edinburgh. His enthusiasm for natural philosophy meshed with a Scottish liking for the fruits of knowledge. In the early 1830s he took part in promoting a steam road carriage project. In the same period, the Union Canal Company (whose line westward from Edinburgh connected to the Forth and Clyde Canal to Glasgow) awarded him a contract to conduct experiments with small iron vessels to ascertain whether the time of travel between Scotland's two principal towns could be reduced. Already a keen member of the Royal Society of Edinburgh (RSE) whose meetings focused on natural philosophy, Russell soon involved himself with the BAAS Edinburgh meeting (1834).[28]

[26] RMSP Minutes (3, 13 March 1840).

[27] Emmerson (1977: 2–7); Smith (2012: 38–9); Smith (2004: vol. 3, 1390–2) (Meikleham).

[28] Emmerson (1977: 7–12).

Russell's investigations into bodies moving through water generated an early paper 'On the solid of least resistance'. Read before the BAAS' Dublin meeting (1835), it prompted the Association to authorize a small committee to develop these researches, at once mathematical, experimental and practical. Russell and Sir John Robison (RSE secretary and son of the late Edinburgh natural philosophy professor) made up the Committee. With BAAS funding, up to 1844 Russell recorded some 30,000 observations on around one-hundred vessels that ranged in size from thirty-inch models to 200-foot steamers. As the vast quantity of data increasingly threatened to overwhelm any analysis, the project provoked criticism from BAAS managers, notably James David Forbes who had beaten Russell in a bid for the Edinburgh chair of natural philosophy. Nevertheless, Russell's active participation in the BAAS resulted in a rapid rise in his scientific credibility outside Scottish academe.[29]

Unsuccessful in obtaining a chair, Russell secured a position as manager of a Greenock engine-building firm. The head of the firm, John Caird, had earned a reputation for reliable marine steam engines since the 1820s. Four of Burns' paddle steamers, for example, received their engines from Caird's works between 1835 and 1839. In 1838, John Caird died in his early-forties, as did a brother not long after. One of his six sons, also John, had spent eighteen months both in the office and in several departments where he acquired the engine-making skills of mechanical drawing and working with mechanical tools. The family deaths, however, 'led him to look to the Church as the profession which he would prefer to all others'. A scholar of liberal Presbyterian views, he later became Principal of Glasgow University. With no other member of the Caird family ready to manage the business, Russell became senior manager.[30]

Russell was no stranger to Greenock. Caird had already provided the engines and Robert Duncan the hulls, of three iron boats for his canal experiments. Originally a partner in the shipbuilding firm of James Macmillan, Duncan began his own shipbuilding yard in Greenock in 1830 with the paddle steamer *Earl Grey* for David Napier (Chapter 2). Six years later Duncan opened a second yard at nearby Cartsdyke in Greenock and it was there that the first Cunard mail steamer, *Britannia*, took shape in 1839. Firm friends with Russell, he also constructed the hull while Caird again supplied the engines, for Russell's first wave-line vessel, the *Flambeau* in 1839–40, the waterline curves of which gave material form to his findings on the form of least resistance for ships.[31]

[29] Morrell and Thackray (1981: 275, 505–08) (Russell's experiments); Marsden (2008: 67–94) (wider BAAS context of engineering science and naval architecture).

[30] Caird (1899: vol. 1, ix–xiii) (memoir by Edward Caird); Smith (2012: 37–8).

[31] Emmerson (1977: 22–5); 'Russell, John Scott (1808–82)', *ODNB*; Ferreiro (2007: 184) (wave-line).

As Caird's manager, Russell had chosen an ideal site for the continuation of his experimental investigations. Close at hand were the Clyde steamers, the new ocean-going steamers for Cunard, traditional sailing vessels of every variety and up-river at Bowling, the entrance to the coast-to-coast Forth and Clyde Canal. He and his wife Harriette (daughter of Lieutenant Colonel and Lady Osborne) also leased Virginia House, the former mansion of a Greenock ship-owner. Russell pursued his investigations into the forms of least resistance using an experimental tank that he constructed in the grounds. As one witness later recalled, 'Scott Russell's office and the adjacent experimental tank were the famed resort of scientific men from all countries'.[32]

Russell was already acquainted with the Wood brothers' shipbuilding practices. 'I must say I never conversed with John Wood without going away instructed', Russell told the new Institution of Naval Architects (INA) in 1861. 'I must say this also, that a great deal of the love of my profession I owe to an early intercourse with John Wood'.[33] John and his brother were also frequent visitors to Virginia House. Russell told the INA that the Wood's 'ships were long the patterns which the builders around them were proud to follow'. He insisted that their reputation was grounded upon 'the combination in a high degree of the perfect practical knowledge of the craft of shipbuilding, with a continual application to the lines of every ship they constructed of careful mathematical calculation of all the properties and qualities of their designs'. A 'long series of successful ships built by them has borne testimony to the value of combining in the same person practical and scientific knowledge', he asserted. The brothers therefore stood as an exemplar of what Russell had written for the *Encyclopaedia Britannica* in 1841 on the importance of the harmony of theory and practice in steam navigation and in the construction of 'a perfect ship'.[34]

According to Russell, the Woods were 'equally distinguished by the liberality with which they communicated that knowledge to the other members of their profession, and especially to the younger members of it, whom they took peculiar pleasure in directing to study and to apply the true principles of naval architecture'. Indeed, even 'to their immediate rivals in business they were ready to lend their plans'. Russell bore personal witness to this practice of sharing knowledge with 'direct competitors ... frequently going to consult them on the vessels they proposed to build'.[35]

Charles in particular had taken responsibility for the design of the aptly named *James Watt*, a 450-ton, 100-hp steamer built in 1820 and running

[32] [Anon.] (1889–90: 192–3) (obituary of Duncan's son, later a famous shipbuilder in his own right).
[33] [Anon.] (1889–90: 12–14); Russell (1861: 145).
[34] Russell (1861: 141–2); Russell (1841: 257).
[35] Russell (1861: 141–2).

between Leith and London. This vessel, Russell claimed, formed, 'for ten or fifteen years, the pattern steam-ship', being designed not for propulsion by mast and sails but by paddlewheels. Thus 'her waterlines were made exactly the same in the forebody as the afterbody, the midship frame being equidistant from the stem and sternpost'. Challenging the conventional wisdom of a 'round and dumpy bow' ('cod's head') and 'long, fine, and slim' after end ('mackerel's tail'), Wood 'determined that the bow should be nearly as fine as the stern, and that the greatest section should not be one-third from the bow, but exactly amidships', thereby introducing for the first time an 'even balancing body'.[36]

The RMSP allocated its contracts in April 1840 to Caird's of Greenock (four), Pitcher at Northfleet on the Thames (four), Scott's of Greenock (two), Acramans of Bristol (two), Whites of Cowes (one) and Menzies of Leith (one). Since Caird's at that time were engine-builders and not shipbuilders, they subcontracted the four hulls to three local yards with high reputations (Thompson & Spiers (one), Robert Duncan (two) and Charles Wood (one)). Similarly, Acramans subcontracted their two hulls to William Patterson, builder of the *Great Western*, in Bristol. Pitcher's four hulls received engines from Miller & Ravenhill (two) and Maudslay (two). Menzies' and White's vessels would be towed to engine-builder Edward Bury of Liverpool.

Scott's subcontracted one hull to their neighbour James Macmillan, the only firm with the capability of building both hull and engines (Figure 7.1). Prices for hull and engines ranged from £45,600 (White and Menzies) to £53,553 (Acramans).[37]

With construction well under way in most yards in the summer of 1840, the Building Committee presented the directors with an ominous report of discrepancies in contract specifications with different builders. These seeming 'clerical errors', however, were only the prelude to the finding of unauthorized changes in design under way in Greenock. Lt Kendall thus told the Directors that he had 'ascertained from Mr Scott Russell of Cairds that they are preparing to place the boilers abaft the engines in the vessels for which they have contracted'. In the opinion of Kendall, and of the Building Committee, this major change was 'inconsistent with the declared object of the Court ... to have all their ships in every respect alike'. In response, Caird's simply stated 'that it was not now in their power to alter the construction of the engines'.[38]

Russell later explained that the Caird decision to alter the position of the boilers, and hence the profile of the RMSP steamers, was related to dissatisfaction with changes to the *Great Western*. Maudslay's had placed the boilers

[36] Russell (1861: 141–3); Smith (2012: 35–6).

[37] RMSP Minutes (4, 7 April and 7 May 1840); Smith (2012: 40–1). See also Nicol (2001: vol. 2, 10–11).

[38] RMSP Minutes (20 , 27 August 1840); Smith (2012: 41–2); Nicol (2001: vol. 2, 11–12).

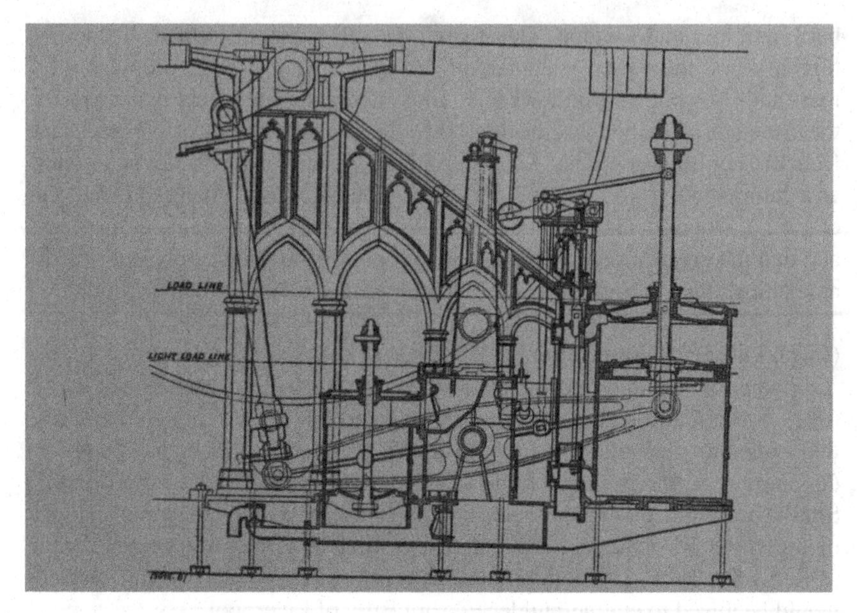

Figure 7.1. Side-lever engine by Scott, Sinclair of Greenock (1840). Note gothic style

(and funnel) forward of the engines, a change that required extra ballasting aft to restore the trim (Chapter 4). Russell was also following the arrangement adopted in the Clyde-built Cunard steamers, the first of which, the Duncan-built *Britannia*, had just entered service on the North Atlantic (Chapter 5).[39]

The 'clerical errors' in the contracts, meanwhile, turned out to carry serious consequences in practice. The drawings and specifications that Lt Kendall had prepared and sent out to the contractors would, if followed to the letter, place the horizontal crank shaft driving the paddlewheels precisely at the level of the main deck, dramatically weakening one of the principal structural strengths of the large hull. Reporting this crisis to the Court, the Building Committee admitted that 'the ships building by Messrs Caird & Co are (with the exception of the intended position of the boilers) of the construction best suited to the service of the Company'. Its unexpected praise for Caird's derived in large measure from that firm's own solution to the main deck problem in the four steamers under their control. A drawing inserted in the Committee's report showed that

[39] RMSP Minutes (26 November 1840); Nicol (2001: vol. 2, 15–16). Russell also seems to have introduced slight variations in the bow lines of Clyde-built RMSP steamers in order to allow their different performances to be observed when in service (Russell (1863); Emmerson (1877: 25)).

Figure 7.2. RMSP's Thames-built *Trent* (note funnel position forward of paddlewheels)

they had simply raised the main deck by two-and-a-half feet above the vertical height from the keel specified by Kendall. The same illustration also contrasted Kendall's specifications with those of the Greenock firm.[40]

Although the RMSP had local superintendents (either a ship's captain or a Royal Navy lieutenant) to oversee construction in each region, these problems all pointed to the Company's failures to control the realization of such a large project. They also testified to the increasing authority of Clyde shipbuilders in general, and of Russell in particular. The change to the boiler arrangement quickly spread from Caird's four vessels to more than half of the RMSP fleet while under construction. All six Clyde-built steamers adopted the altered arrangement. After delays resulting from the bankruptcy of the engine builders, the two Bristol-built vessels not surprisingly followed suit, given Patterson's own experience with the *Great Western*. Sited close to RMSP control, Pitcher's four Thames-built craft followed Lang's design as did White's from Cowes and Menzies from Leith, both of whom depended upon the same Merseyside engine builder (Figures 7.2 and 7.3).[41]

When Russell attended the Court of Directors' meeting in November 1840, he explained that before Caird's hull contractors had laid down the RMSP ships he had consulted 'the most experienced builders in the Clyde'. Now he urged that the Court press for 'the expediency of the decks of the other [non-Caird]

[40] RMSP Minutes (10 September 1840); Smith (2012: 43–4).

[41] RMSP Minutes (30 May, 20 , 27 August and 26 November 1840); Smith (2012: 42–3) (boiler positions).

Figure 7.3. RMSP's Clyde-built *Dee* (note funnel position aft of paddlewheels)

vessels being raised to the same height'. Two days later, the Deputy Chairman Andrew Colvile and Director James Cavan moved a resolution that the decks of the four Pitcher ships, in accordance with a paper signed by him and Russell, be raised to pass over the crank shaft instead of under it (as Lang had argued). They also moved that seven feet above the said deck, there was to be placed 'a spar deck of as slight a construction as may be consistent with due strength and security'. This proposal was RMSP's response to Fincham and Lang's objections to an increase in top-hamper and implied that if heavy guns were ever to be needed, the spar deck would have to be removed first.[42]

The resolution as a whole did not go uncontested. Directors George Brown and Robert Cotesworth objected that whereas the only problem with Lang's solution was a reduction in cargo capacity, Russell's plan 'may involve the safety of the vessels'. Their amendment, however, was voted down and the original motion carried. As a result of Russell's victory, the Court passed two further resolutions. First, it decreed that 'arrangements be made to effect a similar alteration in the other ships provided it can be done at a reasonable charge'. And second, it requested Russell to 'undertake the negotiation with the several builders and to report to the Court the terms on which they would effect the alteration'.[43]

Russell's ensuing round of contractors revealed a variety of problems consequent upon the 'clerical errors'. He and Pitcher had already agreed 'upon the plan of midship section with which to furnish the other builders'. White

[42] RMSP Minutes (3 , 5 November 1840); Smith (2012: 45–6); Nicol (2001: vol. 2, 12–13).
[43] RMSP Minutes (5 November 1840).

consented to make the alterations at a cost of £750 to RMSP and £250 to himself. Menzies, however, had implemented the Lang plan as to the position of the main deck, only to discover what Russell had already found: the paddle shaft would pass right through the centre of the deck. For a sum of £1,285, he consented to correct the problem as far as possible. Scott held out for a much higher sum but eventually compromised at £1,000 per ship. Patterson endorsed the Russell proposal at minimum extra charge, as the two Bristol steamers were less advanced than the others.[44]

Russell's work on behalf of the RMSP Directors ended with an appropriate triumphal flourish. '[I]f at any future time the Directors should find that my services can be of any use to them, I shall be at all times most happy to do all in my power to promote the interests of the Royal Mail Steam Packet Company', he concluded his second report, '& to place at their disposal whatever knowledge the experience of many years in the difficulties of steam navigation & in the practice of the most eminent Ship Builders & Engineers may have enabled me to acquire'. The immediate consequence of the narrowly averted debacle, however, was the swift resignation of a humiliated Lt Kendall and the appointment of a designated marine superintendent, Captain Edward Chappell, RN.[45]

7.3 'The Indefatigable Secretary': Captain Edward Chappell, RN, Takes Command

'I have been constantly supervising steam squadrons since the year 1826', Chappell told the 'Select Committee on Packet Service' in 1849. As agent for the Post Office packets at Milford Haven from 1826, he had assisted Macgregor Laird's African project in the early 1830s (Chapter 3). He subsequently served as agent for the Liverpool packet service around 1836 and three years later presented evidence from his investigations into ships' boats to a parliamentary committee. His appointment as RMSP's marine superintendent at a salary of £600 per annum took effect on 17 December 1840. The directors soon granted Chappell greater managerial authority at the Moorgate Street offices. By February 1842 he had been elevated to the post of RMSP Secretary and later became a joint manager. Until 1856 he directed the Company through its many private and public vicissitudes, often drawing severe criticism for an authoritarian mode of working.[46]

As marine superintendent, Chappell immediately took command of the building programme, even rejecting one of Russell's strong recommendations.

[44] RMSP Minutes (19 , 26 November 1840); Smith (2012: 46–7); Nicol (2001: vol. 2, 15–16).
[45] RMSP Minutes (26 November and 17 December 1840); Smith (2012: 47).
[46] SC (1849), 173, 181; Post Office (PO) (1836), 202; RMSP Minutes (17 December 1840 and 17 February 1842); Nicol (2001: vol. 1, 47; vol. 2, 16–17) (his appointments with RMSP).

Russell had become concerned that the hulls of the two RMSP steamers building at Duncan's yard lacked the same strength as the Cunard Company's recently completed *Britannia*. He advised that the transverse bulkheads be reinforced by means of trusses. The Admiralty's John Edye considered this solution 'highly objectionable' on the basis of negative experience with two large sailing vessels similarly treated. Pitcher's view was also negative. Weighing up these criticisms, Chappell quickly turned down Russell's plan in favour of one suggested by Edye.[47]

Differences in the quality of shipwright practices also preoccupied the marine superintendent during 1841 as the fleet took final shape across Britain. Based on reports received from the surveyor, he approved the quality Pitcher's work but not its progress. Geographical distance between the metropolis and the other yards posed additional concerns over maintaining uniform quality control. Macmillan's *Solway*, Duncan's *Clyde*, Menzies' *Forth*, Thompson & Spiers' *Tweed* and Charles Wood's *Tay* all received praise for the materials and workmanship. But just after the *Clyde*'s launch Duncan contracted typhus and died aged thirty-eight, about three weeks later. Evidence of the subsequent mismanagement soon became manifest in the *Teviot*, where the frames were 'not so well squared as they ought to have been'. Similarly, the bottom plank only touched these timbers at intervals on account of the timbers 'being badly trimmed'. In many parts sap had been left, an invitation to future rot. Sappy timber had also made its appearance in Scott's *Dee* and had to be replaced. In Bristol, he found real problems with Patterson's *Avon* where main deck planks were of pine rather than oak, lower timbers showed sap in many places and several of the upper timbers needed replacing. White's *Medina* at Cowes, although progressing rather slowly, met with general approval.[48]

Chappell also received reports on the engine installation and fitting out stages. In the *Dee* and *Solway* (both engined by Scott Sinclair) the holding down bolts of the engines 'were driven through the bottom planking in a most objectionable manner'. Instead of being raised to clear the shaft, the main deck had been 'bent, so as to give it an unsightly appearance, though it may not be weaker on that account'. Hitherto unproblematic hulls now also gave cause for concern. The *Clyde* had suffered hull strain while aground for the fitting of the machinery. And the *Tweed* had started to leak after the launch.[49]

Other changes to the original plans required Chappell's attention. The controversial seven-foot-high spar deck threatened to alter the sailing characteristics. He therefore took soundings from the masters of three transatlantic

47 RMSP Minutes (14 January 1841).

48 RMSP Minutes (4, 18 March and 1 April 1841); [Anon.] (1889–90: 192–3); Nicol (2001: vol. 2, 20); Smith (2012: 49).

49 Smith (2012: 50).

steamships, including the *Great Western*'s captain, and on the basis of their advice recommended a three-masted barquentine rig, that is, with square sails only on the foremast and fore-and-aft sails on the main and mizzen masts. Such a sail plan was designed to reduce both top-hamper and material costs.[50]

Witnessed by some 60,000 spectators, the *Forth* entered the Firth of Forth at Leith in May 1841. *The Scotsman* spoke of the crowd flocking 'to witness a ... triumph of art and industry, an individual and national enterprise'.[51] After the hull had received the engines from Edward Bury in Liverpool, Chappell arrived to witness the sea trials over a thirty-five-mile run out of the Mersey at an average speed of over ten knots and five psi boiler pressure. 'There cannot be a finer piece of workmanship than these engines', he enthused, 'they worked as smoothly as the machinery of a watch, & were the admiration of every person as to construction & performance'.[52]

This premature judgement gave way to concerns about excessive vibration towards the stern accompanied by clouds of spray abaft the paddle-boxes. Close inspection later revealed that the engine-builders, following the original arrangement of boilers, had been concerned with the trim. They therefore placed the paddle-shaft further aft than originally intended such that the paddlewheels had little room to clear the aft paddle-beam. Equally serious was the 'faulty fitting of brick work round the chimney [funnel]' that caused the boiler casing to catch fire twice during the trials.[53]

With problems addressed, the *Forth* steamed out of the Solent from Southampton on 17 December 1841, the first member of the fleet to depart for the West Indies. Three more vessels left on the following day, each of them bound for the Caribbean to take up initial positions on the scheduled route. At the very beginning of 1842 the departure of the *Thames* and *Tay* from Falmouth marked the official opening of the mail service, although it was Southampton, rather than Falmouth, that was soon to be the home port.[54] Chappell's position altered to that of Secretary. Corresponding to this shift of power, McQueen received the new title of 'Superintendent of the Foreign Agencies' covering the Caribbean region. Evidently marginalized, the one-time General Superintendent resigned around August 1842, less than a year into the Company's operations. At least in the public domain, his departure was to pass virtually unremarked.[55]

[50] RMSP Minutes (31 December 1840); Nicol (2001: vol. 2, 16, 24).
[51] *The Scotsman*, 26 May 1841; Bushel (1939: 14).
[52] RMSP Minutes (18 November 1841).
[53] RMSP Minutes (18 November 1841).
[54] Bushel (1939: 24–5); Smith (2012: 51).
[55] Nicol (2001: vol. 1, 47).

8 'A Most Perilous Enterprise'

Royal Mail Steam Packet's Vulnerabilities

> I still hope it may be saved from that general shipwreck of property, of which this company threatens to present so calamitous an example. I am bound to say that on comparing the great promises (and boastings) of the past, and the similar promises of the present, and seeing how little has been realized, I cannot respond to the somewhat too sanguine expectations of my honourable friend ... The directors are high-minded and honourable men, who have won their way to great commercial distinction by their perseverance and by their talents, but I feel that in a considerable company like this it is impossible to get gentlemen so placed to give that undivided attention to this concern, which is all important.
>
> *Dr John Bowring, MP, voices his grave anxieties to RMSP's half-yearly General Meeting of proprietors in October 1843*[1]

Summary

McQueen's publicly announced *General Plan* had proclaimed that a large-scale system of mail steamship operation between Britain and the West Indies would perform its work efficiently, regularly, frequently and safely. Its guiding principles would be the maximization of useful work and productive labour, as well as regularity and the minimization of wastage of time and materials. The problems of realizing these promises had multiple sources. Contemporary press critics, often informed by disaffected proprietors, highlighted the over-ambitious schedules, the excessive size of the vessels due to Admiralty pressure, the lack of steamship experience and skill within the gentlemanly Court of Directors, the autocratic management style of the Secretary to whom most of the day-to-day operations devolved and his choice of masters and officers whose navigational and seamanship practices they called into question. The total loss of the *Solway* brought these criticisms into public and private focus. When a section of shareholders sought to point fingers of blame at the directors, however, the Court found in McQueen a convenient scapegoat for the conspicuous disorder within the Company's mail steamship system.

[1] RMSP GM (12 October 1843: 18–19). The word 'boastings' has been deleted in the original at some point.

8.1 'Do the Directors Imagine that Such Things will be Long Endured?' RMSP's Bad Press

The original RMSP project attracted gentlemanly proprietors for its promises of regular imperial mail and passenger services and for its expected dividends. Shareholders such as wealthy city merchants and aristocratic military officers were also likely to have business or other colonial interests in the Caribbean region, rendering them prospective cabin passengers on the transatlantic as well as inter-island routes. In his detailed and first-hand account for *The Times* of the rescue of passengers from the wrecked *Medina* in 1842, for instance, the Deputy Adjutant General in the Leeward and Windward Islands, Lieutenant Colonel Thomas Falls, drew attention to the number of high-ranking figures aboard, including 'his Excellency the Earl of Elgin, [that is] the late newly appointed Governor of Jamaica, his lady, and suite, who made their escape from the wreck with his Lordship's despatches only'. Falls also expressed his own confidence in RMSP's promises: 'having some shares in this company, I consider I am entitled to give my opinion on the matter for the benefit of others who may have invested their money in this great national undertaking, and, should even their shares rise above par, I should still hold on firmly with the company'. But not only was the *Medina* a total loss in May 1842, but the *Isis* followed five months later.[2]

Press criticism of RMSP's scheduling, however, had already appeared. The 'Money-market and City intelligence' column of *The Times* cited two examples of the unsatisfactory practice of landing inbound mails at Falmouth before the steamer proceeded to Southampton. In one case, the master arrived by train in London from Southampton, while the mails from Falmouth took a day longer. Complaints of a similar nature issued from the Caribbean where 'the irregularity of the steamers' routes is a source of annoyance to every one. ... At present the complaints on this subject are universal, and the injury which is inflicted is of the most serious nature'.[3] The ships' masters blamed the 'irregularities' on McQueen's over-optimistic schedules. Thus, Captain Miller of the *Trent* reported on 23 April 1842 that 'ship and engine logs prove that none of the Company's fleet' could perform the schedule within the allotted time, the *Trent* only managing the feat by omitting six ports.[4]

'It pains us much to be unable to say anything calculated to repress the growing conviction that this is a most perilous enterprise', began an article in the *NM*, reprinted from the *Civil Engineer and Architect's Journal* in mid-1842, and evidently from the pen of an insider. '[N]ot that there is anything inherent in the nature of the scheme to render it necessarily disastrous, but that the

[2] *The Times*, 11 June 1842; Bushell (1939: 27–9) (*Medina* and *Isis* losses).

[3] *The Times*, 25 April 1842.

[4] Quoted in Nicol (2001: vol. 2, 27).

management is so supremely injudicious as would be fatal to any project, even the most promising'. The author warned that 'men prefer to attribute calamities to any cause except their own ignorance or indiscretion; and a fault once committed, is oftener defended than acknowledged'. And he predicted that should 'any calamity ... fall upon this most spirited and arduous enterprise', it would 'operate most perniciously upon the whole interest embarked in steam navigation, and paralyze the spirit of commercial adventure in the same channel'. The article therefore entreated the 'respectable directors' to look 'to the dangers of their present position'. Neither 'magisterial pomp nor inexperienced precipitancy' could possibly save them from such dangers.[5]

At the core of the critique was the accusation that skill and experience were lacking in an 'enterprise of extraordinary magnitude and difficulty', that indeed it appeared as though those qualities had been 'deemed inessential'. Could the directors 'confidently say', the author asked rhetorically, 'that in the whole compass of their establishment, *there is a single man to be found experienced in the conduct of commercial steam navigation upon the large scale*'? Was such an enterprise 'expected to succeed in the hands of inexperienced persons, even although these persons may unite much general ability with much plausible pretension'?[6]

The article included a summary of widespread press criticism of RMSP's disorderly system. 'There is scarcely a newspaper which does not teem with accounts of mails late – vessels wrecked, or missing – quarrels aboard – passengers almost starved – and vessels following one another *by mistake* – and at an interval of a few hours across the Atlantic', the author insisted. '[And] at other times, the greatest inconvenience and distress are occasioned by merchants being left without their remittances for months together, and correspondents without their letters. Do the directors imagine that such things will be long endured'? He also pointed to the likelihood of unprofitable working. 'One vessel comes in with twelve or fourteen passengers – another with five or six, bringing an income of a few hundred pounds, perhaps, when the expense of a voyage must be several thousands', he noted. 'There is the mail money, 'tis true; but how far will the mail money go towards paying for coal, insurance, and wear and tear?' He therefore set a one-month ultimatum for the implementation of 'reformation' and 'improvement'.[7]

Criticisms also began to multiply from a direction very different from the City and *The Times*. In the wake of RMSP's October 1842 GM, Herapath's *RM* carried much correspondence and considerable editorial comment on the Company's troubles. Early in 1843, for example, 'A subscriber' claimed that

[5] [Anon.] (1842: 419–20).
[6] [Anon.] (1842: 419–20). My italics.
[7] [Anon.] (1842: 419–20).

'the great cause for the failure of the Company is, the expenses are so enormous'. Rather than indict the Company, however, the correspondent attempted to restore some morale to fellow-shareholders 'who imagine their capital is lost'. He instead blamed the Government for 'obliging this Company to build steamers of the size they are' whereas '[s]maller vessels would not only have been very much less expensive, but would have been more suited for the service to be performed'. On the other hand, he accepted that the fleet effectively made available twelve war steamers should trouble arise in the West Indies 'where the French, in June next, will have ten'. Thus he called upon the Government to do everything possible to assist the Company out of its current difficulties and for the managers to 'show that the mails have been regular, [and] the management economical'.[8]

As an advocate of steam, the same correspondent referred to a report in *The Times* concerning a possible return to sailing packets. He noted that the paper had at the same time reported a whole fleet of west-bound sailing vessels awaiting a fair wind to proceed down the English Channel. He therefore argued that '[s]hould the steamers be discontinued', fortnightly mails would be delayed by up to two months. A respondent signing himself 'E' challenged this claim, explaining that departures from Falmouth, with the wide Atlantic in close proximity, would avoid the wait for a fair wind that often beset sailing fleets to the east. He also claimed that sailing brigs could serve the West Indies for £24,000 per annum rather than RMSP's £240,000 contract price. The RMSP had yet to prove itself.[9]

When RMSP's *Clyde* was found to have dry rot in the upper timbers in 1843, the Company deployed the paddle-steamer *City of Glasgow* to fill the space. 'We learn from a passenger who went to Madeira in her, that, so bad a state was the *Glasgow* [in]', the *RM*'s report began, '... several passengers on seeing her refused to go, and demanded their money back'. Worse still, the passenger declared that it was almost a miracle how they ever reached Madeira 'as the men were obliged to be constantly at the pumps'. Had the ship foundered with the loss of the passengers' lives, the *RM* declared, 'we should like to know whether the parties who sent such a vessel to sea, would not have been morally guilty of murder'.[10]

The *Magazine*'s correspondents piled on the pressure in the run-up to RMSP's second GM at the end of March 1843. 'A shareholder' lambasted the Company's 'Honourable Directory' for a third 'New Plan in lieu of the long-looked-for-dividend'. Kept 'a profound secret at that inaccessible of all

[8] *RM* (1843: 7).
[9] *RM* (1843: 42).
[10] *RM* (1843: 206); Duckworth and Langmuir (1977: 198) (*City of Glasgow* built by Wood and engined by Napier in 1835 for the Glasgow to Liverpool service).

unapproachable offices in Moorgate Street', the disaffected shareholder had been 'given to understand' that the latest plan promised a reduction in annual mileage of 100,000 at a saving of £80,000. He expressed himself 'rather sceptical of any of their future high-sounding assertions, seeing how the previous [plans] have [not] been realized'. He therefore warned fellow-proprietors with regard to the forthcoming meeting not to 'be carried away by specious promises'. In the same number, the Editor himself wrote of RMSP's 'credulous shareholders' and its 'opportunities of jobbing'. More broadly, he indicted the fostering of 'these monopolies to the detriment and ruin of private enterprise'.[11]

Assembled in the City's London Tavern, the Court mounted a suave defence in March 1843. Its formal report opened with a confession that the original project 'embraced too extensive a sphere of operation; that its calculations, whether as related to the number of Ships required, to the work they could perform, to the practicability of their visiting all the specified places, to the adequate security of the ports of assemblage, to the cost of building, outfit, and maintenance of the Ships, and to the amount of revenue likely to be realized, – were all of too sanguine a character'. But the Admiralty had since approved two reductions in the original routes which, along with RMSP's control of costs ashore and afloat, augured well for the Company's future.[12]

The Court also neatly implicated the first scheme of routes in the losses of the *Medina* and *Isis* in 1842. Since the revised plan had come into operation, 'none of the Ships have touched the ground, or received the slightest injury, arising out of the navigation they have now to encounter'. It claimed too that those 'acquainted with the construction and outfit of Steam Vessels, their complicated machinery, and the derangements frequently experienced at the commencement of their career' would find the performance 'not only unusual, but perhaps unprecedented, either in a government or mercantile marine'. This accomplishment was especially remarkable, they emphasized, given that the vessels were at least equal, if not superior, in size and power to steamers at work anywhere. And, it noted, the Secretary to the Admiralty in moving the annual estimates in the House of Commons had acknowledged the punctuality of the steamers.[13]

The two-part financial statement (to the end of December 1842) that accompanied the annual report in March 1843 consisted of a balance sheet and a profit and loss account. The former, a statement of assets and liabilities, showed, on the credit side, a 'capital stock' of £831,570, the amount received from the payment in full on shares. Loans amounting to £185,000, together with various bills still payable, yielded a total sum of £1,123,487. On the debit side, the

[11] *RM* (1843: 323, 327).
[12] RMSP *Report* (13 March 1843: 1–2).
[13] RMSP *Report* (13 March 1843: 1–2).

largest sums comprised £669,023 for the twelve steamers delivered, £143,955 for the *Severn* and *Avon* from Bristol yet to be completed, £77,849 for coal and £56,500 of the annual Government contract still outstanding.

The profit and loss account was much starker. The total amount on the debit side stood at £101,405. That figure incorporated the uninsured part of the two wrecked steamers, £15,211 for 'preliminary expenses' incurred in the launch of the service, interest on the loans and a variety of office expenses including salaries. Even after including the mail contract money, the balance on the credit side from the ships' voyages totalled only £21,615. The RMSP's operating loss for the first twelve months of operations effectively came to £79,790. No dividend would be paid.[14]

The *Report* concluded with a resolve 'to continue the same course of rigid economy', including lower consumption of coal, 'resulting from longer experience of its various qualities, and more vigilant supervision of its shipments' (Chapter 9).[15] Sanguine statements such as this, designed to maintain the confidence of the more patient proprietors, provoked Dr Bowring (see above) to caricature the directors' voice as 'that musical sound, under which the proprietors were lulled, that if the directors were left alone there would be a dividend – a most musical word, dividend (laughter), which comprised something like a promise that ere long a dividend would be distributed amongst the shareholders'.[16]

Even more pointedly, the *RM* quoted a fictitious correspondent declaring that he had heard that RMSP directors 'intend holding their annual meeting in future on the FIRST OF APRIL' and that the suggestion had come from one of their members 'who has closely studied human nature, and [who] remarked that at particular seasons men have less objection to be made fools of than at others'.[17] The RMSP was losing its battle to win the approval of the radical press. Three days later, the country received grim news of RMSP's most devastating shipwreck to date.

8.2 'That Moorgate Street Admiralty': Anatomy of the *Solway* Disaster

At 3am on 17 April 1843 at his home in Mayfair, RMSP's Secretary penned a letter to the editor of *The Times*. One hour previously, he reported, the purser of the Company's mail steamer *Solway* had arrived in London 'with the afflicting intelligence of the total loss of that ship' at midnight on 7 April about twenty

14 RMSP *Accounts* (13 March 1843).
15 RMSP *Report* (13 March 1843: 2–3).
16 RMSP GM (12 October 1843: 16).
17 *RM* (1843: 400–1).

miles west of the northern Spanish port of Corunna. Chappell annexed lists of the survivors' names as well as 'those known to have perished'. On the list of survivors were the passengers (twenty-eight), the officers (seven), the engineers (seven), seamen (twenty-two), firemen (twenty-two) and stewards (eleven). The list of those named as lost were passengers (seventeen), officers including the master (three) and stewards (eight). The dead included a family of six and their unnamed servant, together with two other females. Ominously, there appeared to be no casualties among the crew of engineers, firemen or seamen. Later estimates tended to agree on some thirty-five lost passengers and crew, a figure that still pointed to a low death rate among the deck and engine-room crew (seven lost; fifty-one saved).[18]

An anonymous passenger arrived in London some twelve days after the disaster. In his account for *The Times*, this witness testified that during the outward voyage to Corunna the Company had ensured that 'nothing was wanting on their part that could conduce to the efficiency of the service or the welfare of the passengers'. Indeed, the 'elements of comfort were in profusion on board, and the ship was itself a tower of strength and an admirable sea-boat'. The survivor also bore 'testimony to Captain Duncan's kindly bearing to all classes on board his ship, and also [during the subsequent sinking of the *Solway*] his disregard for [his] personal safety'. He had 'evinced a knowledge of his profession, ... while his deportment after the occurrence of the catastrophe proved that he esteemed it his first duty to protect the helpless women and children committed to his care'. In contrast to this uncritical portrait, however, the survivor called for an investigation into whether the ship's officers 'were men of sufficient skill and experience, and whether they had been appointed to their various situations entirely on account of their merit'.[19]

Initially paraphrasing Chappell's correspondence in *The Times*, the *RM* added that the *Solway* 'was a remarkably fine fast-sailing one, Scotch built [and] had lately undergone great repairs'. It further reported that this 'deplorable event will throw the Company's affairs ... into irretrievable difficulty' and that a vessel was to be sent out to St Thomas with mail but no passengers 'to endeavour to stem the current of confusion and disorder now in the West Indies'. It also cited the *Liverpool Courier*'s claim that there were 150 persons on board, of who one-third were lost.[20]

The same issue, however, launched its own editorial tirade against RMSP's managerial incompetence. It began with the assertion that the news 'has for some days created a considerable sensation in the city [of London]. Everyone

[18] *The Times*, 18 April 1843; Bushell (1839: 31) (later casualty figures). For a broader study of Victorian shipwrecks, especially in relation to the vexed chivalric code of 'women and children first', see Delap (2006: 45–74).

[19] *The Times*, 20 April 1843. Also reproduced in *RM* (1843; 424).

[20] *RM* (1843: 418).

exclaims, how is this'? Indeed, it was 'the opinion of the best informed in the city that this *Solway* affair will be the [Company's] death-blow'. All three of RMSP's recent ship losses had occurred on fine, clear nights, prompting the answer that 'it is commonly supposed that there must be something radically wrong in the supreme authority and details of the [Company's] naval department'. Public and shareholders alike had the right to know the answers to further questions: 'what kind of crews and officers have they got – experienced or otherwise – and who has the appointment of all these officers, and to whom are they responsible ... or are they their own masters, and responsible only to some one at home, to whom they owe their appointments?' Indeed, the *RM* believed that 'nearly every one of the Company's ships have [sic] been ashore in different places in the West Indies, and even in some of the finest harbours in that quarter of the world'.[21]

Claiming that it had never seen 'a concern so bunglingly and grossly mismanaged', the *RM* denied that 'all these disasters can be attributed to misfortune or to unavoidable causes'. Something 'radically wrong at headquarters' placed a duty upon shareholders to institute a searching inquiry. Such a duty they owed to themselves, to the Government 'whose money they are receiving' and to the public 'whose lives they are thus jeopardizing, and, in fact, destroying by the score'. Adopting a mocking tone, it suggested that shareholders were 'kind confiding souls' with 'the names ... of wealthy and honourable men for their managers' and would 'patiently await the utter annihilation of their property in due and speedy course'. In contrast, it urged its readers to look at Cunard's line of steamers or P&O's and asked if 'we, in either of these concerns, hear of vessels running on shore in midday, and scarcely leaving a harbour before being wrecked on rocks as well known to sailors as St Paul's to Londoners'? Cunard in particular had 'a much more difficult navigation to encounter', but had not experienced similar losses. And it disputed the 'high encomiums' passed upon the *Solway*'s late captain. Instead, it suggested he 'was by no means equal to the responsibility' and that he lacked authority over his officers who, in keeping with RMSP practice, the Company had selected for him.[22]

A week later the *Railway Magazine* published a letter from 'A Sufferer of the Jobbing Company' that further indicted the *Solway*'s master. He was rather 'well known as an Exciseman in the Isle of Rothsay [Bute]' and had been 'recommended by the Marquis of Bute to the Scotch aristocracy of Royal Mail Company Directors, who, of course, immediately appointed him to a command'. Had he survived the loss, RMSP would have 'promoted him to a lucrative agency, ... it having been their invariable practice in similar cases when

[21] *RM* (1843: 423–4).
[22] *RM* (1843: 423–4, 448).

the recommendations came from such high quarters as the Marquis of Bute, the Duke of Buccleuch, &c'. Moreover, as Chappell's 'brother officer in the revenue service' as well as his protégé, the unfortunate Captain Duncan had had a twofold chance of promotion by virtue of the jobbing practices of 'that Moorgate Street Admiralty'. The 'Sufferer' also quoted from a letter written on the day of departure from Southampton by a Captain Sughrue, 'well-known experienced seaman' and surviving passenger on the *Solway*, that 'such was the confusion and general disorder on board' he had doubted they would 'ever reach the West Indies'.[23]

The *RM* subsequently challenged Irving, Baring and Colvile to set up an investigating committee or, failing that, to 'manfully resign, and show to the world that they are not party to the bungling imbecility or jobbing venality which obviously rules at head-quarters'. It also expressed surprise that members of 'the late Whig Government, under which this scheme was brought forward and fostered, do not make a stir and inquire into the cause of the mishaps perpetually occurring'. And it urged Dr Bowring, MP, as a shareholder who had raised concerns at the March meeting, to call for such a committee. By a 'prompt and complete reformation of men and measures it is still possible to restore the confidence of the public and save the undertaking'.[24]

In a closely related item, the *RM* turned to targeting RMSP's hierarchical management structure in which Chappell held the reins of power. 'No complaint can reach the aristocratic board of the Royal Mail Company, but through Captain Chappell, the secretary', it asserted. 'The noble lords, the directors, are as difficult of approach as his Celestial Majesty'. Unlike subordinate officers in other vessels, moreover, such officers in RMSP's fleet had to make their complaints not to their captains but to 'the gallant secretary'. As a result of bypassing each commander, it concluded sarcastically, 'this must be a good way to promote good conduct and discipline in their vessels'.[25]

In contrast to these allegations, *The Times* published a scientific analysis of the likely causes of the shipwreck in terms of compass deviation. In this letter, London-based barrister Archibald Smith avoided all disputed questions of managerial incompetence. Smith's family had strong ties with navigation. His father, ship-owner and gentleman James Smith of Jordanhill near Glasgow, had persuaded Clyde shipbuilders to co-operate with Russell in his extensive experimental researches into ship performance.[26] Smith senior had a residence at Helensburgh of which town he became Provost between 1828 and 1834. His real love was yachting a pursuit that not only won him many cups, but also

[23] *RM* (1843: 442).
[24] *RM* (1843: 448, 472–3).
[25] *RM* (1843: 449).
[26] Emmerson (1977: 18).

connected with his interests in seamanship and navigation. It further linked to his biblical researches that underpinned his much-read *St Paul's Voyage*.[27] The Helensburgh home placed him, and his family, in close proximity both to Robert Napier's residence a short distance further up the Gareloch and to the Greenock shipyards across the Clyde estuary.

Graduating from Cambridge in 1836 as Senior Wrangler in the Mathematics Tripos, Archibald had been urged by his father to seek the chair of natural philosophy at Glasgow University. But he instead remained in London as a practising lawyer. In addition, however, he began working closely with Colonel Edward Sabine, based at Woolwich, on data resulting from the 'Magnetic Crusade', an Empire-wide research programme sponsored by the BAAS, the Royal Society and the Admiralty into terrestrial magnetism. Smith's particular role involved the analysis of results obtained from a maritime expedition to Antarctic seas in which the effects of the ship's own magnetic character (deviation) had to be taken into account.[28] His letter to *The Times* explored the consequences of that authoritative analysis for practical navigation in the wake of two recent tragedies, one of them the *Solway*.

Smith pointed out that, as a general rule, in northern latitudes a compass needle 'is attracted towards the ship's head'. On a northerly or southerly course, there was no deviation, but on an east or west course the deviation was at a maximum. For sailing ships (with some iron present as bolts or knees, for example), the deviation might amount to half a compass point (five or so degrees). For wooden steamers (with large iron engines and boilers) the deviation could rise to a whole point (ten degrees or more). And for iron steamers the deviation might be as high as two points (twenty degrees or so). Thus a master, ordering a westerly course and dependent on his compasses alone on account of cloudy skies and limited visibility, would be oblivious to the steersman taking the ship on a course much further southward than that shown on the compass card. Anyone, Smith argued, 'may easily satisfy himself by inquiry, that this source of error is almost wholly unknown to shipmasters'.[29]

The danger, he argued, was not manifest during long ocean voyages when the variation of the compass (due to geographical changes which are variable over time in the earth's magnetic field) was determined regularly by astronomical observations. In such cases, therefore, the deviation 'is then not distinguishable from the variation, and is unconsciously corrected'. If, however, a vessel had altered course without making a new determination of the variation or if it simply entered a correction for the variation where the value was well-known

[27] Macleod (1883: 163).
[28] Smith and Wise (1989: 56, 60, 65, 100, 140–4, 276–8, 755–63) (Archibald Smith); Morrell and Thackray (1981: 353–70, 523–31) ('Magnetic Crusade').
[29] *The Times*, 19 April 1843.

(as in the approaches to the English Channel) then the danger was very real. Smith had already written to *The Times* at the end of the previous year soon after the loss of the 1,550-ton East Indiaman *Reliance* on the French coast near Boulogne. According to press reports, this large wooden sailing vessel had been running up the Channel on an easterly course in poor weather while homeward bound from China with 27,000 chests of tea (almost 1.9 million lb) and 116 passengers and crew. After striking the rocks, chaos ensued. The heavy seas swamped the overloaded boats and drowned 110 persons. The shipwreck 'cast a gloom of the most melancholy description among the merchants connected with the East India trade' and 'in a material degree, depressed the feelings of parties in our shipping interest'.[30] From Smith's analysis, on an easterly or north-easterly course the compass needle would be drawn toward the ship's head and once again take the ship further southward, and so on to the French coast. He therefore 'endeavoured to point out my reasons for thinking it probable that the loss of that vessel was caused by the deviation of her compasses, produced by the local attraction of the iron in the ship'.[31]

In the case of the *Solway*, Smith reasoned that the steamer had departed on a northerly course with negligible deviation, but changed to a westerly one in which the deviation would have been at the maximum. 'Her officers had no opportunity of determining her course astronomically', he argued. 'The deviation was in all probability not known and not allowed for, [and] as she had run about 20 miles, [either] a deviation of one point would throw her nearly four miles to the south of her proper course, [or] a deviation of half a point two miles'. At the same time he dismissed reports that the loss could be explained by an in-draught or current from the Bay of Biscay: 'the indraught, I presume, sets along the coast, not on [to] the coast'.[32]

Smith's letters drew forth the mockery of an old guard familiar with wooden-hulled sailing vessels. A self-styled 'Old Mariner' dismissed his scientific analysis as 'scarcely worthy of notice'. It was reminiscent of 'Tenterden Steeple [in central Kent] causing the Goodwin Sands [off east Kent]'. The loss of the *Reliance*, he asserted, was instead occasioned by 'a hurricane, and the indraught of the Bay of Biscay'.[33] Among deep-sea mariners of that kind, ships' magnetism was very much an alien subject. Yet the *Great Britain* was under construction in Bristol and the P&O Company were just embarking on their first experiments with relatively small iron-hulled steamers (Chapters 6 and 14). And on Merseyside by the end of 1842, promoters of iron steamers were concerned about the difficulties of obtaining insurance on such vessels

[30] *Londonderry Sentinel*, 19 November 1842 (with extracts from *The Boulogne Gazette* and *The Times*).
[31] *The Times*, 19 April 1843.
[32] *The Times*, 19 April 1843.
[33] *RM* (1843: 429).

and their cargoes. For example, Liverpool ironmaster Thomas Jevons wrote to Birkenhead shipbuilder John Laird requesting that he attend a meeting 'to devise some means by which the prejudice at present existing against that class of vessel may be removed'.[34]

A different challenge to Smith's analysis came from 'A Commander, R.N.', who asserted his authority not based upon mathematical science but on practical experience 'of that iron-bound coast' gained in a man-of-war packet steamer operating between Falmouth and Lisbon. He offered three explanations for the RMSP losses: that the ships were required to touch at too many ports; that in the *Solway* case there had been 'unpardonable carelessness in shaving the shore too close'; and that the officers' behaviour had 'an ugly look, their finding their way into the boats, when so many poor women and children were left to perish on the wreck'! He reminded readers, however, that the death of the master, who in the merchant service had entire responsibility for the navigation, would most likely exonerate the officers.[35]

The *Solway* disaster also generated wider public questions about the material trustworthiness of steamships. 'A Naval Officer', emphasizing his credentials as a 'sea-officer of some experience', argued that a steamer, due mainly to its power, 'almost invariably damages herself irremediably when she strikes'. With the *Solway*, 'they backed and forced the ship off; when down she went in deep water, not affording time or opportunity to save the lives of all her crew, saying nothing of the helpless and unfortunate passengers'.[36]

Three weeks after the shipwreck Chappell released RMSP's investigation report. The report conveyed a message that the Company was blameless. A 'thorough refit of the ship and her machinery' had preceded departure. The steamer had easily weathered 'some stormy weather in crossing the Bay of Biscay'. At Corunna the ship had taken another eighty tons of coal aboard and then sailed in a fresh west-south-west breeze with a ground swell due to the recent gale. The land, although hazy, was visible and the steamer passed the lighthouse outside Corunna at 10pm.[37]

Much of Chappell's key evidence derived from the purser, Mr Lane, who testified that the captain left the deck only once after leaving port. 'There, west, that will do, Lane', he told the purser after laying down the new course '... that course carries us clear of everything'. After the captain returned on deck, Mr Lane heard him 'order the steersman to keep the ship west'. The purser had made two complete voyages under Duncan's command and had seen him

[34] Jevons to Laird, 1 December 1842, Laird Family Papers (LFP), Merseyside Maritime Museum (MMM). Marsden and Smith (2005: 28–9) (quoting letter); Winter (1994: 69–98) (wider history).

[35] *The Times*, 5 May 1843.

[36] *The Times*, 1 August 1843.

[37] *The Times*, 8 May 1843.

'evince great coolness, skill, and resolution' in the face of a hurricane in the Gulf of Mexico and a storm in the Atlantic. Deck officers, on the other hand, offered little personal testimony in support of Duncan and confined themselves mainly to the events after the *Solway* grounded.[38]

Lieutenant W. G. Hemsworth, RN, Admiralty Agent on board the *Solway*, had agreed 'to furnish the directors with his impartial and competent opinion of this disaster'. In contrast to Lane, he placed the blame fully on the master, 'his unfortunate "error in judgement"' arising from 'his great anxiety to fetch up the loss of time'.[39] The indictment fulfilled the earlier prediction of 'A Commander, RN' that the death of a merchant service captain would exonerate the officers. The Company now had a drowned master who could be praised for his noble conduct, blamed for his error of judgement, and trusted not to ponder other disturbing questions such as those raised by Archibald Smith.

Reproducing Chappell's report in full, the *RM*'s editor denounced it. 'With the intention, we presume, of glossing over, if possible, the revolting affair of the loss of the *Solway*, Captain Chappell ... has appeared frequently in the columns of the *Times*', it asserted before accusing the Secretary of inventing 'a haziness over the land, of which those who were present say nothing'. As the person who possessed the appointments to the ships, the *RM* pointed out, he was responsible for their inadequate conduct and thus 'naturally very *desirous* to envelop in thick "haze" the whole affair'. Chappell's 'very transparent anxiety to stifle inquiry convinces us that the affair ought to be sifted to the bottom, and ... if it is, we shall have matters rise to the surface which will be no very pleasant subjects to behold'. The *Railway Magazine* then alleged that he had not only built up a heroic image of the master after the grounding, but had failed to examine important witnesses including the helmsman, the pilot at Southampton and most of the passengers.[40]

Changing its style, the *RM* now waxed satirical over what it described as 'the endless variety of ways the Company's paragons of seamen seek to get rid of their vessels'. One had attempted 'to cut a West India Island in twain' at noon; another, 'Don Quixote like, makes an onslaught on a rock four miles diameter and thousands of feet high'; a third goes 'miles out of his way to run foul of a rock'; a fourth would 'sink his vessel in harbour, perhaps to save the trouble of taking her out to do it'; while the new *Avon*, 'not finding a rock in the middle of the ocean, runs down a sailing vessel'. Never, it concluded, had there been 'such an odd set of adventurers in the world ... and we are half induced to ask wherever they could have picked up so ingenious a set of fellows in misfortune'. Another biting satire in the *RM* assumed the persona of 'Sam Slick'

[38] *The Times*, 8 May 1843.
[39] *The Times*, 8 May 1843.
[40] *RM* (1843: 494–5).

(Judge Haliburton's creation (Chapter 4)) in which he mercilessly mocked the Court as 'a Scotch aristocracy' characterized by nepotism and jobbery rather than by professional seafaring knowledge. Taken together, the press attacks were devastating. In the space of a single week, the Company's share price fell from £17 on 13 May to £12 by 20 May.[41]

Presented to RMSP's October GM, the Secretary briefly announced the directors' formal conclusion to the *Solway* investigation. '[T]here is every reason to believe, [that] it arose from no want of professional ability or zeal on the part of the Captain', it ran, 'but from an error in calculation or in judgement to which the most experienced navigators are occasionally liable'.[42] Several proprietors, however, challenged the statement. Captain Sweeney, RN, pulled no punches. Had those circumstances happened to him, as a naval officer while in command of the same good vessel with the men on deck, 'I should stand before you self-accused of manslaughter'. He added that many friends were choosing traditional merchantmen under sail for their passage west. Admiral Austin also condemned the directors' judgement, citing a naval colleague's view that the disaster demonstrated 'either want of skill, or the want of attention'. An unnamed proprietor echoed these concerns with respect to the West Indies where the report 'is capable of doing the company serious injury, for the hue and cry [there] is general'. Moreover, the directors' removal of blame from the captain 'will take away a great deal of that confidence which we should have in them'.[43]

Responding, the chairman (Irving) acknowledged the doubts but deferred to Captain Chappell as a 'professional man' who had 'taken his own means of enquiring into the whole subject'. Chappell explained that 'the calamity arose from the captain's error in not allowing for the variation'. The directors' conclusion, thus, was not to say there was no fault with the captain. Indeed, they 'conceive there was great fault, but were willing to draw as delicate a veil (hear, hear) as possible over the misfortune of a family which is left in great distress by this calamity'. Chappell also countered press criticisms of Duncan's credentials. 'He came to us in the first instance through being chief officer of a Trans-Atlantic steam ship going from Liverpool; he was recommended to us by two captains, Captn Miller, who commanded the [Cunard] Acadia, and Captn Favrer [RN] who commanded the [RMSP] Forth'. The Secretary therefore summed up: the late captain was not ignorant of the variation but between leaving his cabin and going on deck he had made the error in calculation.[44]

[41] *RM* (1843: 491, 495, 515, 524).
[42] RMSP *Report* (12 October 1843: 1).
[43] RMSP GM (12 October 1843: 21–2, 27, 44–6). It is likely that Admiral Austin here was Horatio T. Austin (chapter 11)
[44] RMSP GM (12 October 1843: 46–9).

Relying largely on the accuracy of the purser's recollection of the master's words, Chappell simply assumed that Captain Duncan had not taken account of the variation and that, had he done so, the *Solway* would have cleared the land. The Secretary did not, however, put any figure on the variation. Archibald Smith's point had been that without astronomical or terrestrial checks, adding in a previously known figure for variation ignored the unknown deviation which could in the *Solway*'s case have amounted to at least ten degrees. But for the Company's proprietors, the loss of the *Solway* was just one part of a deeply distressed public image which their directors tried desperately to redeem.

8.3 'The Dastardly Assassins': Finding a Scapegoat in James McQueen

'Men ask, with incredulity in their countenance, if the John Irving, the Chairman of this Company, is the highly respectable merchant of London, the talented partner of the eminent house of Reid, Irving, and Co., one of our best and most highly esteemed men of business', teased the *RM* in June 1843, 'and when they are told that it is the veritable and true John Irving, they lift up their eyes in wonder and astonishment'. A month earlier, a sarcastic editorial in the same journal had maliciously suggested that the 'trumpeting forth' of Irving's image as 'a very upright honourable man' of high character was 'the magnet by which their [the shareholders'] money was drawn from their pockets'.[45]

Taking the chair at the October 1843 GM, Irving pleaded indifferent health and delegated to Robert Cotesworth the role of Court spokesman in the absence of more eminent directors such as Baring and Colvile. Better known as a director of several joint-stock railway companies, Cotesworth unfolded a narrative of increasing success that included a reduction in debt from an original £260,000 to £120,000 and a rhetorical declaration that 'never was anything more beautifully perfect than the working of your establishment the whole of this year'. Promising to hide nothing, he could not avoid the uncomfortable question of insurance costs. Underwriters, he stated, had originally set the premium 'at a higher rate than that paid by any other shipping company per annum' as the underwriters judged that the service 'embraced vast difficulties and vast dangers'. After the loss of *Medina* and *Isis*, he admitted, insurers were very reluctant to take on the risks at the high premium of 8 per cent. He made no mention of the likely detrimental effects of the *Solway* disaster, but advocated future self-insurance funded by annually setting aside 5 per cent of each vessel's value.[46]

[45] *RM* (1843: 567, 472).
[46] RMSP GM (12 October 1843: 2–14).

Cotesworth's soft words, however, provoked a Bowring-led group of proprietors to launch an attack on RMSP's management. Bowring hailed from a family of wealthy Exeter wool merchants. His Unitarian pedigree placed him outside the Anglican establishments. Well versed in political economy, he had been a close friend and neighbour in London of the late Jeremy Bentham who had appointed him political editor of the radical *Westminster Review*. A staunch promoter of free trade, an opponent of slavery, advocate of popular education and a one-time active supporter of Greek independence from Turkish rule, Bowring was indeed an enthusiast for radical and often romantic causes. He had served as radical MP for the Scottish industrial town of Kilmarnock (1835–7) and then for the Lancashire manufacturing town of Bolton (1841–9).[47]

Bowring's critique contrasted RMSP's earliest 'halcyon days of prodigality' with the subsequent realization that the original calculations as to expenditure had been erroneous. He recalled how, at the October 1842 GM, he and others felt 'very considerable anxiety respecting the disastrous and dilapidated state of this concern'. They had therefore suggested an investigation into all the affairs of the Company by men chosen from the proprietors rather than from the Court. He then parodied Cotesworth's response: 'for God's sake do not inquire – do not throw any obstacles in the way – we are negotiating with the government – true we are in poverty now, but paradise is at hand – let everything go on in our hands, and we shall get you through all your difficulties (laughter)'.[48]

Two fellow proprietors picked up the gauntlet. A Mr Poynder complained that 'there was not that efficient and active superintendence' demanded of the Company's affairs. He reminded the GM that a year before he had urged 'taking [on] some gentlemen as directors who were considered to be more minutely acquainted with naval affairs than any general body of merchants'. And he expressed disquiet on hearing that the Secretary and Captain Charles Mangles (Chapter 9) were to constitute a new Board of Management.[49]

Proprietor Ridgway sought access, on behalf of the shareholders, to the Company's accounts. 'I look upon this company as an individual company', explained Ridgway, 'and every one of us has a right to know how every farthing of his money is laid out'. Stung by this challenge to the Court's authority, Irving responded as though with hurt feelings: '[d]oes the honourable proprietor think we come here destitute of the feelings of gentlemen? If he does, I tell him he is mistaken, and that we would not submit to degradation [sic] (hear hear, and applause)'. Bowring, however, rose to defend Ridgway. He insisted that there was nothing whatever 'at which the honourable chairman ought to

[47] 'Bowring, John (1792–1872)', *ODNB*.
[48] RMSP GM (12 October 1843: 15–19).
[49] RMSP GM (12 October 1843: 31–3).

take offence'. Turning to the directors, Bowring expressed the hope that to any enquiries 'directed to them by a proprietor decorously' they would be happy to provide answers. 'Undoubtedly', was the directors' muted reply.[50]

To add to Irving's troubles, a proprietor raised the matter of the *RM*'s damaging criticisms, and their potential consequences for the value of RMSP's shares. In reply, the chairman denied ever reading a single word of 'that Magazine' and affirmed that he always treated 'those kinds of subjects in such publications with the utmost indifference'. He admitted, nonetheless, having heard extracts read within his hearing and proceeded to infer that 'some persons in the employ of the company' had communicated with the anonymous authors. He stated that every honourable man, whether 'in an inferior or superior station', would consider such communications to constitute a breach of confidence and deserving of immediate dismissal. But Irving's 'stringent enquiry' among Company officers for anyone 'guilty of so great an impropriety' failed to discover the culprit. At this point, the minutes noted, the chairman 'rested from indisposition, amidst signs of applause'.[51]

Recovering his superior poise, Irving without warning turned his fire on McQueen. 'I really can hardly trust myself with language sufficiently strong', he began, 'to describe what I feel to be due to the conduct of that individual, who was the originator of the plan, or scheme perhaps it might be called'. For the Company, it was nothing less than 'a misfortune that we trusted so much to him ... the evil arose from so much confidence that was given to Mr McQueen'. Irving indicted the former general superintendent for unreliable calculations, for his misrepresentation of RMSP during his negotiations with the Admiralty and now for his attempt at self-justification in his writings. Pointing to McQueen, the Chairman ended with the words 'That is the man'.[52]

McQueen rose to 'repel with scorn and indignation what has been insinuated against me'. He challenged Irving to 'point out one single thing in my calculation whether in regards to the cost of the ships, the cost of the coals, the supply of coals, the consumption of coals, which has not been borne out, except where people would not do their duty'. And unlike 'the dastardly assassins' (Chappell and his agents) who wrote under a cloak of anonymity, he had put his name to everything he had written. Moreover, if he had so misrepresented the Company to the Government, 'would they have sent me away with every proof of their confidence'? Tellingly, he laid the *Solway* disaster at the feet of the captain and condemned the Company for trying to hold responsible its former General Superintendent. He further cited 'a communication from the chairman himself who said the blame [for wider failures of the service] was

[50] RMSP GM (12 October 1843: 42–3).
[51] RMSP GM (12 October 1843: 51–2).
[52] RMSP GM (12 October 1843: 52–4).

entirely to be imputed to your captains, who were not prepared to carry out the details'. Implicit in the argument was the fact that Chappell had initially joined the Company as marine superintendent in December 1840, with responsibilities for overseeing the appointments of captains, officers and crew (Chapter 7). Although Cotesworth countered with a second attack on McQueen, the scapegoat received little further opportunity to respond and the meeting came to an abrupt end.[53]

The surviving verbatim record of the October 1843 GM, together with the *RM*'s devastating critiques, provides a rare glimpse into the secretive world of this metropolitan joint-stock enterprise. With a public reputation as the 'Moorgate Admiralty' directed by a 'Scotch aristocracy', it functioned in clear contrast to the intimate partnership of Cunard's line of steamers (Chapters 5 and 6). As RMSP directors and proprietors dispersed until the next GM, the Court had found its scapegoat in McQueen. Captains Chappell and Mangles now had full day-to-day management of this 'most perilous undertaking'. What changes they implemented, what practices they introduced and what internal and external contingencies threatened the orderly working of the system will be the subject of the next chapter.

[53] RMSP GM (12 October 1843: 54–7).

9 'In Highly Creditable Order'

RMSP's New Board of Management in Action

> For at least a mile along the edge of the reef, inside the breakers, nothing was to be seen but *wreck*! Piled up several feet in awful confusion; – timbers, planks, doors, crushed boats, beds, trunks, baggage, barrels, seamen's chests, &c. – and all that remained in the surf, of the once proud *Tweed* ... was the port side, from the sponson to the figure head, over which still stood the bowsprit and jib-boom, bending, as each sea covered it, like a reed. The [paddle]wheel was still attached to the sponson, and the paddle-box boat covered the paddle-box, all held together by the machinery and the shaft. ... [W]hen, to our dismay, no land could be seen on daylight appearing, we began to conjecture that we had struck on the all-dreaded 'Alacran[es] reefs'.
>
> *John Black Cameron, RMSP official and passenger on the Tweed's last voyage, describes the disintegration of the mail steamer in his 1856 account for SPCK*[1]

Summary

The wreck of the *Solway*, and the consequent criticisms of Company management in the press by proprietors and at the Admiralty, placed exceptional pressures on the new Board of Management to deliver on RMSP's promises of a regular mail system. As the managers constituting the Board, Captains Chappell and Mangles now imposed a more rigorous discipline upon RMSP's masters, officers and crews. With restless proprietors demanding dividends, the managers directed special scrutiny to the dangerously large coal consumption. Encouragement of engineers and masters to economy, experimental trials with coal from different mines, investigation into the handling and shipment of fuel and attention to engine performance all formed part of the tighter regime. As a model of such disciplinary virtues, the *Tweed* nevertheless strayed from the proper course, meticulously plotted on paper, to break up on an isolated Caribbean reef with the loss of seventy-two passengers and crew. Against this bleak event that threatened to shake the Company's reputation to its foundations, an eyewitness invoked a transcendent perspective embodied in Christian practices, civilized community and Church of England theology.

[1] [Cameron] (1856: 21–3).

9.1 'Efficient and Active Superintendence': Disciplining Company Operations

In mid-August 1844 RMSP's *Tay* grounded on the Colorado Reef off Cape St Antonio, Cuba. Refloated, the ship proceeded to Havana and arrived at Southampton a month later. Captain Chappell travelled down from the head office in London and went off to the ship at the moorings. As he recorded in his 'Daily Minutes', he there ascertained that the steamer's master, Captain Hayden, had gone ashore 'without waiting to see her safe moored which, owing to the gale of wind then blowing, was not accomplished for two hours after'. The Secretary then dispatched a messenger demanding Hayden's immediate return that he might answer for this serious breach of trust. But when the captain arrived back on board to take his ship to the Thames for extensive repairs, Chappell observed that 'his manner was so excited as to show it would not be safe to trust him with the charge, even with the assistance of a pilot'.[2]

The RMSP's marine superintendent, Captain Barton, RN, volunteered to accompany the ship. In order to emphasize the seriousness of the episode, Captain Mangles arrived the next morning. After discussion, he and the Secretary issued written instructions to Hayden 'to obey all orders given by Captain Barton as if issued by the Court of Directors'. They also authorized Barton 'by a separate order to even assume sole command and charge of the ship if necessary'. Upon the ship's safe arrival at Blackwall, Captain Barton reported that Hayden had exhibited 'the same negligence on the voyage to the River [Thames] as appears to have characterized his conduct previously'.[3]

The extent of the *Tay*'s injuries quickly became apparent. As soon as the engines were stopped, the leak 'increased so fearfully upon the ebb tide as to leave only 18 inches of her watertight bulkhead uncovered'. Extra pumps and hands arrived and 'by unremitting exertions night and day the ship was got into [dry] dock'. The Secretary then informed Captain Hayden, as well as the second and third officers, that the Court of Directors 'had dismissed them from their service'. In contrast, the Court voted the chief engineer Mr Wilkie, its thanks and a gratuity of £50 'for his zeal, resolution, perseverance & ability' throughout. Other engineers received the Court's thanks and a gratuity of one month's pay subject to Mr Wilkie confirming each man's 'uniform good conduct' throughout the ordeal. They would, moreover, receive preference for re-entry into the Company's service.[4]

Hayden had already been the subject of at least two management investigations in the course of the previous twelve months. On the first occasion the managers looked into 'differences between Captain Hayden & his officers'.

[2] RMSP Daily Minutes (DM) (14 August and 16 September 1844).
[3] RMSP DM (14 August and 16 September 1844).
[4] RMSP DM (14 August and 16 September 1844).

But while they deemed that 'his being on bad terms with all his officers is far from creditable', they admitted that he had commanded his ship for almost two years 'without the slightest damage either to the lives or property entrusted to his charge'. They therefore recommended that the Court strongly caution him 'to avoid in future all cause of personal altercation and disagreement with his officers, which only lead to the formation of cabals, the propagation of injurious reports, the dissatisfaction of passengers and various other evils injurious to the interests of the Company'. Instead, Chappell and Mangles blamed the chief officer, Mr Mitchell, 'the principal promoter of the cabal against his captain'. Mitchell had admitted to them that not only had he been 'dismissed the naval service', but that he had also been 'imprisoned in the Marshalsea for misconduct'. The managers thus concluded that he was 'an unfit person to be retained in the Company's service'. They also recommended the removal of the third officer to one of the other ships of the fleet as his manner had been 'disrespectful to Captain Hayden during the inquiry'.[5]

Up to this point, the managers had given the master the benefit of doubt on all counts. But Captain Hayden was effectively on trial. 'As the *Tay* would thus proceed next voyage with an entirely fresh set of officers', they concluded, 'means would be afforded to test more thoroughly whether the former differences have arisen from the fault of the subordinate officers, or from the infirmity of the Captain's temper'. Further complaints followed, this time from passengers. But the Admiralty Agent on board, Lt Brereton, gave testimony 'highly in favour of Captain Hayden' and the managers bowed to the absence of proof against him.[6] The subsequent grounding of his vessel, however, destroyed his reputation as a trustworthy master. Above all, three former captains, Chappell, Mangles and Barton, had witnessed at first hand his unreliable behaviour.

Hayden was far from being an isolated offender. Chappell and Mangles frequently summoned masters to account for perceived sins of omission and commission that threatened the orderly working of the Company's system. In July 1843, for instance, the 'Daily Minutes' recorded that the managers had interrogated Captain Sharp of the *Tweed* 'as to several points that had occurred during his absence from England'. He had 'got the ship aground upon two occasions'. He had stood 'too great a distance off shore when at Madeira'. He had apparently not remained long enough at one port of call for freight. He had brought home a quantity of Madeira wine 'without permission'. He had 'occupied the ladies cabin'. And he had not 'transmitted home abstracts of the *Tweed*'s log as ordered'.[7]

[5] RMSP DM (21 October 1843).
[6] RMSP DM (21 October 1843 and 29 February 1844).
[7] RMSP DM (17 July 1843).

As Secretary, Chappell not only had to manage discipline within the fleet, but he also had to contend with the Company's most significant external power, the Admiralty, upon whose inspection and patronage the line's viability depended. The *Tay*'s arrival in Blackwall coincided with an uncomfortable one-to-one meeting between the Secretary and Rear Admiral Bowles at the Admiralty. Sir William Bowles had recently taken office on the Board of Admiralty headed by Lord Haddington. He had an impeccable pedigree and formidable aristocratic connections. Born in 1780, his mother was an admiral's daughter. He was also related to the Earl of Malmesbury. After a distinguished seagoing career, much of it spent in North and South American waters, he married the sister of his friend Henry John Temple, Viscount Palmerston. While comptroller-general of the coastguard service, he sought to restrict coastguard recruitment to men who had served in the Royal Navy and later argued for Admiralty control of the coastguard in times of national crisis.[8]

In Sir William's august presence, Captain Chappell risked losing his usual aura of authority. As the interview turned to the *Tay* affair, the Admiral 'expressed his astonishment at the gross negligence on board that ship'. The Secretary assured him that the Captain and both officers had been dismissed and that not a single officer involved in any of RMSP's lost steamers had been retained in the working fleet. Sir William nevertheless pushed for much greater Admiralty involvement. He 'was of opinion that naval officers of higher rank than Masters, properly recommended by the Admiralty, would be best adapted for the Company's service'. He expressed his intention to recommend that 'the Admiralty should co-operate with the Company in the selection of officers, by sending a List to the Directors of thoroughly competent persons, that the Directors might select from such a list, and dismiss them again at their pleasure'. Any misconduct 'would be noticed by the Admiralty, as well as by the Company'.[9]

The Rear Admiral's opinion mirrored his earlier views on Admiralty control of coastguard manning. Chappell did not, however, allow him an uncontested victory. Agreeing to communicate Sir William's suggestion to the directors, he refused to anticipate their views but pointed out some rather uncomfortable facts. '[N]aval officers had been tried by the Comp^{y}', he stated, 'and the *Medina*, *Isis*, *Medway* and *Tay* had all been run on shore while in command of officers of the Navy; while in the *Trent*, *Teviot*, and other ships such officers had been above the work of attending to passengers and mercantile matters'.[10]

With all the authority of a high aristocrat, Sir William retorted that 'this arose merely from not consulting the Admiralty as to their [the commanders']

[8] 'Bowles, Sir William (1780–1869)', *ODNB*.
[9] RMSP DM (20 September 1844).
[10] RMSP DM. See also Lindsay (1874–6: vol. 4, 294).

previous character, and from employing the class of masters who were apt when placed in command, to degenerate into what they sprang from'. Deftly, Chappell ended the 'interview' with a 'pledge that the Court of Directors would not appoint any Naval Officer hereafter, without referring to the Admiralty for an opinion as to his character and capability'. He made, however, no such commitment with respect to appointment of officers without a naval pedigree. And he made a note in his 'Daily Minutes' that 'as in all former conferences with the Admiralty, it was best to place the matter upon record while it was fresh in his recollection'.[11]

In the wake of the three steamship losses and the seeming waywardness of masters, the Secretary issued instructions to tighten up masters' practices. Each captain received orders that upon arrival in Southampton he was to submit his ship's books and papers to the marine superintendent's office there. These documents were to be forwarded without delay for inspection to the London headquarters. Each commander also had to regularly complete a printed form regarding all the freight brought home. Chappell further instructed masters to attend at the London office on the Wednesday of the week prior to departure when they would appear before the Shipping Committee, now scheduled to convene at noon precisely on that day. Captains would remain overnight to 'take leave of the Court, as usual, on Thursday'.[12]

Chappell also addressed the wider problem of crew discipline and sobriety. When the crew of an outward-bound vessel signed the articles, he decreed, they would be summoned by the Admiralty Superintendent to attend muster aboard ship at eleven o'clock on the day before departure. Anyone failing to appear would be regarded as a deserter and barred from service on the Company's ships. Policed by the master and officers, no crew member would be permitted to return ashore thereafter.[13] Enforcement of regulations was made easier by the practice of mooring the RMSP steamers off Southampton.

All these more rigorous practices were exemplified in the departure from Southampton of the *Tay*, fresh from restoration in Pitcher's Thames dockyard and under new command. Aboard were sixty-three male and seven female passengers, one child and two female servants, along with 1,195 bottles of quicksilver for the Mexican silver mines, £7,356 worth of freight and 661 tons of coal. Although the steamer was damp from the incessant rain, the 'engine room & machinery of this vessel appeared in highly creditable order'. Acting under instructions from London, the superintendent (Captain Liot) 'caused all the officers to be assembled in the Captain's Cabin, and addressed them with the view of more forcibly impressing upon their minds the paramount necessity

[11] RMSP DM (20 September 1844).
[12] RMSP DM (11 July and 25 October 1844).
[13] RMSP DM (20 July 1844).

of observing the Company's Regulations, & being zealous & attentive in the discharge of their various duties'.[14]

The managers were never slow to interrogate instances of bad practice as steamers prepared to depart. At the March 1844 inspection, Mangles found the *Avon* 'not in that state of cleanliness & order' expected. The deck and some cabins were dirty, while unoccupied cabins 'were lumbered up with furniture which appeared to have been thrown into them with an utter disregard for its preservation'. The saloon steward, too, was 'dirty in his dress & he and the establishment of servants seemed to be a day at least behind in their work'. On the other hand, he acknowledged that 'the appointments of the table were clean & neat and the dinner was good – the cook was a Frenchman and judging from the dishes and the table, efficient'. Moreover, the *Avon* had arrived late from the West Indies and had required repairs such that 'some allowance may be found in this circumstance for the want of order &c on board the ship'.[15]

It was the *Severn*, however, under the command of Captain William Vincent that earned the most consistent praise. In February 1847 Captain Liot bore testimony to the vessel showing 'undoubted signs of having been beautifully kept'. Handicapped by a fractured cross-head in the starboard engine, the quick passage home was 'in a great measure due to the zeal and intelligence of the Commander in setting all additional sail possible upon the vessel'. And in April 1850 Captain Mangles, witnessing the ship's departure, remarked that it 'was, as the *Severn* always is, in the highest order throughout'.[16]

As the managers implemented the Company practices, it seemed as though the early problems had diminished or even disappeared. The directors' *Report* in April 1844 attributed the Company's 'heavy charges arising from accidents', with the exception of the *Solway*, to its first year of operations. A combination of 'the less dangerous navigation' and 'the greater experience of the Company's Officers' augured well for a non-recurrence. Indeed, not a single casualty had 'occurred during the last twelve months'. By the October meeting, the directors claimed that 'the expectations formerly held out have been fully realized'. The Company's debt, they announced, would be paid off early the following year; in the half-year to June freight and passage money stood at almost £55,000, some £17,000 up on the same period in 1843 and coal costs had fallen by nearly the same amount. Unknown to the directors, however, the 650-ton feeder vessel *Actaeon* had been wrecked on the coast of New Grenada (Colombia) just two days before the meeting.[17]

[14] RMSP DM (1 November 1844). Quicksilver (liquid mercury) was vital to the extraction of gold and silver.

[15] RMSP DM (1 November 1844).

[16] RMSP DM (17 February 1847 and 2 April 1850).

[17] RMSP *Reports* (11 April 1844: 2); (10 October 1844: 1); Haws (1982: 31) (*Actaeon* loss).

9.2 'The Great Expenditure of Coal': Curtailing Fuel Consumption

In early 1844 the *Avon*, commanded by Captain Strutt, left Vera Cruz for Havana with only 120 tons of coal aboard. The managers commended him for showing 'nerve, energy and judgement' throughout the voyage, especially as any time spent taking on more coal would have 'thrown the Service into great confusion, caused great clamour in the West Indies, as well as expense to the Company'.[18] At other times, however, masters would be censured for allowing their vessel to run low on fuel and thereby put at risk the safety of the ship, its passengers, mail, freight and the reputation of the line.

From its creation in mid-1843, the Board of Management was desperate for any method of mitigating a very costly, and at times dangerously large, consumption of coal. In the Company's first full year (1842), the cost of coal ran to almost £78,000. In the first half of 1843, the cost was £55,390, pointing to a figure well in excess of £100,000 for the full year and representing over 50 per cent of the target annual expenditure of £200,000. Between 1844 and 1849, the available working accounts consistently showed lower full-year coal shipments at between £77,000 and £80,000. These costs approximate to an annual quantity of 100,000 tons at a price of fifteen shillings per ton delivered to coaling stations in the West Indies.[19] Given, however, the much lower price for Welsh coal shipped only to Southampton, this annual tonnage was probably very much higher. Moreover, it must be remembered that all the mail steamers were sail-assisted.

In comparison, the other major annual disbursements were wages (between £37,000 and £45,000) and provisions (between £54,000 and £61,500). Total expenditure in the same period (between £217,000 and £299,000) far exceeded the promise of expenses remaining at £200,000 per annum, but receipts for both freight and passengers showed a generally steady increase to ensure that the surplus on the working account reached £129,300 by 1849. Looked at from another perspective, however, without the £240,000 contract the surplus would become a loss of over £100,000 on passengers and freight. After 1850, the annual coal bill rose from some £88,000 to peaks of well over £200,000 per annum between 1853 and 1858.[20] Even without the positive effects of the Crimean War on trading patterns and profits, revenues rose to compensate for larger, more powerful and more expensive vessels and extended routes.

In the 1840s, the managers pursued a number of options in an attempt to reduce consumption of coal, including the instruction to selected masters to

[18] RMSP DM (17 February 1844).

[19] RMSP *Reports* (1843–9); RMSP DM (18 March 1844); William Schaw Lindsay Papers (WSL) LND 35/2, 9–10. Freight costs for West Indies coal shipments recorded by Chappell typically show around fifteen shillings per ton. Lindsay cites seven shillings for the pithead cost.

[20] RMSP *Reports* (1843–9).

undertake trials with fuel from specific locations, testing the calorific value of coal from different mines, encouraging the seagoing engineers to use expansive working, attending to better means of selecting coal suited to climate, and weighing up the economy of oscillating *versus* side-lever engines. The *Clyde*'s Captain Symons, for example, submitted a statement on the quality of coals from the Risca Black Vein mine in South Wales in August 1843. He reported that 'they burnt very well, making but little clinker, and averaging about 25¼ tons per day' compared to the ship's previous average of thirty-four and a half tons per day. A few months later, superintendent engineer Mills pronounced the *Avon*'s engines likely to consume twenty-eight or twenty-nine tons per day.[21]

In early August 1843, Chappell received from Mills a report on 'the results of the trial of the [Risca] Black Vein Coal at Woolwich Dock Yard'. The experiment deployed one and a half tons of coal (3,360lb) to evaporate 30,720lb of water distributed in sixteen separate tanks. Each pound of coal therefore turned 9.142lb of water into steam.[22] While the figure appeared to agree exactly with Welsh coal from a different colliery, the report noted the results of a similar trial with 'Lower Black Vein Coal' apparently showing a higher figure of 9.795lb of water turned into steam per pound of coal. But the quantity of coal (1,176lb), with 11,520lb of water in six tanks, prompted Mills to conclude that 'equal reliance cannot be placed on the experiment'.[23]

In a second investigation later that year Mangles travelled to Newport, South Wales, to witness the working of coal transfers between pithead and loading jetty. As a site of coal shipments overseas, Newport handled 7,256 tons in 1840, but in 1851 loaded some 151,668 tons reflecting the overseas demand from steam navigation companies. In 1874 some 1,066,000 tons of coal left the berths.[24]

Mangles, a thorough-going sailing ship master and Surrey gentleman, seemed an unlikely candidate for such a task. Brought up in a family deeply entwined with the EIC, Charles had risen from midshipman to commander in the course of seven round voyages on EIC's Cape route to India. Around 1830 he had joined a brother in taking over their father James' Thames mercantile business that had included shipping convicts to New South Wales. James had served as mayor of Guildford and High Sheriff of Surrey, as well as sitting as an MP. From 1834, the company began trading to the Swan River in Western Australia. Less than enamoured with its intrusion, the local press angrily

[21] RMSP DM (4 August 1844 and 18 March 1844).

[22] RMSP DM (4 August and 9 November 1843).

[23] RMSP DM (August 1843); Smith (2013a: 513–16).

[24] Daunton (1977: 4). Newport was second only to Cardiff, which by 1874 was shipping over 3.75 million tons.

declared of the Mangles' company in 1836: 'Are second hand Wapping merchants to make a monopoly of Swan River? They never shall!'[25]

The tone of Mangles' report on Newport coal certainly carried a strong flavour of gentlemanly disdain towards, and distrust of, the lower orders. Mangles identified several points at which the fuel might suffer a reduction in quality and recommended corresponding remedies. First, a loss of quality occurred in the transfer of the coal between the smaller trams (used to bring it to the pithead) and the larger trams (employed to carry it alongside the ship). He therefore urged hand-picking or 'screening' of the coal to separate out the unwanted small coal and shale. Second, Mangles pointed out the need for more rigorous checks on tickets issued at the weighing house immediately prior to arrival of the trams at the loading jetty in Newport. Suspected fraud, he noted, occurred when the gatekeeper received a 'gratuity' in exchange for a second ticket showing a greater tonnage than the official measurement.

Third, he objected to the great height, exacerbated by the large tidal range in the Bristol Channel, from which the coal was thrown down by means of shoots from the trams into the vessel, leading to broken coal. Fourth, he found that 'at present stevedores of low character are in the habit of meeting the ships on their way up the River to Newport, and offering their services at low rates'. In accepting their terms, masters received larger cargoes, but with a higher proportion of broken coal making up the additional weight by filling the void spaces. He thus urged the use of stevedores approved only by RMSP's agent. Finally, he advised the stocking of coal at Newport as 'it would place Mr Russell [the pit owner] in a more independent position in regard to his unruly pitmen; and would secure a supply for the Company always readily available, in the event of a strike in the mines'.[26]

Although RMSP had no control over shipments of coal across the Atlantic, the number, small size and variety of the chartered vessels ensured that the Company was unlikely to suffer from a failure in delivery. In March 1844 the Secretary accepted the coal freights shown (Table 9.1).

Within a month, a further five sailing vessels had been accepted for Bermuda and the West Indian stations. But the appetite for coal was far from satiated. On a single day, less than a week later, Chappell ordered freights to be obtained for St Thomas (1,500 tons), Jamaica (1,000 tons), Havana (800 tons) and Bermuda (800 tons).[27]

The RMSP mail steamers did on occasion run short of fuel while on a passage. In June 1843, for example, Captain Boxer was called before the managers

[25] SC (1837), v, 86–7; Statham-Drew (2003: 1, 38–9, 46, 50, 238, 307 (quotation from the *Swan River Guardian* 10 November 1836), 340). My thanks are to Philip Boobbyer for this source.

[26] RMSP DM (9 November 1843).

[27] RMSP DM (18 March, 3, 8 April 1844); Smith (2013a: 514).

Table 9.1. *RMSP coal freights (March 1844)*

Lord Canterbury	599 tons at 16/- for Bermuda
Ann	190 tons at 16/- for Bermuda
Trial	158 tons at 15/- for Bermuda
Marie Antoinette	160 tons at 15/- for Havana
William	240 tons at 15/- for Grenada
Cuba	291 tons at 15/- for Grenada
Demerara	288 tons at 14/7 for Jamaica

to report 'how the *Trent* became short of coals on the passage home from Bermuda'. In July the following year, Captain Strutt passed Fayal in the Azores without taking on coal and reached Southampton with only two tons remaining in the bunkers. In the managers' opinion, this omission was tantamount to 'running unnecessary & unjustifiable risk, which might have produced very unpleasant consequences'.[28] The RMSP steamers were by no means alone in facing this hazard. In January 1849, P&O was caught out when the prestigious *Bentinck* exhausted all the coal while on a passage from Madras to Calcutta (Chapter 14).[29]

At the behest of the Admiralty, the Company arranged rigorous steaming trials for their mail steamers, usually after a major engine overhaul. The six-month-old *Avon* prepared for such a formal trial in early August 1843. Mr Mills attended the Court of Directors in person to report the machinery, 'upon trial, to be in a thorough state of effectiveness; and that the consumption of coal did not exceed one ton [per] hour'. This performance roughly equated to five and a half pounds of coal per horsepower per hour or twenty-four tons per day. Six days later, the Secretary recorded a 'trial of the *Avon*'s engines and boilers round the Isle of Wight in which the performance was very satisfactory not only to the Managers, but to Messrs Maudslay, Field & Pitcher who were present' as experienced witnesses. Also aboard was Mr Kingston, the Admiralty Surveyor, who 'expressed himself well satisfied with the performance of the machinery and boilers'.[30]

The Admiralty also insisted on steaming trials for second-hand tonnage to be employed on Caribbean feeder services. In October 1843, for instance, the recently purchased Caird-built *Actaeon* underwent trials in the Thames. Such trials tended to focus on meeting the Admiralty's stringent standards with respect to the safety and seaworthiness of the boilers, engines and hull, while

[28] RMSP DM (20 June 1843 and 12 July 1844).
[29] RMSP DM (5 January 1849).
[30] RMSP DM (3, 9 August 1843).

also ensuring that the machinery delivered of its best results. Accompanied on the trip by the Admiralty surveyor and Messrs Maudslay, Field and Company, the Secretary recorded that the 'new boilers gave an abundant supply of steam and a Diagram taken by the Indicator shewed a most efficient performance of the [side-lever] engines'.[31]

Towards the end of 1843, the managers instructed Mills to 'adopt some means of lessening ... [the] consumption of fuel' by the *Severn*, last of the original mail steamer fleet to be completed. By altering the furnaces and 'chimney' (funnel), the engineer expected to diminish the draught of the furnaces and thus effect 'a considerable reduction in the consumption of fuel, with acceleration of speed'. As an additional measure, Chappell ordered the shipment of sixty tons of patent fuel (on top of the 620 tons of coal). The patent fuel was 'to be wholly consumed between Southampton and Madeira; and ... a minute account be transmitted, from the latter place, as to the consumption per day, generation of steam, &c &c'.[32] The results of the changes were scarcely encouraging. By March 1844 the Secretary recorded the 'expenditure' of coal on the *Avon*'s second voyage to be 'at least 8½ lbs per hp per hour'. The *Severn* also appeared to suffer a similar tendency. He thus ruefully reflected that if this statement were correct 'it would really appear a serious consideration if it would not be a better plan to withdraw these ships altogether, or only to employ them in the shorter voyages'.[33]

Attempts to obtain the best performance from machinery also continued while RMSP ships were in service. Chappell reported in September 1843, for example, that the managers had questioned the *Dee*'s master, Captain Hemsley, 'as to sundry delays on his voyage and as to what attention had been paid to working the steam expansively' (Chapter 7). The managers had also called 'the attention of Mr Mills to the great expenditure of coal in the *Dee*' – a massive thirty-nine and a half tons per day over the first eighteen months in service, although this had been reduced to thirty-four and a half tons per day for the most recent voyage.[34] Some conception of this rate of consumption can be gained by observing that every ten days a mail steamer such as the *Dee* would require one or even two sailing-ship loads of the best Welsh steam coal.

Deeply conscious of this vexed question of fuel supply and consumption, the managers suggested that they be authorized to reward captains and engineers 'where striking cases occur, evincing unusual care and economy in the management of stores or expenditure of coals'. Captain McDougall of the *Medway* was a case in point. The managers recommended presenting a reward to him

31 RMSP DM (25 October 1843); Smith (2013a: 514–15).
32 RMSP DM (13, 25, 30 November, and 18 December 1843).
33 RMSP DM (22, 30 March 1844).
34 RMSP DM (25 September 1843).

'for his great attention to the expansive working of the steam and other means of economizing fuel [and] that this be made known to all the captains and engineers'. The managers had recently received from the *Medway*'s engineers a complaint against Captain McDougall. Upon investigation, they discovered that 'Mr Hurst Chief Engineer had absconded having been detected in extensive smuggling & that Mr Jukes 4th Eng^{r} had also left the ship [while] Mr McLaren 3rd Engineer died at Havana'. They nevertheless wanted all RMSP engineers to be informed that Mr Hurst would have shared in the reward 'had he not disgraced himself subsequently by an act of smuggling & been compelled to abscond to avoid the consequences'.[35]

The directors' *Report* in April 1844 admitted a loss on the coal account 'arising chiefly from waste and deterioration by climate since the commencement of the service'. Coal stocks held for a time in the West Indies, they believed, lost a measure of their calorific value. They nevertheless assured shareholders, somewhat vaguely, that a recurrence was unlikely since 'the Foreign Agencies and Coal Stations are now placed on a more systematic and economical arrangement' and because 'experience has enabled the Directors to select the coal least liable to such deterioration and waste'.[36]

Towards the end of that year, Chappell directed Thames shipbuilder William Pitcher, in conjunction with Mr Elsmie (RMSP superintendent shipwright), to prepare the drawings for the hull of a wooden steamer of about 1,050 tons in order to make the ship more manageable in Caribbean waters than members of the original fleet. The vessel had 'to be fitted so as to mount heavy ordnance at a short notice with light poop and forecastle that can be removed immediately when required'. The hull was to 'be built of the best materials [and] copper fastened' from the keel to just above the waterline. All the requirements followed to the letter the demands of the Admiralty. With regard to the design of the 350-hp engine, there were two main options: 'oscillating' or beam (side-lever) engines, each with either tubular or 'common' boilers.[37] The promoters of 'oscillating' engines, in which the cylinder moved in accordance with the rotation of the crank, claimed reduced friction of moving parts with consequent economy. The engines had recently become the talking point of the London marine engineering community and the engine of choice for a new generation of P&O passenger steamers (Chapter 14). The 895-ton steamer that Pitcher delivered in 1847, the *Conway*, was designed and commissioned not for the ocean mails, but as a West Indies feeder. The only new-build for RMSP since the completion of the main fleet, the ship was fitted with oscillating engines.[38]

[35] RMSP DM (26 September 1843); Smith (2013a: 515–16).
[36] RMSP *Reports* (11 April 1844: 1).
[37] RMSP DM (8 October 1844).
[38] RMSP DM (9, 10 December 1844).

Circumscribed as it was by the legacy of fourteen original mail steamers each with two side-lever engines operating at or below 6 psi, RMSP had little opportunity to experiment with other engine designs during the 1840s. At the same time, these tried-and-tested simple engines were evidently the least problematic and arguably the most reliable part of the Company's complex system, both in terms of individual vessels and on the larger geographical scale. For its managers in this period, economy leading to a reduction in coal consumption, and thus to an improvement in the Company balance sheet, focused not on changing the engine design but on improving the calorific value of the fuel. For them, it was a value that could vary significantly with local conditions: the character of the particular mine and mining region, the handling of the coal between pithead and jetty, the subsequent shipment and storage of the coal under specific climatic circumstances, the zeal and skill of chief engineers and their boiler room crews and of course the experience of the ships' deck officers and crews to take advantage of favourable winds.

The relatively stable annual cost of coal for the fleet between 1844 and 1849 contrasted with rising expenditure on wages and provisions. But freight revenue in the same period rose steadily from a meagre £30,000 to £72,200, while passage money increased from £82,000 to £116,000. As a result, the annual surplus remained in the region of £130,000. With the elimination of early debts, the Company began in the mid-1840s to pay modest dividends to its long-suffering proprietors: £1:10:0 per share in April 1846 rising to £1:15:0 in October and to £2 in April 1848.[39] This new-found confidence coincided with the absence of any further losses within the fleet of transatlantic mail steamers between April 1843 and February 1847. But soon after came the shock news from the Caribbean of the loss of the *Tweed* and many lives.

9.3 'Under Providence': John Black Cameron's Narrative of Deliverance

A week before Christmas 1846, the mail steamer *Tweed* departed from Southampton bound for Vera Cruz in Mexico with intermediate calls beginning at Madeira. From there, Captain George Parsons, RN, dispatched news to the Line's Moorgate head office that the weather during the ten-day passage had been 'more severe ... than on any former occasion'. For two whole days the dining saloon had been ankle-deep in seawater as the vessel rolled in the heavy swell. And as though to demonstrate the veracity of the master's report, his message took more than three weeks to reach Captain Chappell in London.[40]

[39] RMSP *Reports* (c. 1843–9).
[40] RMSP DM (18 January 1847).

The *Tweed*'s course crossed the Atlantic to the West Indian islands of Barbados, Grenada, Trinidad, Grenada again, Jamaica and Cuba where, at Havana, the steamer took on board 480 tons of coal (some of which was for delivery to HMS *Hermes* in Mexico) and exchanged mail with the homebound *Medway*. Departing deep-laden for Vera Cruz on 9 February, the ship had on board sixty-two passengers, eighty-nine crew and 1,115 bottles of quicksilver valued at £18,000. The master set a safe track on the chart midway between the northern coast of the Yucatan peninsula of Mexico and the extensive Alacranes (Scorpion) reef some seventy miles further north.[41]

On 5 April *The Times* reported 'the melancholy intelligence of the loss of the *Tweed*, and 60 lives'. The grim news had followed a tortuous route, beginning with the wreck of the steamer on the north-east end of the Alacranes reef on 12 February, three days out of Havana. It travelled first by frail ship's mailboat to the Yucatan coast, thence by Mexican schooner from Campeachy to New Orleans with publication in the *Picayne* newspaper on 1 March, shipment of that paper aboard the Red Star Line sailing packet *Baltimore* to Le Havre, and across the Channel with the South-Western Railway steamer to Southampton on 4 April with onward travel by train to London. The following day *The Times* recorded that this news 'caused great excitement in the city; and anxious inquiries were made during the day at the company's offices in Moorgate-street'. It commented, however, that the account received from Le Havre 'is too circumstantial in its details to be untrue in fact', the alleged locality of the grounding agreeing 'so exactly with the supposed position of the ship at the date quoted, that little, if any doubt, is entertained of her loss'. Two days later, the arrival at Liverpool of more American papers indicated the scale of the disaster.[42]

At this very time, RMSP's *Avon* was in the Western Approaches to the English Channel, homeward bound with the surviving officers, crew and some passengers. A pilot boat informed the *Avon* that 'rumours were abroad, for the last three days, of the loss of the *Tweed*, but that it was not generally credited'. In gathering darkness, the *Avon* hove to in Cowes Roads inside the Isle of Wight, but failed to secure a competent pilot for the final passage up Southampton Water. During the night one of the passengers, John Black Cameron, accompanied the Admiralty Agent's boat with the mail but failed to catch the London mail train from Southampton at 2am. Aided by the Postmaster and the 'ever-active correspondents of the *Times*, *Herald*, and *Daily News*', he obtained a seat on a special train that reached Nine Elms at 6am and thus became the first witness to bring confirmation of 'the melancholy tidings to London'. As Chappell tersely noted in his 'Daily Minutes' that same day, '[p]articulars were received by the Avon of the loss of the Tweed upon the Alacranes on the

[41] [Cameron] (1856: 5–6).
[42] *The Times*, 5, 6, 7 April 1847.

12th February, upon which occasion 31 of the passengers and 41 of the crew perished'.[43]

Cameron was no ordinary passenger. Voyaging 'for the benefit of his health', he was in fact a Company servant. Back in 1841, his name appeared in RMSP Minutes as a £120-per-annum clerk in the Moorgate office. In the spring of 1843, the *RM* intriguingly asserted, perhaps as another insinuation of jobbery against the Company's 'Scotch aristocracy', that the latest route plan 'is the production of another of their thousand-and-one superintendents named Cameron'. The day after his return to London, *The Times* printed his account of the disaster entitled 'The wreck of the West India ship *Tweed*' and attributed the text to an anonymous eyewitness. The SPCK tract (1856) revealed the author's name. Judging by the speed of publication, Cameron had crafted *The Times* version during the *Avon*'s four-week passage home. Both versions followed a very similar structure. The SPCK text, however, expanded the account in a number of key places to enhance, but not radically alter, the original moral and spiritual perspectives.[44]

At first sight Cameron's account reads as a simple tale of survival and heroism. But his narrative offered Victorian readers a skilfully woven fabric constituted of integrated practical, moral and spiritual components. Rather than lay bare the Company's shortcomings, the author represented the shipwreck as a lesson that combined a direct experience of God's awesome power manifest in a storm at sea with a personal witnessing of divine providence at work. On the one hand, the narrative illustrated the insignificance of human achievements in the face of omnipotent governance, while on the other hand it highlighted the Supreme Ruler's care for individual souls who consecrated themselves and their actions to Him. Framed within a broad reformed Protestant (Church of England) Christianity, Cameron's account used material objects with symbolic resonance to show how officers, crew and passengers recovered Christian civilization previously expressed by and within the noble *Tweed*.[45]

Most notable among these civilizing objects that washed ashore from the *Tweed* were the Bible, the *Book of Common Prayer*, the magnetic compass, clothing, bread and wine, candles and matches and a set of ships' signal flags, all of which served to restore light, order and life in the face of chaos and darkness. Persistent references to these objects allowed the author to demonstrate how RMSP's values of disciplined work, co-ordinated by the officers, directed the survivors to reconstruct civilization on the reef from the fragments bequeathed to them by the tempest. Concomitantly, he evoked strong

[43] [Cameron] (1856: 74–5); RMSP DM (8 April 1847).

[44] *RM* (1843: 323); *The Times*, 9 April 1847; [Cameron] (1856: 5, 12–13, 18).

[45] This analysis draws heavily on Courtney and Smith (2013: 183–203). I thank Stephen Courtney both for his permission to use this work and for his invaluable analytical insights.

resonances with Biblical episodes concerning providential deliverance very familiar to a nineteenth-century Christian readership back home.[46]

The reader first joined the *Tweed* as a 'Norther' comes to blow when it was bound from Havana for Vera Cruz. 'At midnight it was very black, with thunder and much lightning, and the wind came round suddenly to the north', Cameron wrote. On that day and on the following day 'no observation of the sun was obtained by the captain' due to 'it being cloudy and obscure'. As a result, Captain Parsons had to trust entirely to dead reckoning. The steamer's conformity to the track on the paper chart could not be confirmed from the heavens. Already there were echoes of the voyage of St Paul from Acts: 'And when neither sun nor stars in many days appeared, and no small tempest lay on us, all hope that we should be saved was then taken away'.[47]

Order and civilization nevertheless prevailed, embodied in the strength of the steamer, the trustworthiness of the officer on watch and the observance of regular timekeeping that offered support to dead reckoning as to distance run. All was quiet, 'except the boisterous waves now and then dashing against the sides of our noble vessel, the tread of the officer keeping watch on deck, and half-hourly the sound of the bell proclaiming the time'. Company discipline, moreover, insisted that the officer be stationed on the forecastle while occasionally going aft on the spar deck to check on the proper course. The regulations also required that he be assisted by a midshipman, that two lookout men be posted forward at each bow, that one or more men (depending on the weather) be at the wheel and that a quartermaster be close by the wheel to see that the compass course was held. On the main deck, seamen of the watch would be resting but dressed and ready for making or taking in sail and any other urgent duties. In the engine room, apart from the firemen and coal trimmers at work, two engineers would be standing by the engines.[48]

Such rational order suddenly yielded to total destruction of the little world that was the *Tweed*. In pitch darkness, and under both steam and sail, the vessel struck the reef, swung broadside to the seas and broke her back before the aft section disintegrated. The similarity to St Paul's experience would again strike Victorian readers: '[t]he hinder part was broken with the violence of the waves'. Timber from the stern supported Cameron and fellow survivors through the surf to the calm of the reef. From there, he witnessed all along the edge the dismembered parts of the *Tweed* (see Epigraph). Because the Alacranes were covered at high water, the first priority was a raft-like platform to allow survivors to remain dry at all stages of the tide.[49] But it was from the civilizing

[46] Courtney and Smith (2013: 186–8); [Cameron] (1856: esp. 27, 29–31, 39–40, 42, 47).
[47] [Cameron] (1856: 6–7); Acts 27: 20.
[48] [Cameron] (1856: 7–8).
[49] Acts 27: 4; [Cameron] (1856: 12–23).

objects lying among the disordered remains that a new moral and spiritual economy took shape.

Chief among the manifold objects was Cameron's Bible salvaged from his cabin trunk. Reading and rereading Psalm 107, he insisted, provided him with the assurance of deliverance, 'that the Lord would bring us unto the haven where we would be; that we should yet live to praise Him for his goodness and declare the wonders that he doeth for the children of men' (Chapter 1). He also reported how one of the passengers had, on leaving his cabin, snatched 'a much valued prayer book'. Underpinned by the King James Bible, the Church of England's *Book of Common Prayer* had shaped the spiritual practices and values of Protestant England. It now became the means by which the survivors consecrated their efforts to restore Christian civilization and perhaps thereby secure the means to return home.[50]

At first sight, all the ship's boats appeared to have been damaged beyond repair. Led by the chief engineer and chief officer, however, a dozen of the crew took on the challenge of repairing the small mailboat with improvised tools, including a copper bolt as a hammer. Others searched the wreckage for nails, oars and odd sails. The chief officer picked up 'one of the ship's compasses: wonderful to relate, it was *uninjured*, although it had been torn out of the binnacle, washed on to the reef, and was found lying among the mass of wreck and stones, with not even the glass broken'! The survivors, Cameron declared, 'hailed it as sent by heaven to assist in our deliverance'! Here was a crucial civilizing object, resonant with material, moral and spiritual meaning for the Victorian reader at home.[51]

As the chief officer, compass on his knee, took command of the mailboat with nine other men aboard, the community of survivors all gathered for a service of consecration using the prayer book. '[W]e offered up our united thanks to Almighty God for our wonderful preservation, all joining in the responses with the greatest fervour, solemnity, and emotion', Cameron wrote. 'We besought our Father in heaven to watch over our frail little bark, and conduct her in safety to some friendly port'. In *The Times* version, following the reading of the prayers 'each had a little wine, and a little oatmeal'. Once again, the resonance with the narrative of St Paul's shipwreck was evident: 'he took bread, and gave thanks to God in presence of them all; and when he had broken it, he began to eat'.[52]

For the remaining company, Christian practice went hand-in-hand with consolidation of civilized community. Its first act at daylight was 'to offer up our

[50] [Cameron] (1856: 26–31, 47); Psalm 107: 23–31.
[51] [Cameron] (1856: 27–30); Courtney and Smith (2013: 187); Winter (1994: 69–98).
[52] [Cameron] (1856: 30–3); *The Times*, 9 April 1847; Acts 27: 35; Courtney and Smith (2013: 187).

grateful thanks to God for our preservation during the night'. This prayer was the 'The third collect for grace' from the *Book of Common Prayer*:

> O Lord our heavenly Father, Almighty and everlasting God, who hast safely brought us to the beginning of this day, defend us in the same with thy mighty power, and grant that this day we fall into no sin, neither run into any kind of danger; but that all our doings may be ordered by thy governance, to do always that is righteous in thy sight, through Jesus Christ our Lord. Amen.[53]

The officers then took charge of different working parties, 'every one as orderly and obedient as when the unfortunate *Tweed* proudly ploughed the ocean'. The division of labour yielded results: '[o]ne party was sent along the reef to look after stray provisions of any sort; another to collect pieces of timber, and materials to enlarge and strengthen our raft; and another to carry stones, with which to build up a breakwater around to fortify it; the few who were too much injured to do such hard work remaining on the raft to spread out the wet clothes ... stow the provisions in the best manner, and make the raft as snug as possible'. A box of matches, when dried, provided fire and candlelight, symbolic of hope. With the ship's signal flags recovered, one distress flag now flew from a spar in the hope of attracting notice from a passing vessel. The men also 'picked up a few fish, one or two fine lobsters, three bottles of wine, and a can of oil'. Civilization had been restored. And within a short time the Spanish brig *Emilio*, arriving with the chief officer aboard, conveyed the survivors to Campeachy (Figure 9.1).[54]

After RMSP's Court received news of the shipwreck, it declared that 'the energy and firmness of the Passengers, as well as of the Officers and Crew, were most praiseworthy, and tended much, under Providence, to the ultimate safety of the survivors'. Its investigation predictably placed blame for the disaster on Captain Parsons, who was dismissed. But it hailed 'the very satisfactory conduct of Mr Cameron and the able arrangements made by him on the service after the loss of the *Tweed*'. It awarded him the princely sum of £200 and duly confirmed his appointment as Company Superintendent at St Thomas in the Virgin Islands. In contrast, it awarded £200 in total to the twenty-five widows and fifty children of crew members lost. Always ready to submit a request for an increase in salary, Cameron's annual income from RMSP reached £1,000 per annum by 1854. His narrative had provided notable enhancement to both his own status and RMSP's much-depleted spiritual and moral capital.[55]

Neither Cameron's *Times* letter, nor his later SPCK tract, mentioned any of the more disturbing features of the shipwreck. There were no survivors from

[53] [Cameron] (1856: 37); 'Morning prayer', *Book of Common Prayer*, 57.
[54] [Cameron] (1856: 37–40, 42–5); Courtney and Smith (2013: 187–8, 200n).
[55] RMSP *Report* (8 April 1847: 2); RMSP Minutes (15, 22 April and 21 May 1847).

Figure 9.1. Rescue in sight for the shipwrecked survivors of RMSP's *Tweed*

among the four women and two children aboard. Breakdown of officer and crew numbers shows that two-thirds of the officers, quartermasters and seamen survived, while more than three-quarters of the thirteen coal trimmers made it home. In contrast, half the passengers had drowned.[56] Yet Cameron's account provided Victorian readers with an eyewitness narrative of the powers of deliverance at work for the *Tweed*'s survivors and placed the tragic event within the perspective of a Christian culture whose practices, values and beliefs transcended everyday material concerns.

[56] [Cameron] (1856: 76–81); Courtney and Smith (2013: 188, 200n).

10 'She Was One Mass of Fire'

Reading the Maiden Voyage of the Royal Mail Steamer Amazon

> We prefer speed to safety. Competition hastens men to premature enterprises. There is a rage to be first. Machinery is put to its stretch. Vessels are used too soon. Captains will graze the rock to save a few minutes in the voyage. ... Haste makes waste. Speed is a god that must be worshipped, although, like the ancient [Canaanite idol] Moloch, with living and human victims. If we will pass through the fire, we must be burned. ... No laws can be broken with impunity, whether natural or revealed. If we commit the trespass, we must pay the penalty.
>
> *A Congregationalist minister preaches at Islington Chapel on 11 January 1852 on the moral and spiritual lessons from the loss of RMSP's* Amazon *one week previously* [1]

Summary

Control of both RMSP's extended geographical system and the steamships that ploughed its tracks was easily lost. The tragic consequences of an unaccountable, all-consuming fire as the first of a new class of mail steamers crossed the Bay of Biscay on the maiden voyage were bad enough. But these were arguably exacerbated by the duty engineer's obedience to the Company's strict rule of service that only the master could authorize the engines to be stopped. In a rising gale, the *Amazon* ploughed on, rendering futile the safe launch of the boats. Controlling the fallout at home proved equally problematic, as inquiry witnesses reported widespread presentiments of disaster and an eyewitness clergyman preached literal and metaphorical messages of hell and sudden destruction. Restoration of confidence included construction of the line's first oceangoing iron steamer *Atrato*, but such measures were insufficient to placate anxious shareholders for whom the line's superintendent engineer represented all that was flawed in RMSP's system of management.

10.1 'It is a Rule of the Service': The Trials and Tragedy of RMSP's New Steamer *Amazon*

Two weeks before Christmas 1851 the RMSP Secretary, superintendent engineer (George Mills) and superintendent shipwright (George Elsmie) joined

[1] Hollis (1852: 24–5).

engine-builders Seaward and Capel on board the 2,200-ton *Amazon* moored on the River Thames off Blackwall. They prepared to bear witness to the initial trials of the largest wooden-hulled mail steamer constructed since the *President* in 1840 (Chapter 6).[2] The first of five new RMSP ocean steamers under construction at Green's yard in Blackwall, Pitcher's in Northfleet, two vessels Wigram's in Southampton and Patterson's in Bristol, it was also one of the largest and, at over £100,000, one of the most costly merchant vessels anywhere in the world.

Financial guarantees for the vessels went hand-in-hand with a revised mail contract awarded in 1850. Now worth £270,000 per annum, the contract included a long-projected service to Brazil and Argentina. It also specified a new 'mainline' system between Southampton and Chagres on the eastern side of the Isthmus of Panama, where a projected railroad would link to shipping lines on the Pacific coast (Chapter 11). Before long, there might be a trans-Pacific line to Australia. A call at St Thomas in the Virgin Islands would allow passengers to transfer to and from 'branch' steamers in the West Indies.[3]

The Company had invited tenders from ten shipbuilders for wooden steamers, from six shipbuilders for iron-hulled vessels and from eleven marine engine-builders for steam machinery. The Admiralty's recent directive, however, that iron vessels in combat risked more lives than wooden vessels weighed decisively in favour of timber hulls (Chapter 14).[4] The *Amazon*'s builders, Richard and Henry Green, had an an impressive pedigree extending back to the days when the Blackwall Yard was turning out high-quality East Indiamen and Blackwall frigates. In 1843 they had split from their former partners, the Wigram's (Chapter 14), to inherit half of the famous Yard. But while the Wigram's added iron shipbuilding to their repertoire of skills, the Green's continued to build wooden ships, both steam and sail, including deep-sea sailing vessels that they operated themselves. As recently as 1848 they had completed one of the earliest British tea clippers, *Sea Witch*, and while the RMSP steamer rose on the stocks, Green's commenced work on a second tea clipper, *Challenger*, on a neighbouring slipway.[5]

As the *Amazon* took shape, favourable reports flowed from experienced witnesses. 'Captain Mangles visited Mr Greens yard at Blackwall', the Secretary recorded in late August 1850. 'Considerable progress has been made in the construction of the new ship: 12 pairs of frames erected, the entire length of

[2] RMSP DM (10 December 1851).
[3] Bushell (1939: 63–4); Haws (1982: 14); Forester (2014: 11–14) (mail contract); Nicol (2001: vol. 2, 52) (cost).
[4] RMSP DM (21, 22 May and 25 July 1850); Bushell (1939:63–5); Tyler (1939: 170) (Admiralty experiments comparing wood and iron under projectile fire); House of Commons Order (HCO) (1851), 1–3 (Order to P&O and Cunard not to build iron mail steamers).
[5] Banbury (1971: 181–6).

the keel completed and other preparations in a state of forwardness'. Little over a week later, the Secretary himself testified to the erection of more than twice that number of frames. As Mangles continued on a tour of the other yards, he found that none matched the pace of work achieved by Green's. On another visit at the end of October, he reported that very soon the forebody would be closed in. 'The *Amazon* is in a more forward state than any of the other new ships', Chappell wrote, 'and having got the lead, Messrs Green state that they are determined to keep it'. And whereas the *Orinoco* at Pitcher's Northfleet yard displayed timber that 'appeared of good quality throughout and the workmanship unexceptionable', the *Amazon*'s 'materials and workmanship appear of the very best description throughout'.[6]

Late in 1851, the Bristol-built ship *Demerara* for the same line grounded while being towed down the River Avon on passage to Caird's at Greenock to receive the engines. With 1,200 tons of ballast amidships in lieu of its engines and boilers, the hull was so badly strained as the tide ebbed that the vessel was written off as a total loss. Minor delays beset the other three vessels, especially as two of them also had to be towed from the Thames and Southampton to the Clyde for the installation of their machinery.[7]

With the *Amazon*, however, the managers intended leaving nothing to chance. When, for example, the builders moved the ship's hull to Seaward and Capel's engine works on the Isle of Dogs upriver from Blackwall, the owners sought to avoid strains on the hull if left to dry out with the ebb. They therefore appointed their own pilot 'to see that the ship is hauled into the stream at the proper time and that she is well secured at her moorings'.[8]

In mid-November, one of their most valued captains, frequently commended for the superior working condition of the *Severn* (Chapter 9), arrived at Blackwall with the Secretary to give 'directions ... respecting entering a crew, undocking the ship, effecting sundry necessary alterations &c'. Captain Vincent had recently docked his ship at Southampton after bringing forty-four passengers and over £27,000 worth of diamonds safely from the Brazils. Among the crew entered for the *Amazon*'s maiden voyage to the West Indies was a midshipman, Captain Vincent's seventeen-year-old son William, who seven years earlier, had been granted permission to accompany his father on the *Severn* and who had already served three years with RMSP.[9]

By mid-December, the steamer was ready for preliminary trials on the Thames. The ship was well-trimmed, drawing about one foot more water aft than forward in contrast to the ships of the first fleet that were usually lower by

[6] RMSP DM (29 August, 9 September, 1, 14, 15 and 30 October 1850). See also Bushell (1939: 66).
[7] RMSP DM (10 November 1851); Bushell (1939: 67–9).
[8] RMSP DM (2 July 1851).
[9] RMSP DM (14 November 1844, 13, 24 November 1851); *Amazon* Report (*AR*) (1852, 62).

the head. There were only routine concerns: '[b]oth paddle shaft bearings also heated and could only be kept cool by pumping a continual stream of water upon them'. Below Gravesend the Thames widened and straightened into the Long Reach with its measured mile. The *Severn* had on board 550 tons of coal, 918 tons of machinery, including filled boilers, and forty-five tons of fresh water. At 13 rpm, the *Amazon* managed a speed of 10¼ knots, slowed by the fact that there was 'not more than three feet [of] water between the keel and the ground'.[10]

On board were more than 'a hundred Gentlemen [who] sat down to luncheon in the saloon'. *The Times* reported the following day that 'a number of officers and scientific persons were on board to witness the trial'. They included several RMSP directors, several captains (Chambers and Appleton, both RN, and Claxton, former managing director of the now-defunct GWSS), representatives of the new General Screw Shipping Company with a fleet of auxiliary steamers under construction for Indian and Australian services, Thames shipbuilders (including Green and Mare), the chief engineer at the Woolwich Dockyard and his assistant, P&O's chief engineer at Southampton (Andrew Lamb), chief engineer at the steam factory in Bombay (Ardaseer Cursotjee) and Isambard Kingdom Brunel (most probably with his *Leviathan* project in mind). The reporter noted that 'her powerful engines worked with great perfection, and were the admiration of all on board', being 'fitted in a framework independent of the vessel' such that 'no perceptible vibration is felt when standing on deck over them'.[11]

Less than a week later the *Amazon* undertook six formal trials in the deeper waters of Stokes Bay near Portsmouth. Here the ship averaged 11¾ knots at 14½ rpm with a consumption of fifty-three hundred weight of coal per hour, or a daily consumption of about sixty-four tons. On 2 January 1852, the *Amazon* slipped the mooring off Southampton. During passage down Southampton Water and as far as the Needles at the western tip of the Isle of Wight, the *Amazon* averaged 10½ knots at only 10½ revolutions of the paddlewheels. Aboard were 1,128 tons of coal, fifty passengers (comprised of five ladies, thirty-eight gentlemen, four artisans, one infant and two servants of unnamed gender), and freight worth £862. The press reported the departure with enthusiasm and pride, fully endorsing the ship's performance. As this triumph of British shipbuilding and marine engineering disappeared from view, an air of satisfaction pervaded the RMSP officials who left the *Amazon* as the steamer neared the open waters of the English Channel and prepared to set a course for the Atlantic and the West Indies.[12]

[10] RMSP DM (10 December 1851).

[11] RMSP DM (10 December 1851); *The Times*, 11 December 1851. *Leviathan* was the provisional name for the *Great Eastern* (Chapter 15).

[12] RMSP DM (17 December 1851 and 2 January 1852).

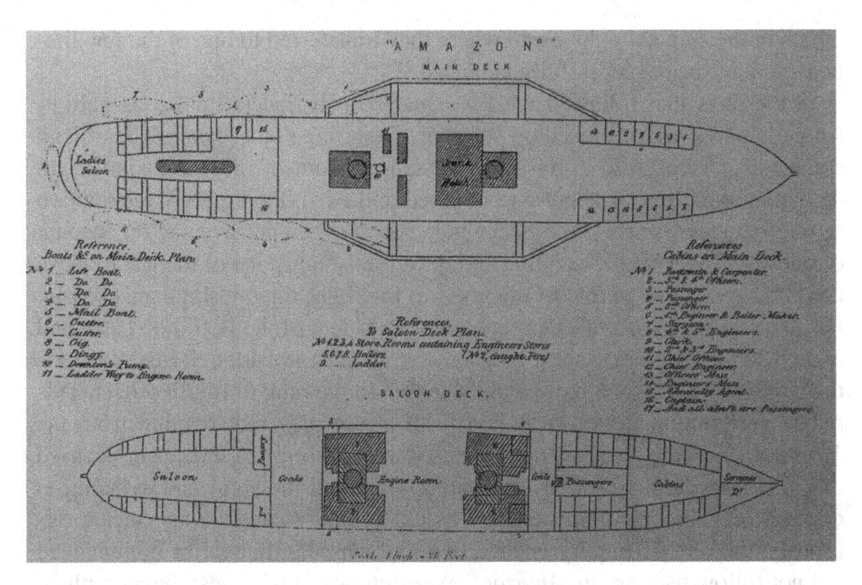

Figure 10.1. Deck plans of RMSP's prestigious *Amazon* (1852)

'Destruction of the Steam Ship *Amazon* by Fire: Great Loss of Life', ran a sensational headline in *The Times* on 7 January 1852. Early on the previous morning the brig *Marsden* of London, bound from Cardiff for South Carolina with a cargo of railway iron and under the command of Captain Evans, had landed survivors at Plymouth. The initial account, most probably drafted by Liverpool merchant and passenger Robert Neilson, reported that the vessel 'has been totally consumed by fire, and of the 153 persons who were on board her ... it is feared only 21 have been saved'. When the *Amazon* was about 110 miles west-south-west of the Scilly Isles, fire had broken out below decks on the starboard side just ahead of the forward funnel. With half a gale from the south-west, and the ship steaming out of control at perhaps 8½ knots, '[m]any were burnt in their berths, others suffocated, and a great many were drowned in the lowering of the boats' (Figure 10.1).[13]

Since none of the senior officers had survived, the duty of providing a trustworthy account fell upon Midshipman Vincent. Rendered all the more harrowing by its measured tones, focused primarily on matters of navigation and seamanship, Vincent's narrative in *The Times* told readers of how he had been on watch on the quarterdeck when, shortly after midnight on 4 January 'fire was observed bursting through the hatchway foreside of the fore funnel'. Every exertion to extinguish the fire 'proved ineffectual'. Vincent escaped with four

[13] *The Times*, 7 January 1852.

others in the ship's dinghy, from where they transferred to one of the few lifeboats to have been successfully launched.[14]

In the days that followed the fire, small Dutch traders known as galliots (of perhaps 120 tons) picked up survivors from three other boats. The official death toll decreased to thirty-seven passengers (out of fifty) and sixty-eight crew members (out of about one hundred and two). Finding any redemption through the heroism of the officers and crew proved controversial on account of questions over their discipline and duty. Neilson, as one of the last to leave the blazing steamer, wrote to the press to highlight the midshipman's valour. He also reported Vincent's courage to the President of the Board of Trade and urged RMSP directors to show him some mark of approbation. By late January, the Company had promoted the midshipman to the rank of fourth officer. This public recognition, however, called forth a very different reading from the *Morning Chronicle*. Suggesting a breakdown of discipline, the paper accused him of forsaking his post 'by prematurely leaving the ship' and evincing 'a disregard for the fate of those whom he left behind'. Questioning Vincent rigorously, the official Inquiry wanted to know why the lifeboat he commanded did not follow the burning *Amazon*. 'We kept away'; the midshipman replied. '[S]he was one mass of fire'. The Inquiry, however, fully exonerated Vincent from the *Chronicle*'s charge of prematurely abandoning ship.[15]

Anna Maria Smith, a young Dublin woman en route to Puerto Rico to take up a position as a family governess, acquired the rare accolade of heroine, achieved through defiance of gentlemanly advice. Written anonymously in novel form for *Bentley's Shilling Series* a couple of months after the disaster, an awakened Miss Smith ran from her berth in her nightdress only to be 'met by some gentlemen who conducted her back to her cabin' where they 'besought her not to be alarmed, as the danger was but transient, and the flames [would be] soon subdued'. Resolving, however, to save her own life if possible, she refused the assurance and 'with coolness and intrepidity' hastened on deck, took hold of a rope attached to the bulwarks and swung herself into a lifeboat. 'Was not this the courage of a true heroine?' the author asked rhetorically.[16]

Preaching to a non-conformist congregation in Southampton, the Revd Thomas Adkins explained that he had received from Miss Smith's own lips the statement that 'though constitutionally very timid, she was blessed at that awful moment with perfect self-possession, to plan and execute the most likely means of her escape'. The Saviour's words, that God's providential care extended to the hairs of the head and even to sparrows, 'enabled her, a weak

[14] *The Times*, 7 January 1852.

[15] *The Times*, 9 January 1852; *Morning Chronicle*, 19 and 20 January 1852; *AR* (1852 xi, 62–8). I thank George Mitchell for the *Chronicle* reference.

[16] [Anon.] (1852: 16–18); Courtney and Smith (2013: 192).

and timid female, to sustain the drooping, desponding, companions' in the boat. The Revd R. B. Hollis (see below) also mentioned the heroic action of Mrs Eleanor Roper McLennan who saved her eighteen-month-old infant.[17]

Presided over by Captains F. W. Beechey and W. H. Walker, the official Inquiry began in late January. Captain Beechey had served first with Franklin and then with Parry on the Arctic expeditions of 1818 and 1819. His subsequent naval career focused on charting the coasts of North Africa, the Bering Straits and the Irish Coast, the latter on board steam survey vessels between 1837 and 1847. He became superintendent of the newly established marine department of the Board of Trade (BoT) in 1850. Walker had served the EIC up to 1833, after which he commanded in turn three of the 'finest and largest ships trading to India' that he himself part-owned. Retiring from the sea in 1847, he joined Beechey as a senior officer in the BoT's marine department. [18]

With just a hint of potential weakness in an otherwise first-rate material and moral system, Beechey declared the *Amazon* to have been 'the finest of her class, constructed by one of the first builders, and fitted with machinery by one of the first engineers, owned by one of the most respectable companies, and, *it may be supposed*, in every respect well manned and well looked after'. He also stressed, perhaps with RMSP's controversial record in mind, the independent character of the Inquiry. Altogether, it interviewed some thirty-seven witnesses under oath. Apart from three passengers (including one navy lieutenant), all of them were either surviving crew or those involved in the ship's design, construction or management.[19]

The most experienced eyewitness was second engineer William Angus, whose late brother George had been the *Amazon*'s chief engineer. 'I have been brought up all my life in steam-ships', he told the presiding officers. He also stated that he had been on board the vessel for some twelve days before departure. Asked about speculation concerning witness reports of heated paddle-shaft bearings, Angus replied decisively that they were 'no more than is common with new engines, both on land and on shore'. He instead suggested that painters might have thrown inflammable material into a narrow space separating the boilers from the forward coal bunker bulkhead.[20]

What followed in the interview, however, was especially significant. 'If I had received orders from the deck when the fire first broke out I could have stopped the engines;' he told the Inquiry, 'but it is the rule of the service that you must not stop the engine upon your own account except anything is wrong with the machinery'. Company discipline thus precluded the one action that would have

[17] Atkins (1852: 25–6).
[18] *AR* (1852, xxi); 'Beechey, F. W. (1796–1856)', *ODNB*; [Anon.] (1871: 196).
[19] *AR* (1852, xxi). My italics.
[20] *AR* (1852, 32–7); *The Morning Chronicle*, 19 January 1852.

enabled the crew to secure the safe launch of more boats and perhaps also to have bought time by reducing the draught to the boiler room. Instead, Angus 'ran aft to get on deck' where he found 'nothing but smoke and confusion, and people running about'. By the time the captain gave the order to stop the engines, it was too late: 'we could not go below for smoke and flame'. In first following the rule of the service, he then refused to obey the orders of the captain to return below.[21]

The Inquiry also pursued with Vincent the question of a wider breakdown of discipline. He explained that when Captain Symons came on deck he called the men to try and get the fire under control but found some of them already in the boats. Only about ten minutes after the fire had been detected, Vincent recalled, the captain had 'given it over'. In its findings, the Inquiry noted that deck officers alone appeared to put duty before personal survival but largely failed in their role as enforcers of Company discipline. 'Confusion and dismay appear to have prevailed', it concluded. Given the shocking speed with which the vessel had been transformed into an uncontrollable conflagration, however, the report largely exonerated the officers and crew.[22]

The Inquiry chose as its principal scientific authority Thomas Graham, FRS, professor of chemistry at University College London. His remit concerned 'questions connected with spontaneous combustion, both as regards the immediate cause of the fire ... and also as respects the propriety of keeping separate certain stores [such as oils and greases] used by the engineers'. Graham focused a large part of his report on a known 'tendency of coals to spontaneous ignition ... increased by a moderate heat, such as that of the engine-room'. Conversely, he noted that the most obvious precaution against such ignition was to take the coal on board 'in as dry a condition as possible'. He accepted that the Welsh coals had been shipped in a dry and dusty state a month or two previously and that no strong odour had been remarked. He therefore rejected this explanation. The professor, however, did not address the possibility that the new ship's bunkers were no longer a dry store for the 1,100 tons of bituminous coal on account of the hull taking in seawater. Nor did he consider Angus' insistence that he had seen a 'gleam of fire' in the bilges followed by an effect 'more like an explosion than anything else'. As a result of Graham's report, the Inquiry reached no certain conclusion as to the origin of the fire.[23]

[21] *AR* (1852, 34, 36).
[22] *AR* (1852, 62–8, v–xi).
[23] *AR* (1852, iii, xii) (origin of fire), (xv–xvi) (Graham and coal); 32–3 (Angus). Angus' description was very reminiscent of the explosive and much-feared 'firedamp' that afflicted deep coal mines.

10.2 'My Brethren, Rush from the Flames of Hell'! The Authority of Personal Experience

'Did any persons converse with you as to their feelings with respect to the ship?' Captain Beechey asked Vincent at the Inquiry. 'Not any passenger', he replied with characteristic precision. 'There was a presentiment among the officers'. Coming as it did from someone who had hitherto spoken clinically on matters of navigation and shipboard discipline, Vincent's revelation was unexpected. 'Tell me the nature of it?' Beechey persisted. 'I cannot tell you;' came the twice-offered reply. 'We went to sea with the feeling that something would happen before the voyage was over'. 'You sailed on a Friday?' Beechey tried again. Vincent denied the superstition: 'That had nothing to do with it'. Moreover, he endorsed a view that nothing in either the physical or the disciplinary aspects of the vessel had generated the anxieties.[24]

Co-survivor Neilson had retired to his berth 'under the apprehension of danger, though not from fire'. Three minutes later, he heard the alarm. Beechey also invited passenger Frederick Glenny to communicate anything regarding 'what has been referred to as an impression among the passengers that something was going to happen to the ship, something that created a feeling of uncertainty and apprehension as to the vessel'. 'I believe there were great forebodings among many of the passengers'; he answered. 'I had the same foreboding myself'. Glenny, however, denied that his own anxieties arose from 'any vulgar superstition', suggesting instead that they arose from an awareness that the new steamer had not undertaken the conventional trial trip to Ireland or Scotland.[25]

As a naval officer, Lieutenant Grylls, RN, might have been expected to deny personal fears and forebodings. He confessed, however, to having 'had some sort of feeling, – I cannot describe it; something impressed me with the idea before I went on board the ship that something would happen to her'. This unaccountable apprehension had, he explained, nothing to do with the steamer ('the most beautiful ship I ever saw') but had come over him before he arrived at Southampton.[26]

Uttered under oath by four very different survivors, these remarks suggest that in this period the boundaries between visible, material nature and an unknown, perhaps spiritual realm remained permeable. It was no coincidence either that Grylls had spent long hours in an open lifeboat, followed by some ten days at sea aboard a tiny Dutch merchant sailing vessel commanded by a devout Calvinist master, Captain Gruppelaar, and in the company of an increasingly zealous evangelical clergyman, the Revd William Blood.

[24] *AR* (1852, 67).
[25] *AR* (1852, 26, 77–8).
[26] *AR* (1852, 25–6).

Blood had been a Church of England priest since 1844 and held a living worth £146 per annum, but without a parsonage, in the small English village of Temple-Grafton (population 403) near Stratford-upon-Avon. He was also chaplain to the Marquis of Hertford, who was probably patron of the parish. Seeking funds towards a parsonage, he travelled to Paris to lobby a nobleman acquaintance and while there had fractured his leg, developed pleurisy and, after bleeding, leeches and blisters, only just recovered. Granted a year's leave of absence by his Bishop to recover his health, he chose a voyage to the 'sunny islands of the West Indies'.[27]

The clergyman's traumatic personal experiences of escaping from the burning steamer formed the raw materials for a series of sermons delivered in churches in Plymouth, Southampton, Bath, Bristol and elsewhere. Word of his providential survival attracted very large congregations – estimated at 4,000 souls for his first address in St Andrew's Church of England in Plymouth. Taken down in shorthand, several of the sermons appeared as tracts that also recorded the preacher's emotional pauses. The printed version of the Plymouth sermon, for example, apparently had a run of 12,000 copies 'distributed all over the kingdom' and an edition printed in French. Each sermon had two components: the first on his personal, providential experiences and the second on the spiritual lessons to be drawn.

Although both Cameron's and Blood's narratives concerned providential deliverance, the former's genteel, mainstream account contrasted dramatically with the Puritanical intensity of Blood's preaching. His sermon at All Saints in Southampton caused, in the words of the *Hampshire Independent*, 'that vast edifice to be crowded in every part, and numbers could not gain admittance to that church'. The Revd Blood began with 'a thrilling and affecting narrative of the events of that awful night ... [that] possessed a deep interest as coming from the lips of one who personally passed through the frightful catastrophe, and who was providentially snatched from the jaws of death'.[28] In another published address, he grounded his vivid providential narrative on God's direct assurance to the prophet Isaiah: 'When thou passest through the waters, I will be with thee; ... when thou walkest through the fire, thou shalt not be burned; neither shall the flame kindle upon thee'.[29]

Every detail of the voyage provided evidence of special providence at work on Blood's behalf. Prior to departure, for example, Captain Mangles had arranged for his transfer from a cabin just forward of the engine room (where he would most likely not have survived) to a cabin next to the saloon aft.

[27] Courtney and Smith (2013: 193).

[28] [Blood] (1852a: 5, 1852b: 30) (end paper, including extract from the *Hampshire Independent*). See also Kurihara (2011: 1–16) for contemporary sermons in the United States, which the author neatly links to different denominations.

[29] [Blood] (1852c: 3) (title page); Isaiah 43:2.

There, on the night of the fire, 'a most unaccountable feeling' persuaded him to lie fully dressed on the bed. Awakened by the stewardess' cry of fire, he narrowly avoided boarding one of the boats that was quickly swamped. Here again 'Divine interference was manifested on my behalf'. A fraught launch of a second boat was watched over by 'the All-seeing Eye ... still upon me'. Once clear of the burning steamer, the prospects looked dark but 'the divine ruler of our destiny had not deserted us; for as the Sabbath morning's light dawned, the wind abated, and the sea became comparatively calm'.[30]

The master of the Dutch sailing vessel assigned the thirteen survivors, including Lt Grylls, Revd Blood, William Angus, three stokers and two ordinary seamen, a cabin six feet by four and a half feet. There, Blood instituted 'a regular form of daily worship' for the benefit of the captive audience. 'Every day to us was a Sabbath', he told the Plymouth congregation in his first sermon, 'and our little barque our church, within which our feeble voice of prayer ascended, and our notes of praise mingled with the music of nature, in the whistling winds and roaring billows, while the great God of nature – of the storm and the tempest – heard our supplications and sheltered us in his pavilion'. On deck, Captain Gruppelaar also '[f]requently ... in the day had ... his faithful [Dutch-language] Bible with him, thoughtfully perusing it'.[31]

Unlike the *Tweed* survivors' calm thankfulness expressed through the *Book of Common Prayer*, Blood's evangelical message was stark: that 'we may be saved from the wreck here and the deep abyss of eternal woe hereafter, [only] through the merits of our Lord Christ Jesus'. By the 'blood' of Christ alone would these sinners be 'washed and made clean'. Powerful emotions, in the form of 'heartfelt sighs and sobs' swept round the cabin. An anonymous survivor, however, appeared to resist the clergyman's exhortations. 'We want no religion here', was the cry of this impenitent sinner, a man according to Blood, 'almost ruined by former low and vile habits, yet at times even he seemed to relent'.[32]

A slightly later address went further in merging human experience with revealed vision. Blood now recalled his providential retreat from the first, stricken lifeboat when he 'returned to the blazing pile. What a picture of Hell'! He at times seemed to conflate the *Amazon* conflagration with hell itself when he gave thanks for the providential kindness 'in snatching me from the burning element and saving me from the excruciating torture which might have followed'. By the end of the sermon, the second lifeboat had become a metaphor for Christ: 'Oh, Jesus! Thou art the Life-boat of the soul. ... Thou art the life of the sinner in danger of perishing. My brethren, rush from the flames of hell'!

30 *Morning Chronicle*, 20 January 1852.
31 [Blood] (1852a: 15–16).
32 [Blood] (1852a: 14); [Blood] (1852c: 21–4).

In this passionate evangelical theology that brought the preacher to tears, a fallen, sinful world tempted 'the vengeance of heaven'. The omnipotent God, who manifested His power in nature's mountains, fire and flood, expressed His true Power through the Gospel: 'when God gives energy to the word of His Grace, no darkness is too dismal for it to dispel, no obduracy too hard for it to subdue'.[33]

Blood had unique clerical authority by virtue of his eyewitness experience, but he was by no means alone in his preaching on the *Amazon*'s sudden destruction. Indeed, some clergymen acted with remarkable speed. Even before the Dutch galliot had landed Blood, the Revd Hollis preached with considerable passion in his Islington Congregationalist Chapel. He pondered the loss of a quarter of a million pounds of 'trading capital, destined to do great things for society' that was now 'irrevocably gone'. He reminded his congregation that it was 'human intellect and heart which fabricated these wonderful works, the vessel and the freight, or which made gold "current coin with the merchant"'. In the larger scheme of humankind's eternal destiny, however, 'if but *one soul* be saved as the result of this dire catastrophe, a quarter of a million of money will not be thought too vast a sacrifice'.[34]

Christ the 'Great Teacher', Hollis asserted, 'seized passing events, as well as the facts of history, to awaken the attention, to inform the judgment, to quicken the conscience, to save the soul'. Following His example, the minister would do likewise with the *Amazon* narrative. Thus the disaster was 'not the end designed; it was a means to an end'. This 'unusual calamity', he claimed, 'must contain some instructive lessons; it must be the voice of God to man, bidding him "prepare to meet Him"'. Even more powerfully, it anticipated not simply the death of individuals, but the end of all things as foretold in the scriptures: 'it must be a miniature of the day of judgment, of that "day of the Lord which so cometh as a thief in the night"'.[35]

For the benefit of merchants and ship-owners, Hollis included a lesson on human imperfection and the need to seek God's blessing on all their doings. We should be thankful 'for our modern associations for insuring property and life; for our precautionary measures against danger and death; for sanatory [sic] commissioners and life-boats', he observed. But all these, he warned, tended towards 'scepticism, prayerlessness, and self-sufficiency'. Forgetting God, let alone defying Him, he implied, was at the heart of the kinds of ingenuity built into the *Amazon*. With lifeboats enough for everyone aboard and provisions against every kind of contingency, 'any gentleman at Lloyd's would readily have taken the risks at a moderate premium'. Yet, he affirmed, 'when

[33] [Blood] (1852c: 7, 28–9, 40–8).
[34] Hollis (1852: 5–6).
[35] Hollis (1852: 16–18). (Hollis was paraphrasing 2 Peter 3:10.)

God's angel of death is commissioned to do his fatal work there will always be some door left open for him to enter ... Invent and provide as we may, there will always be sufficient interstice for the finger of God'.[36]

Hollis dedicated the tract to Midshipman Vincent 'in admiration of his magnanimity and skill ... with an ardent wish that a life so spared may be distinguished by eminence in his profession, [and] above all, by consecration to the glory of God, his preserver'. Vincent duly became a living embodiment of Hollis' belief that '[t]o work, and to work hard; to watch, and to watch diligently, are of divine ordination'. The minister proposed that the congregation present him 'with a handsome copy of the Pictorial Bible' with an inscription expressing 'our ardent hope and prayer for the prosperity of his future career in this world, and above all, that when death shall call him away – it may be suddenly, to the same watery and undistinguished tomb as his last companions, – [that] he may have a "sure refuge", ... and "eternal life"'.[37]

If Vincent stood for the values of discipline, duty and diligence in the ship's officer, Hollis challenged the aspirations of both contemporary society and the Company. Modern society and commerce, he observed, seemed to have forgotten that 'the laws of gravitation and projectile power, and all the various laws in mechanics and chemistry are an expression of the wisdom and will of God'. In this voluntarist world, violating the laws of nature, as well as moral laws, brought penalties. He thus suggested that a great public sacrifice of victims had been made at the 'shrine of *speed*' (see Epigraph).[38]

The strident preaching of the Revd Blood was in sharp contrast to Cameron's earlier *Times* narrative of Christian heroism and providential deliverance that resonated with the Company's values (Chapter 9). In January 1852 the Court of Directors resolutely rejected Blood's request for financial assistance, informing him that it was not Company policy to make any return of passenger fares 'in the case of accidents to the ship'. Instead, 'passengers wish[ing] to secure themselves against pecuniary risk ... must insure their passage money and effects'. Not for Blood's pains and preaching, then, RMSP's generosity that was afforded to Cameron. Moreover, the Company's lack of effort to appease Blood may indeed relate to his, and Hollis', damning association of the *Amazon* with divine vengeance against a sinful and arrogant humanity.[39]

[36] Hollis (1852: 18–19).
[37] Hollis (1852: 20–3).
[38] Hollis (1852: 23–5). He probably grounded his condemnation of 'speed' on the letter from 'Quaestor' in *The Times*, 8 January 1852. There was no evidence that the *Amazon* was steaming especially fast that night.
[39] RMSP Minutes (22, 29 January 1852); Courtney and Smith (2013: 195).

10.3 'Not a Practical or Scientific Engineer': Misfortune Versus Mismanagement

The RMSP directors faced a crisis of public and investor confidence. Within days of the fire, a special meeting of the Court authorized Chappell to communicate with Messrs Burns and Cunard regarding the purchase of their wooden-hulled mail steamer *Arabia*, recently launched from Steele's yard in Greenock and now receiving the side-lever engines from Robert Napier in Glasgow (Chapter 2). The Court also instructed Chappell, Mills and Elsmie to proceed there to examine the vessel. Very close to the length and tonnage of the *Amazon* but with 1,000-hp engines, the steamer, renamed *La Plata*, cost RMSP almost £120,000 and entered service in August 1852. Self-insured, the Company was paying a high price for the disaster.[40]

Managers also acted within weeks to introduce changes to both material and social practices aboard their vessels. They charged their superintendent engineer with the task of investigating the fitting of metal linings in engine rooms. They ordered alterations to the remaining three ships of the Company being fitted out, shifting the engine room bulkheads further away from the forward and after boilers, lining these bulkheads and the adjacent internal sides of the hull with zinc, removing storerooms for oils away from the engine room, fitting mechanical communication between officers on deck and engineers below and enhancing firefighting pumps and hoses. In the wake of the Inquiry, they also promised to take steps with regard to anyone involved with the 'clandestine shipment of combustibles'. The Court further approved a recommendation to confine the largest vessels to the ocean link between Southampton and St Thomas (Virgin Islands) and thus reduce the risks of stranding.[41]

A few weeks before the *Amazon*'s loss, the Court had received a letter from the Admiralty approving the construction of iron ships for its mail services. The RMSP had already been communicating by letter with Caird and Company for an iron-hulled mail steamer to take the two side-lever engines completed in Greenock but never fitted to the stricken *Demerara*. Worth £30,000 for the pair, they were allocated to the new vessel, duly named *Atrato*.[42] After Russell's departure for London in 1844, James Tennant Caird had taken over as manager and in the same year led the firm into iron shipbuilding following the purchase of the yard of Thompson & Spiers in Greenock. Caird's early iron vessels ranged from 100-ton steamers for Firth of Clyde services to two vessels for

[40] RMSP Minutes (14 January and 29 July 1852); Bushell (1939: 72–3); Haws (1982: 37–8).

[41] RMSP Minutes (5 February, 3 June and 2 September 1852); Bushell (1939: 73–4); Courtney and Smith (2013: 198).

[42] RMSP Minutes (18 December 1851).

Figure 10.2. RMSP's iron paddle-steamer *Atrato* (1853) built by Caird's of Greenock and shown at speed in the Firth of Clyde near Ailsa Craig

G. & J. Burns, including the ill-fated 899-ton *Orion* in 1846 (Chapter 1). Later that year Caird's constructed their first P&O steamer (Chapter 14).[43]

James Caird had been born in 1816 in the cotton manufacturing and calico-printing village of Thornliebank near Glasgow. His mother died soon after his birth and his father, also James, never returned from America. His paternal grandparents, John and Mary Caird, brought him up in a devout and intense Presbyterian culture. Beginning his apprenticeship in 1831 in the Greenock firm at a wage of seven shillings per week, he gained further engineering experience at Glasgow's St Rollox Locomotive Works and then spent a further two years in Glasgow at the mill machinery business of Randolph, Cunliff & Company (Chapter 16). Russell, as engineering manager and natural philosopher, and Caird, as principal draughtsman with practical knowledge, worked together on changes to the design of the four Caird-engined RMSP steamers around 1840–2 (Chapter 7).[44]

Caird's *Atrato*, however, was on an altogether different scale from their previous vessels and became, for a time, the largest mail steamer in the world (Figure 10.2). Carrying the same beam as the *Amazon*-class but with over thirty

[43] Duckworth and Langmuir (1977: 129–98); Shields (1949: 125).

[44] *Glasgow Herald*, 31 January 1888 (Caird obituary).

feet more length overall, the new vessel had a keel formed of nine lengths of iron with stem and stern posts each cast in one massive piece. Iron bulkheads divided the hull into seven watertight compartments. Passengers numbered 224 in first class, twice that of the *La Plata*.[45] Entering service after delays due to engineering strikes, the iron steamer acquired a much improved reputation over predecessors. Proving capable of 14 knots, the *Atrato* well outpaced the *Amazon*-class with the same horsepower, and was at least equal to the higher-powered *La Plata*. As the Chairman (Colvile) told proprietors in April 1854, 'I know she was faster than the [*Amazon*-class] *Magdalena* ... She is the fastest vessel we have had'. But he also admitted that the introduction of faster ships under the conditions of the new mail contract had increased the coal consumption. 'No body could prevent that increased consumption, and the increased cost that was consequent thereon', he explained fatalistically. Moreover, he attributed the enormous coal costs for 1853 (£218,507) compared to 1852 (£164,169) and 1851 (£115,595) to higher freight rates and colliery strikes.[46]

The spring 1854 GM generated exceptionally heated debate between directors, determined to resist uncomfortable proposals for a Committee of Inquiry into the Company's affairs, and with disaffected shareholders unwilling to take assurances at face value. Proprietors challenged the Court's repeated insistence that all of the Company's shortcomings and problems were to be attributed to external causes outside the managers' control. On the contrary, critics identified, through specific episodes, the failings of officials and directors to admit openly their own blunders, especially in relation to shipbuilding and engineering practices. Dr Alexander Beattie (also a P&O shareholder) began by taking the meeting through the finances to show that the Company was 'in debt to the extent of between £200,000 and £300,000'. Against Beattie's quantitative analysis, the Chairman's repeated phrases that the Company was 'wholesome and sound', 'perfectly sound', and 'sound and wholesome' served only to increase shareholders' anxieties.[47]

At the core of the 1850 Government contract was the requirement for greater speed. But the remaining three *Amazon*-class steamers of a 'superior description' were proving highly problematic. In particular, directors admitted that the final vessel, the Wigram-built *Parana*, required strengthening. Dr Beattie seized on this case to open up wider questions of management competence. 'Is it the fact', he asked, 'that the *Parana* after making [her first] three voyages was found to be in so shattered a condition that it was not only found requisite to reduce the size of her large paddle wheels but that when her engines were being repaired it was also found necessary to introduce new beams and new fastenings to strengthen

[45] Bushell (1939: 75–6); Haws (1982: 38).
[46] RMSP GM (April 1854: 5–7); *Report* (April 1854: 3).
[47] RMSP GM (April 1854: 11–12, 31–2, 53–4).

Figure 10.3. *Amazon*'s sistership *Orinoco* departing Southampton for the Crimea (c. 1854)

the ship[?]' If the steamer had been constructed 'strictly in accordance with the directions expressed in the builder's contract', he persisted, how has the ship 'proved so complete a failure'? Beattie extended his criticism to the other two 'superior' steamers. The *Orinoco* had been taken up for Government service as a troopship (Figure 10.3) and so the extent of repairs 'in consequence of being fitted by these enormous wheels' was not yet known. The *Magdalena*, however, had been refitted at an (unspecified) cost 'far beyond anything that would have been required had they been properly constructed in the first instance'.[48]

Captain Sweeney then rose to implicate the Company's engineering superintendent in what now looked like a major debacle. Sweeney alleged that Mills had, largely on his own authority, ordered a substantial increase in the diameter of the paddlewheels at a cost believed to be £12,000 per vessel. This change in specification produced a 'result [that] proved that the engineer was wrong'. Moreover, the captain told the meeting that he had 'heard it stated that he [Mills] was only a foreman to a factory before he was engaged by us, and not a practical or scientific engineer'. Sweeney reminded the meeting that the directors had also appointed Mills to manage the shore-based RMSP works in Southampton and expressed his opinion that 'the selection of Mr Mills was not a wise selection on the part of the directors'. The Chairman, however, strongly

[48] RMSP GM (April 1854: 14–18, 44).

defended Mills, claiming that the Admiralty itself had even tried to recruit him to a more remunerative post. One proprietor wryly responded that, given the Admiralty's own record in government yards, this was 'no very great recommendation'. Later in the meeting, Captain Shepherd (deputy chairman and EIC chairman) chastised Sweeney's attacks made upon Mills 'whose character is his stock in trade'.[49]

Sweeney's assault on Mills' reputation lacked named sources. All that changed when another proprietor, T. R. Tufnell, intervened with devastating evidence of the engineer's incompetence. Tufnell claimed to represent 'the opinions of a body of proprietors who hold stock to the amount of £200,000 in this Company'. Moreover, Lady Anne Tufnell had performed the naming ceremony of the *Orinoco* at Pitcher's Northfleet yard in May 1851. Tufnell now described as a misstatement the directors' assurance that, during the building of the *Orinoco* and the other new vessels, free communication was allowed between shipbuilders and engine-makers. Instead, 'an intimation was sent to the builders from the Board [of Management] stating that all communications between the ship builders and the engineer were to go through Mr Mills'. As a consequence, one of the builders, 'knowing the incompetency of Mr Mills absolutely refused to go on with the contract unless he had free communication with the engineers who were to make the engines'.[50]

Tufnell's most damning evidence, however, arose from two recent cases. First, Pitcher had built the hull of the 800-ton *Derwent* in 1850 while Mills constructed the boilers at Southampton. When the hull arrived to receive the boilers, they were found to be much too large. Mills had blamed the ship but Tufnell placed the drawings before the GM to show that 'there was no error whatsoever in the dimensions of the ship'. The second case concerned a major refit to the *Avon*. Pitcher completed the hull alterations to schedule, but the vessel languished for many weeks while Mills finished work on the engines. These instances alone, Tufnell argued, showed both Mills' incompetence and that 'a committee of enquiry is necessary'.[51]

Shepherd was a principal opponent of any committee of inquiry. He claimed never to have known instances where such a motion was proposed 'except where bad management & not misfortune were complained of'. Such a motion, he insisted, 'would be a suicidal act' damaging the property of the Company. An ally of Tufnell's, Mr Davies, seized on this political blunder. 'I hope & trust that we shall not be led away by our respect for the very respectable gentlemen who have for so many years managed the business of the Company', Davies

[49] RMSP GM (April 1854: 19–23, 29, 37–8, 43, 55–6).
[50] RMSP GM (April 1854: 47, 54–5); Bushell (1939: 66) (Lady Anne Tufnell).
[51] RMSP GM (April 1854) 47–9.

pointedly remarked. 'It is clear that there has been gross mismanagement'.[52] The Court's tendency to hide behind a cloak of misfortune was being challenged.

In a strategic move to build momentum for changes at Court, another proprietor, Mr Payne (Chapter 8), spoke of the advantages to be obtained from an infusion of new blood into the Board. In almost the same breath he referred to Mr Tufnell as 'a practical man and a man of business'. Payne pulled no punches. 'I believe in all our arrangements with respect to our vessels', he told the meeting, 'we are at least ten years behind the time (hear hear)'. He justified his assertion on the grounds that 'we have not yet adopted the screw but confine ourselves to the old fashioned paddle wheels – and that so many of our vessels have broken down'. Shepherd hotly disputed the last claim. '[O]ur vessels have never broken down (hear! hear!) – and that is where we say so much credit is due to Mr Mills our superintending engineer'. 'But I still say we are very much behind the age with our vessels', Payne insisted.[53]

As the meeting drew to a contentious conclusion, both the proposal for a committee of inquiry and the election of Tufnell as a director failed. Early in 1856, however, Chairman Andrew Colvile died. His successor, Captain Mangles, chaired the April GM at which the disaffected proprietors' rejection of the Report prompted the resignation of the entire Court. In the subsequent reconstitution, Tufnell was elected along with several other newcomers. In the same year, Chappell retired as Secretary.[54]

The old era was closing. But any new era for the Company was still uncertain. Chairman Mangles was hardly representative of a younger generation and many of the older directors resumed their seats on the Board. The question as to whether or not RMSP, under pressure from long-suffering proprietors, had addressed the problems of its seemingly accident-prone mail steamers and perhaps begun to reshape its fleet remained unanswered. For some passengers, nothing much had changed. Returning from Australia in 1859, the young political economist William Stanley Jevons boarded the aging *Medway* at Chagres for a seven-day passage to St Thomas. It was 'a period of extreme discomfort caused by the bad accommodation of the steamer, the tropical heat of the weather and the perpetual pitching of the ship against head wind & waves', he wrote. 'The officers too from Captain down to Midshipman might be described as insolent conceited puppies who thought of nothing but their gold lace'. And for someone who had voyaged contentedly from Britain to Australia and from Australia to Peru by sailing vessel, Jevons was eminently qualified to judge.[55]

[52] RMSP GM (April 1854) 44–6, 58–9.
[53] RMSP GM (April 1854: 60).
[54] Bushell (1939: 112); Forrester (2014: 36–7).
[55] Black (1973: vol. 2, 380, 397). See also Trollope (1985 [1859]: 2–3) (Trollope's retrospective verdict on the *Atrato*).

11 'An Uncompromising Adherence to Punctuality'

Pacific Steam from Valparaiso to Panama

> We then swept round within a few fathoms of the mole-end, and passed between a line of boats, with bands of music and decked with flags, lying off the beach ... [and] thence threaded our way in and out the different merchant-vessels, which were ornamented with flags, receiving their cheers, and around the sterns of the men-of-war, ... the shores and every house ... being lined with spectators, who kept up a continual cheering and waving of handkerchiefs; in short, this reception far exceeded our most distant or sanguine expectations, and I only regret that the worthy projector of the enterprise, Mr Wheelwright, was not present to witness this gratifying display of enthusiasm; all the public offices were closed, shops shut, and business suspended.
>
> *Captain Peacock tells of the arrival of the first two PSNC ships at Valparaiso, Chile, in 1840*[1]

Summary

The PSNC project represented an unusual mix of United States vision, British capital and post-colonial enthusiasm from local merchants and leaders not long 'liberated' from centuries of Spanish rule. William Wheelwright initially secured patronage from the influential and wealthy Scarlett family. The PSNC's promoters then used print cultures (including pamphlets and prospectuses, maps, statistics and newspaper and magazine articles) to persuade investors, Governments, shippers and the wider travelling public that the enterprise was trustworthy. As commodore, Captain Peacock seized on contingent events to highlight his skill at turning seemingly inevitable defeat for the project into hard-won victory. Able at first to fund only two steamers for a system running for thousands of miles along the relatively unknown coastline endowed with little local fuel, few repair facilities, a frequent threat of war between rival states and a Court of Directors a world away, PSNC struggled to survive its first five years in the face of loss-making services.

11.1 'The Worthy Projector of the Enterprise': Wheelwright's Manifesto of Promise

On 18 June 1836, His Majesty's Consul General in Peru addressed, on behalf of the British Government, a communication to British merchants and residents based in the capital Lima and the principal port Callao. He invited his countrymen

[1] Quoted in Wardle (1940: 40–1). This chapter draws extensively on Smith (2013b: 82–112).

to furnish him 'with every information in their power respecting the probable utility, practicality, and the most convenient and least expensive method of effecting a periodical intercourse between the several ports of the Pacific and Panama'. He also specified that their submission should provide an estimate of expenses and of their defrayal, along with the likely volume of correspondence expected to flow through the new channel. Similar British Government communications reached Consuls at the ports of Guayaquil and Islay, British merchants at Arica and Payta and the Commodore Commanding HM Naval Forces in the Pacific. The Consul General in Chile received the same message for onward transmission to British merchants and residents in the port of Valparaiso.[2]

The public meeting held at the British Consulate in Lima set up a Committee consisting of nineteen merchants and residents chaired by the Consul General. Within two months, it issued a full report. Given 'the very inadequate income to be derived from the postage of letters compared with the heavy expenses attending such an establishment', the Committee declared the 'total impracticability' of establishing such a line of packets on the Pacific coast were it to be funded solely or jointly by 'Native, British, or Foreign residents'. It also asserted the 'insufficiency of sailing-vessels for the proposed object' on account of the 'general prevalence of southerly winds, currents, and calms' producing delays in delivering mailbags at the principal ports. Such delays, it concluded, meant little saving of time over the usual route from Europe round Cape Horn.[3]

While the Government call had not mentioned steamers, the Lima Committee now raised 'the possibility of establishing steam-navigation in the Pacific'. Only William Wheelwright had submitted a project for packet communication since the June invitation. But the New Englander had already been granted 'exclusive privileges' by the Governments of Chile, Bolivia and Peru for a period of ten years from the commencement of just such a steamship project. The Committee thus recognized that everything depended on Wheelwright's 'willingness to render his project available for that object'.[4]

Wheelwright duly undertook 'to recommend to the projected "Pacific Steam-Navigation Company" an article into its constitution, binding it to convey all letters, addressed to and by British subjects, from any one point to any other within the limits of navigation of the company'. The rates would be standardized at sixpence for single letters and one shilling for double letters, two shillings per ounce for packets of letters and sixpence per pound for newspapers of all countries. He further insisted on a corresponding liberality

[2] [Wheelwright] (1838: 14–15).
[3] [Wheelwright] (1838: 16–18 (public meeting), 18–24 (Committee Report on 19)).
[4] [Wheelwright] (1838: 19–20).

in establishing the rate of postage by HM packet from Britain to Chagres for transfer across the isthmus to Panama. His Majesty's Consuls at the various ports would arrange acceptance and delivery of mails. The Committee then recommended these proposals to the British Government.[5]

Born in 1798 at Newburyport, Massachusetts, Wheelwright had a distinguished pedigree both in faith and in works. His seventeenth-century English ancestors included the Revd John Wheelwright, a dissenting Lincolnshire clergyman who preached salvation by faith rather than works. The later Wheelwrights worshipped in the Old South Church in Federal Street, Newburyport, the Meeting House of the First Presbyterian Society and famous in the eighteenth century for its associations with the Wesley's friend George Whitefield. Baptized and later married there, William shared an evangelical Christianity with his family, including his brother Isaac Watts (named after the famous hymn writer) who preached for the Bible Society of America. Later in life, William himself generously funded the translation of the Gospels into Turkish as a means of offering Protestant Christianity to the Ottoman Empire.[6]

Wheelwright's maternal grandfather, William Coombs, was a New England sailing ship-owner. His own father, Ebenezer, commanded one of these vessels trading to the West Indies. William went to sea at sixteen, survived shipwreck and rose by the age of nineteen to be master of a barque trading to the east coast of South America. While he was in his mid-twenties, the barque ran ashore in the River Plate, but all on board survived. From Buenos Aires, he took a passage to Valparaiso and from there to Guayaquil in the new state of Ecuador, where he set up a trading company and served as United States consul. The business failed but he acquired a small trading schooner and built up a fleet of sailing packets that earned a reputation for both comfort and reliability.[7]

In the early 1830s, Wheelwright began to promote steam navigation in his adopted Pacific region. A meeting, held at the house of British merchant Joseph Waddington in Valparaiso in June 1835, took place a few months after HMS *Beagle*'s protracted visit to Valparaiso under the command of Robert Fitzroy. The *Beagle*'s departure was delayed by Charles Darwin's illness following an expedition into the interior. Fitzroy, too, was unwell but his subsequent publicized judgements gave credibility to the project (see below). Attended by Chilean statesman Diego Portales, the meeting addressed the project in concrete detail. When Portales revealed that it would have the Chilean Government's blessing, Wheelwright obtained monopoly privileges and important exemptions from harbour dues and customs duties from both Chile

[5] [Wheelwright] (1838: 201–21).
[6] 'Wheelwright, John (1592?–1679)', and 'Watts, Isaac (1674–1748)', *ODNB*; [Alberdi] (1920–2).
[7] Wardle (1940: 13–14); Fifer (1998: 7–24).

and Peru for a ten-year period from the commencement of the service. He was thus in a uniquely strong position to win the support of British and local merchants when the British Government issued its calls in mid-1836.[8]

Wheelwright supplied the local committees with estimates of both the annual expenses and receipts for three steamers of around 450 tons, each powered by two 80-hp steam engines. The expenses were divided into the costs of officers and crews on the one hand and fuel and other running expenses on the other. On the recommendation of the Lima Committee, he subsequently added a fourth vessel as a reserve. The Committee also noted that several items had been costed at the highest price and that therefore, actual running costs under 'economical arrangements' should be lower. For each steamer, he estimated the annual costs as follows:

Table 11.1. *PSNC projected operating costs (1838)*

	Labour costs
Captain	£400
'Head' engineer	£ 300
First mate	£200
Second engineer	£200
Second mate	£100
Clerk	£100
Steward	£70
Cook	£70
Stokers (4)	£192
Sailors (8)	£288
Servants (4)	£120
Total per vessel:	**£2,040**
	Other costs
Provisions	£1,871
Coal	£3,733
Agencies	£1,200
Wear and tear	£2,037
Post Office costs	£200
Insurance (5%)	£1,000
Interest capital (6%)	£1,350
Commission (5%)	£1,556
Total per vessel:	**£12,947**

[8] Wardle (1940: 15–20); Fifer (1998: 34–42); Keynes (1988: esp. 249–64).

For the four vessels (including one in reserve with only insurance and interest) the total annual expenses came to a little over £47,300.[9]

Coal consumption was by far the largest single item of annual expenses, averaging over £300 per month for each active steamer. The Peru to Chile route, he asserted, would consist of twenty-four passages per annum, each passage occupying fourteen days of steaming at a rate of twelve tons of coal per day and requiring 4,032 tons annually. Similarly, the Peru to Panama route consisted of twelve passages taking fourteen days and consuming 2,016 tons per year. Rounding up the total coal tonnage required to 7,000 tons left 952 tons for contingencies. At one pound twelve shillings per ton, the annual fuel would therefore cost around £11,200.[10]

Both the Lima and Valparaiso Committees endorsed Wheelwright's estimates and accepted that the venture would yield profits to investors. The former justified the reserve vessel by highlighting the 'wide distance of space and time between the Pacific coast and European resources'. The Lima Committee felt the rates were 'extremely moderate compared with existing charges, and well calculated, therefore, to encourage and increase the intercourse, by steam, between the different ports of the Pacific'. And it recommended that at least 5 per cent of all receipts be paid to the principal managing agent, Wheelwright, for his central role in the project.[11]

The Valparaiso 'Commission', consisting of seven British merchants, added a warning that 'want of capital and experience in such undertakings in these [South American] countries would render it difficult to procure a sufficient number of subscribers'. At the same time, it pointed out that the privileges granted to Wheelwright had been conceded 'on the understanding that the natives of the respective countries would be invited to participate in the enterprise'. In order to balance these desiderata, the Commission urged that the Company be formed either in the United States or in Europe, with 'a sufficient sum ... to carry the project into full execution', and that the proposed constitution of the Company be transmitted to the different West Coast states 'with instructions to advertise in the public papers ... for subscribers'. The Commission's full report received the support of representatives of some thirty-five merchant houses of various nationalities based in Valparaiso.[12]

Wheelwright turned first to the United States. In the words of an observer, both in America and in Britain Wheelwright had 'great difficulty in forming a company ... Unfortunately for the projector, the extreme pressure of the money-market at that time, coupled with the distance of the intended scene

[9] [Wheelwright] (1838: 32).
[10] [Wheelwright] (1838: 32, 35).
[11] [Wheelwright] (1838: 21–4).
[12] [Wheelwright] (1838: 26–31).

of operations, the want of confidence in the grants of South American states, and the political changes to which they were exposed, all conduced to impede the enterprise'. Then, 'when his capital was nigh wrecked', he had the 'good fortune to meet with the late Lord Abinger, who, together with the noble members of the Scarlett family, warmly espoused the undertaking'. They and their friends, acting as Wheelwright's patrons, came together to form the PSNC in London.[13]

Cambridge-educated James Scarlett (Baron Abinger from 1835) built his reputation as a highly persuasive barrister with a peak income of some £18,500 in one year. Although a committed Whig MP from 1819, he served as attorney general under the liberal Tory ministry of George Canning, whose views he sympathized with. As foreign secretary in 1824, Canning had memorably declared that 'Spanish America is free, and if we do not mismanage our affairs sadly, she is English'.[14] This statement implied that Britain now had a golden opportunity to trade with the new South American republics. As a personal friend of the family, Canning advised Scarlett to encourage his youngest son, Peter Campbell, to follow a diplomatic career. After serving as an attaché in Rio de Janeiro, the young Scarlett set out across the continent in 1835 and met Wheelwright upon arrival on the Pacific coast. Meanwhile, Scarlett senior was appointed Lord Chief Baron of the Exchequer in late 1834 under the Whigs and held the post for more than nine years.[15]

Early in 1838, Peter Scarlett's account of his travels caught the attention of the metropolitan press.

> The most important part of the work ... consists in observations on the possibility of establishing a communication between the eastern and western coast of the American continent, by cutting a canal across the isthmus of Panama, or by forming railroads, and, in plans and statements annexed to the second volume, for establishing steam navigation on the Pacific.

The latter section included 'statistical tables and statements collected with great industry, and calculated and arranged with accuracy and care'. As the work of Wheelwright, the section was also printed as a pamphlet in October to accompany the PSNC prospectus.[16]

In his introduction, Wheelwright left prospective investors in no doubt as to his credentials. The 'extent and minuteness of my calculations' bore witness to his careful assessment, extending back over four years, of the prospects for

[13] [Peacock] (1859: 84–7) (excerpt from Hadfield's work on Brazil and the River Plate (1852)).

[14] 'Scarlett, James, first Baron Abinger (1769–1844)', *ODNB*; Hinde (1973: 368); Knight (1999: 122). Knight evaluates Canning's claim in relation to subsequent Latin American history.

[15] 'Scarlett, Peter Campbell (1804–81)', *ODNB* (1804–81); Wardle (1940: 20).

[16] *The Times*, 6 February 1838; Scarlett (1838); [Wheelwright] (1838: 3–4).

Pacific steam. 'Disinterested parties', armed with 'probity as well as practical knowledge of the subject', offered 'the best assurance that my data have been fully and fairly examined'. The British Government, furthermore, had confirmed its support for the project with the promise of a Royal Charter. And the Governments of the Pacific states had granted 'exclusive and valuable privileges for the navigation of their coasts'.[17]

Recognizing that loans amounting to millions of British capital had been made to unstable South American states, Wheelwright insisted that only 'the development and improvement' of those countries would bring about 'an amelioration of their domestic and international affairs'. In particular, the absence of prompt communication resulted in ineffective efforts to suppress rebellions. Regular steam navigation would therefore 'strengthen the executive authorities, ... promote the industry of the people, and ... contribute to an improved state of public and private credit'. And, crucially, it would improve the currently blighted prospects for a return on British investments.[18]

Wheelwright projected a system that would link to a steam packet service between Britain and the Caribbean (Chapter 7). Together these systems would reduce communication between Britain and Chile from more than four months to not more than forty days. Such an 'accelerated and easy communication' would to a great degree obviate, through this means of 'transmitting frequent and regular advice', the 'uncertainty and fluctuations which attend all mercantile operations with those now distant markets'. A future trans-Pacific system also beckoned. Incorporating a letter and chart from Fitzroy, Wheelwright argued that sailing vessels could take advantage of prevailing westerly winds between Australia and Callao, from where passengers and mails would travel north to the isthmus aboard PSNC steamers and thence join West India steamers for Britain. Such a system would also advance 'the civilization of the inhabitants of the numerous islands of the Pacific, to which the Missionary Societies have, for a considerable time past, been directing much of their attention'.[19]

Local trade too 'would derive great advantages from an accelerated communication between the several ports in the Pacific'. Thus 'many voyages would be performed in forty or fifty hours, which now occupy twenty or twenty-five days'. As evidence, he included a chart of the West Coast with respective tracks showing daily runs for both sailing and steam vessels. The former's quest for wind power took them far westward into the open Pacific, while the latter steamed on direct coastal courses between scheduled ports (Figure 11.1).

[17] [Wheelwright] (1838: 4–5).
[18] [Wheelwright] (1838: 7–8).
[19] [Wheelwright] (1838: 8–9).

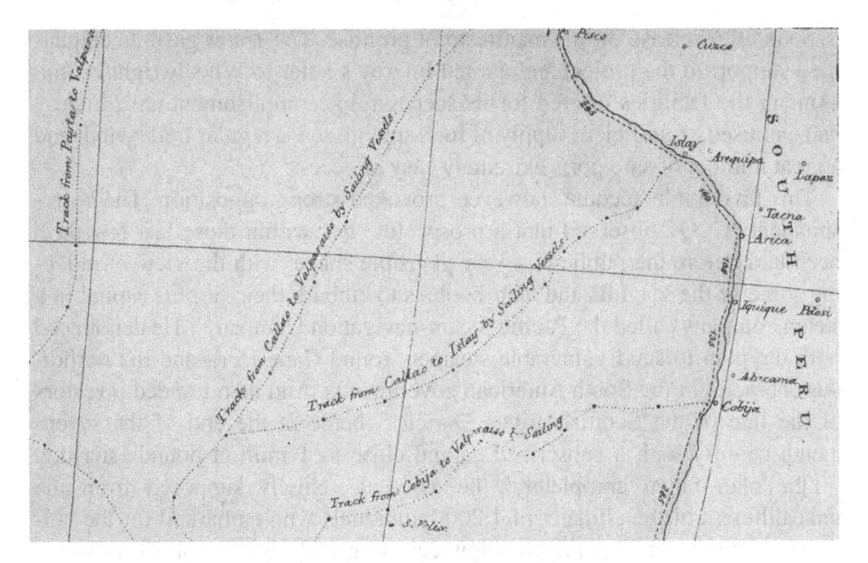

Figure 11.1. Wheelwright's chart of the West Coast of South America (extract). Steamer tracks are those close to the coast
Source: By kind permission of the Syndics of Cambridge University Library.

Reflecting the saving in time, the projected fares showed a typical reduction of about 50 per cent on current rates.[20]

In the remaining part of Wheelwright's publication, he identified five components for fulfilling these promises. First, he claimed that there existed a great abundance of cheap coal in Chile, together with the option of shipping it from England ('at a moderate price') or from Australia (where the supply was both abundant and cheap). Second, he asserted that the Company would be afforded greater security because their vessels 'will sail under the British flag, be under the protection of a British squadron, and possess the special guarantee of the separate local governments'. Third, he assured investors that Guayaquil in Ecuador possessed 'an excellent arsenal, and [is] particularly favourable for the repair of steam-vessels; while some of the ports of Chile offer in this respect almost equal advantages'. Fourth, while Chile and a Peru–Bolivia confederation were presently at war, he emphasized that the project's neutral character, together with its privileges, rendered groundless any fears of interference from either side. And fifth, the different states had approved the use of offshore hulks 'for the reception of coal, provisions, water, and cargo'.[21]

[20] [Wheelwright] (1838: 5–7, facing 36 and 42).
[21] [Wheelwright] (1838: 9–13). See Fifer (1998: 42) (war).

Soon after release of this manifesto of promise, *The Times* gave its unqualified support to the project and quoted Fitzroy's letter to Wheelwright in full. 'Among the facilities offered for its successful accomplishment are', Fitzroy had enthused, 'a sufficient supply of fuel, smooth sea, a regular trade wind, and a great number of safe ports extremely easy of access'.[22]

This favourable account, however, provoked strong opposition. The correspondent 'C. Q.' observed that a prospectus 'has within these last few days been laid before the public in a very plausible shape, with the view of inducing some of the seagulls and land-boobies to embark their surplus capital in a steam company called the Pacific Steam-navigation Company'. He denounced both the plan to send vulnerable steamers round Cape Horn and the perfidious character of the South American governments. And he reminded investors of the fate of the Scottish-driven Darien Scheme at the end of the seventeenth century with a subscribed capital close to 1 million pounds sterling. '[T]he plan failed completely', he asserted, 'chiefly supposed from the unhealthiness of the climate: of 1,200 individuals who embarked for the colony not one above 30 survived shipwreck, war, and disease – a fine prospect for those who now contemplate embarking themselves and their capital in the South Pacific Steam navigation Company'! Both Wheelwright and an anonymous merchant challenged 'C. Q.'s' gloomy predictions within days.[23]

The PSNC's prospectus named eight directors including the Hon. P. Campbell Scarlett, Captain Horatio T. Austin, RN, George Brown and J. Todd Naylor. Scarlett's older brother Robert, as an MP, had been instrumental in winning Government approval for the Company's Royal Charter. Brown was at the same time an RMSP director, while Austin had commanded some of the Royal Navy's earliest steam vessels (see below). All of the directors except for one at this stage were London-based gentlemanly merchants and investors. Only Liverpool's Todd Naylor represented outside interests. His family firm of Naylor, Boardman and Company, merchants, were strongly connected with South America and Frederick Boardman had been closely involved in promoting the project in Valparaiso.[24]

In late February 1839, the Company launched a press advertising campaign. Advertisements appeared in the morning papers (*The Times*, *Morning Chronicle*, *Morning Post* and *Herald*), in the evening press (*The Courier*, *Standard* and *Sun*) and in the weekly *Railway Times*. Wheelwright, meanwhile,

[22] *The Times*, 8 November 1838. See also [Anon.] (1837c: 255–6) (Wheelwright's project endorsed), (24–31, 97–100, esp. 30) (report of the *Beagle*'s voyage and Fitzroy's comments on the opportunities for steam navigation).

[23] *The Times*, 16, 21 November 1838. The reference to the 'South Pacific' probably recalled the infamous 'South Sea Bubble'.

[24] [Wheelwright] (1838: 1–2) ('Prospectus'); PSNC Minutes (14 December 1838); *RM* (1843), 1160 (doubts over Charter); Wardle (1940: 22–4).

had been promoting shares in Liverpool and the northern provinces. In the immediate aftermath of the granting of the Royal Charter, the capital of the Company stood at a mere £5,000 compared to the £250,000 sought. Directors nevertheless agreed to the Construction Committee (Wheelwright and Austin)'s recommendation that the Company place an order worth £17,400 for two pairs of 90-hp steam engines with the Thames firm Miller, Ravenhill 'of known experience and high reputation'. The Committee had apparently negotiated with Merseyside's Thomas Wilson for two iron hulls at £9,000 each, but the directors, wary of such a break with tradition, instead contracted with Curling & Young for two wooden hulls, half of the projected fleet.[25]

11.2 'The Greatest Confidence in Steam': Self-Fashioning a Master of Steam Navigation

Early in 1840, PSNC appointed its first commander, Captain George Peacock, at a salary of £500 per annum plus a further sum of about £100 per month promised out of the ship's earnings. Back in 1835, Peacock had served under Austin's command during HMS *Medea*'s Mediterranean duties (Chapter 3). In a report to the Admiralty, Austin listed Peacock's valuable skills: 'a thorough practical and scientific knowledge' of navigation, a 'zeal or aptitude ... in making himself acquainted with the pilotage of foreign coasts', a perfect proficiency in astronomical observations, and an unrivalled attention to 'the general principles and localities' of 'the various systems of winds and currents'. His knowledge of coasts ranged from the English Channel ('most perfectly') to the Brazils, West Indies, Portugal, Spain and the Mediterranean ('a good knowledge'). Austin emphasized especially his officer's skills in pilotage as '[m]ost trustworthy in every way – I have never hesitated to enter a port without a pilot, which he had visited before'.[26]

Born in Devon in 1805, Peacock served an apprenticeship aboard his father's sailing vessels that traded as far as the eastern Mediterranean and the east coast of South America. By the mid-1820s he was chief mate aboard the family-owned brig *Fanny*. Three years later he became executive officer on the steam naval vessel *Echo* (a ten-gun brig) deployed in surveying the Thames Estuary and providing data for William Whewell's theory of tides sponsored by the BAAS. On each voyage he maintained a meticulous log. He later recorded that he had made the acquaintance on board of Francis Pettit Smith with whom he 'conversed on the subject of screw propulsion'. He also met one of the Maudslay's who came aboard to survey the *Echo*'s engines, and

[25] PSNC Minutes (28 February, 4 May and 1 October 1839); Wardle (1940: 24–6) (iron hulls).

[26] PSNC Minutes (30 January and 6 February 1840); Austin to Peacock, 24 August 1835, George Peacock Papers (PP), LRO.

gained hands-on experience working a lathe and assisting in the construction of marine engines at Maudslay's Lambeth works.[27]

Peacock was quick to fashion himself as someone with the combined skills of marine engineer and surveyor. As he told Captain William Oldrey, RN, 'I would take the liberty of strongly urging on His Majesty's Government to have a correct survey made of this important place [Victor Cove on the east coast of the Isthmus of Panama], which may in future times, *when steamers cover the seas*, become the high road to China and Australia'.[28]

As the semi-official imprint of the Hydrographic Office of the Admiralty, the *Nautical Magazine* was the perfect forum for Peacock to publish on matters of pilotage, seamanship and navigation. An early contribution concerned his attempt to propel the *Medea* in Malta using the crew rather than steam to turn the paddlewheels. The vessel achieved a speed of two knots despite the crew suffering the after effects of influenza. The system, he concluded, was 'simple and economical, and I hope may be found serviceable'. Other contributions during his Royal Navy career included an account of the salvage of a sunken sailing vessel near Quebec, a survey of Pictou Harbour in Nova Scotia and the evaluation of a new type of ships' anchor. He also attracted the *NM*'s admiration for his rescue from drowning of a boatswain's son at Sheerness, for which action he received a Royal Humane Society Silver Medal.[29]

In the winter of 1839–40 the Admiralty declined Peacock's request for a China station posting. Austin then advised Peacock to come up to London and call on Macgregor Laird with a view to becoming master of the new *President* (Chapter 6). Failing that, he should 'wait on Mr Wheelwright at the office of the Pacific Navigation Company ... who will be happy of your services'. Under a misapprehension, Peacock offered Wheelwright his services 'for the command of one of your Transatlantic Steam Ships'. Clarifying the PSNC system, Wheelwright explained that the steamers 'will be confined to the coasts of the Pacific. The Royal Mail [RMSP] will meet us at Chagres'. Refused leave of absence by the Admiralty to deliver the PSNC's new steamer *Peru* to Chile, he resigned his commission in February 1840 and turned with characteristic zeal to work on the imminent launch of the Wheelwright project.[30]

[27] Peacock, 'Log of the brig *Fanny*'; 'Log of the *Echo*'; 'Statement of the services'; '[Death of Sir Francis Smith]', (from the *Shipping and Mercantile Gazette* 24 February 1874), 387 PEA 1/1, 1/14 and 3/22, PP, LRO; Morrell and Thackray (1981: 425–7, 515–17); Marsden and Smith (2005: 25–6) (Whewell and BAAS); Wardle (1940: 29); 'Peacock, George (1805–1883)', *ODNB*.

[28] Peacock to Oldrey, 14 November 1831 in [Peacock] (1859: 21–7). My italics.

[29] Peacock (1837: 730–2, 1838: 776–8, 1839: 146–8, 1840a: 447–8, 1840b: 599–600). The Award is recorded in *NM* (1838: 411, 627).

[30] Robinson to Peacock, 6 January 1840, 387 PEA/B2, PP, LRO; 'Statement of the services', 387 PEA 1/14, PP, LRO; Wheelwright to Peacock, 6 January 1840, in Wardle (1940: 30–1); [Peacock] (1859: 45).

The *Peru*'s 'experimental trip' down the River Thames and back was less of an experiment than a spectacle staged for the great and the good of Empire. Among the nearly 200 invited guests, according to *The Times*, were the Earl of Wiltshire, Sir Thomas Cochrane, RN, PSNC chairman George Brown, vice-chairman the Honourable Peter Campbell Scarlett and 'many merchants of London connected with the steam navigation of the Empire'. Hospitality included an 'elegant *déjeuner à la fourchette*' aboard and dinner at the West India Tavern upon return to Blackwall. *The Times* welcomed the event as 'an era in the history of steam navigation' that would act 'in no little degree to civilise the inhabitants and restore good government' in the region. No reference was made to the war between the two states after which the twin steamers took their names.[31]

In September 1840 *The Times* carried Peacock's account of the passage from Plymouth to Rio – 'a most prosperous voyage, answering in every respect my most sanguine expectations'. He reported particularly on trials with sail alone during which the vessel attained speeds of up to eleven knots in the trade winds. The funnel, hinged at the base, had been lowered to a horizontal position to allow clear working of the mainsail. Under steam the average coal consumption conformed to Wheelwright's projection of twelve tons per day. And the passengers daily expressed themselves 'very contented, happy, and comfortable' (Figure 11.2).[32]

After a passage lasting a little under a month, the *Peru* rendezvoused with the *Chile* at Rio and proceeded to the Straits of Magellan. As circumstances permitted, Peacock again ensured the despatch of sanguine reports back to London. For the benefit of the investing public in particular, Peacock's reports not only reassured readers of the project's viability, but of its triumphant inauguration and high symbolic promise as the bringer of British cultural and commercial values.

At Port Famine (Puerto del Hambre), part way through the Straits, they anchored not only to take on water and wood fuel, but also to stage a ceremony coinciding with the anniversary of Chilean Independence. Peacock seized the opportunity to celebrate simultaneously the political occasion and the advent of practical steam navigation on the Pacific. 'I erected a beacon 25 feet high on the height of Santa Anna, depositing underneath it a manuscript parchment roll, descriptive of the particulars of each vessel', he reported. 'Length of passage, consumption of fuel, &c., together with several British coins of the present year'. They then 'hoisted the Chilian [sic] ensign, and saluted it with three cheers; at sunset we lowered it, gave three cheers for the Pacific Steam Navigation Company, and returned on board, leaving the beacon a conspicuous

[31] *The Times*, 8 July 1840; Wardle (1940: 34–6).
[32] *The Times*, 7 September 1840; Wardle (1940: 38).

Figure 11.2. PSNC's first paddle steamer, *Peru* (1840)

seamark for the harbour, and a firmly fixed monument, commemorative of the triumph of steam in this part of the world'.[33]

Encountering northerly gales, the *Peru* arrived at Talcahuano some ten days after leaving the Straits. Beset by 'the utmost anxiety and uneasiness' over the fate of their consort, Peacock rejoiced at the *Chile*'s safe arrival one week later. Following a refit, he ordered steam raised using a small quantity of local coal obtained from the foot of a precipice. Judging it inferior to Welsh steam coal, he nevertheless reported no difficulty in keeping up the steam. At the same time he arranged for the ship's surgeon and the British consul to accompany a Polish geologist, Lololscky, in the service of the Chilean Government, to examine coal formations in various other locations along the coast. Their report claimed Peacock 'establishes the fact of the existence of coal suitable in every respect for steam navigation'.[34]

Heading north for some twenty-five hours, the two steamers were met in the approach to Valparaiso by a boat carrying a letter from Naylor, Boardman and Company who had been appointed PSNC agents in the Chilean port. 'We beg leave to inform you', the message read,

> that his Excellency the Governor of Valparaiso, being desirous that your entrance into this port should be as conspicuous as possible, has commissioned the bearer of this

[33] *The Times*, 29 January 1841 (letter dated Valparaiso 17 October 1840).
[34] *The Times*, 29 January 1841.

> [letter] ... to request that you ... [stand off] until a signal be hoisted at the lighthouse on the point; whereupon a procession of two lines of boats, decked with flags, will leave the harbour to receive and welcome you, between which [two lines] you may enter.

The bearer also provided Peacock with 'a plan of the bay and shipping, and programme of our proceedings'.[35]

Peacock eagerly seized the opportunity for a grand pageant (see Epigraph). While awaiting the signal, he hoisted the Chilean ensign, honoured it with a twenty-one-gun salute and 'performed various evolutions with both vessels'. Proceeding towards the breakwater, the two steamers passed in line ahead formation between the British naval flagship HMS *President* and the Chilean flagship *Chili*. The crews of each vessel occupied the rigging in traditional fashion as further 'royal' salutes took place from the guns of various ships.[36] But a 'gratifying display of enthusiasm' alone was no guarantee of a steamship line's long-term prosperity. Without skill and knowledge, regularity and safety were elusive commodities.

'You may imagine the anxiety I felt in leaving Valparaiso without a soul on board who had ever been at the intermediate ports [between Valparaiso and Callao] before, and at a period when the coast is hidden by fogs the greater part of the day and all night', Captain Peacock told *The Times* readers early in 1841. 'I found Captain Fitzroy's ... [sailing directions] of essential service, as it pointed out the correct latitude and longitude of each place, and having found the ship's place by observation, and shaping my course, and running the distance by patent log accordingly, I managed to hit every port to a nicety'.[37] Not every officer, however, could claim such safe arrivals.

Less than a year into PSNC's service, the *Chile* struck the Quintero Reef in the approaches to Valparaiso while under the command of an inexperienced master. The ship's quick-acting engineer immediately switched the engines to draw condensing water from the bilges instead of the sea, allowing the steamer to be safely beached in a severely damaged condition. Leaving the *Peru* in the hands of his first officer, Peacock now took charge of the *Chile* and organized the temporary repair of the hull. By rigging sheerlegs, he arranged for the removal of the machinery. Having sealed all the openings above the waterline, he used the sheerlegs to heel the vessel first one way and then the other in order to raise the damaged portions as high as possible above water. With 'a preparation of mashed tallow candles, wipings and ashes', he organized a plastering of the splintered hull planks and the construction of a watertight internal bulkhead

[35] Naylor, Boardman and Company to Peacock, October 1840, in *The Times*, 29 January, 1841.
[36] Wardle (1940: 42).
[37] *The Times*, 23 February 1841.

at the forward end of the vessel from keel to deck. Finally, he oversaw the reinstallation of the machinery.[38]

Such an effort had been made imperative both by the absence of a dry dock on the coast and by the small tidal range at Chilean ports. More than 2,000 miles away to the north, however, Guayaquil offered a gridiron made practicable by a tidal range sufficient to allow a vessel to dry out at low water. The *Chile*'s voyage was equally dramatic, especially as rocks were apparently still embedded low down in the hull. A cargo launch, taken in tow to serve as a lifeboat should the steamer founder, was itself lost during the first night at sea. Reaching the destination safely, he recorded that the ship was 'thoroughly repaired under my own eye'.[39]

The *NM* published Peacock's account of the methods he employed to heel the *Chile*. The conclusion to his letter addressed a wider audience. He proclaimed PSNC to be 'in a very flourishing state'. The *Peru* under his command had steamed more than 50,000 miles since departure from Plymouth and earned about £24,000 in revenue from mail, passengers and freight. This 'most excellent beginning' overlooked the fact that PSNC had come close to losing half its fleet in the single accident to the *Chile*. But Peacock was not someone to reflect on such negative episodes. 'I think we shall do even better yet', he affirmed, 'but we want more steamers, and larger ones. It is one of the soundest speculations ever proposed, and although we had a thousand difficulties to contend with, things are now looking well, and people have the greatest confidence in steam. It is a delightful coast for steamers'.[40]

11.3 'With a Due Regard for Economy': Pacific Steam in Practice

'It is therefore evident', PSNC's board minutes decreed in mid-1842, 'that the necessity for an uncompromising adherence to punctuality in the departure of the vessels on the days on which they have been previously advertised to sail is paramount to every other consideration'. The primary reason for this decree concerned the lack of sufficient hotel accommodation at ports to which passengers travelled long distances from inland locations. No Company official, from the chief superintendent down, was thus permitted to delay a vessel on his own authority, since such delays would only lead to 'a general derangement of the service'.[41]

This declaration against the 'derangement' of PSNC's system represented one attempt to exert Court control over its officials on the West Coast. As the

[38] Peacock to the editor of the *Shipping and Mercantile Gazette* (*SMG*) (12 May 1880), 387 PEA 3/28, PP, LRO (published in the *SMG* 15 May 1880).
[39] Peacock to the editor of the *SMG* (12 May 1880).
[40] Peacock (1842: 413–15).
[41] PSNC Minutes (10 August 1842).

personification of trustworthiness both as shipmaster and gentleman, Peacock had earned the directors' gratitude for his 'zealous and extraordinary exertions in promoting the interests of the Company on every opportunity'. But even he could be chastised on occasion. After reporting in 1843 on a Board of Council meeting in Lima, for example, the Court ascertained that no actual meeting had been held. 'Such irregularities', the Court declared, 'were calculated to shake the confidence which the Court was disposed to place in their Officers'. Wheelwright came in for even more censure when he apparently ignored special orders from the directors to hold a vessel at Panama to await the arrival of Colonel N. H. Nugent en route from Britain to take over the accounts at Callao.[42]

The Court also insisted on controlling information that reached Britain relating to the operations of its steamers. Directors demanded a continual flow of abstracts of bridge and engine room logs. When the *Chile* grazed a rock in the approach to Valparaiso in mid-1843, the damage this time was not serious. But the Court criticized the first officer for not immediately including the 'particulars in the abstract of his log'. Instructing masters to implement this practice, the Court explained that 'the absence of such communications ... [is] productive of the greatest evil'. Thus 'in such cases, it almost invariably happens that, either through the public newspapers, or through private channels, exaggerated statements are transmitted, which the Court, when not in possession of official information ... is utterly unable to correct or contradict'. Such distorted press accounts back home were deemed to threaten passenger and investor confidence.[43]

The PSNC's system remained especially vulnerable to the instabilities of the West Coast states. Good relations between Company and state depended upon the British line being seen to stand above partisan politics and to maintain strict neutrality and even-handedness in times of local conflicts. In 1841, for example, a Bolivian force took over the hitherto Peruvian port of Arica. Peru imposed a blockade. The PSNC's Board of Council in Lima sanctioned the conveyance of Peruvian military officers from Callao to join their squadron. The PSNC Court subsequently decreed that its Board's decision 'seriously affects the neutrality of the Company's vessels'. By October 1843, the Court formally reiterated to its agents and officials the importance of preserving 'the neutral character of the Company's vessels, as upon this depends the security of the Company's property'.[44]

Trust between Company and local states was also consequential for the gradual build-up of shore establishments and coal hulks. The dearth of repair

[42] PSNC Minutes (12 January 1842; 7 June and 30 May 1843).
[43] PSNC Minutes (20 December 1843).
[44] PSNC Minutes (11 May 1842 and 11 October 1843).

facilities on the West Coast threatened the stability of the entire system. The PSNC established an engineering works at Callao in late 1843 with a twofold function. First, the 'Factory' enabled the engineering work that could not be carried out at sea to be done on shore. And second, it had 'the object of keeping employed in the service of the Company a sufficient number of engineers so as to guard against casualties in this department by which the voyages of the steamers might be interrupted'. The Chief Engineer, with immediate charge of the establishment, received £280 per annum (plus £30 for maintenance) which compared favourably with about £200 for a first engineer, £270 per annum for the master or £120 for the first officer aboard the *Peru* (it may be noted that these rates of pay were considerably less than those projected by Wheelwright in 1838). Engineers not required aboard the ships assisted in the work of the establishment which could take on outside work, the profits from which would be divided among the employees ashore and afloat. The Factory also served as a depot both for coal and spare machinery. In its original privileges granted to Wheelwright, the Peruvian Government agreed not to impose duty on materials imported for the purposes of operating the service. Now it agreed to the Factory and received rent for the site, but soon complained that the works were 'calculated to afford facilities for smuggling'. Despite political uncertainties and threats to its existence, the Factory was nevertheless deemed to be at work in December 1843.[45]

Coal was the *sine qua non* of the PSNC system. Although Wheelwright failed to locate suitable coal around Callao, Talcahuano in Chile promised a plentiful supply. 'The coal mine I am working is yielding beautifully', Peacock told *NM* readers in 1842. 'We have now several thousand yards of galleries, and got out nearly 4,000 tons, which answers our purpose very well, as the coal does not cost us more than three dollars [about 12s.] per ton'. Drawing on knowledge acquired from an earlier sojourn in Sunderland, Peacock taught the local inhabitants 'how to work it, for no mine had ever been worked there before'. He also left as foreman one of his stokers, while Peacock himself oversaw the construction of a pier linked by railway to the mine 'so as to save carriage by bullocks, which is very expensive'. From Lima in April 1841 Wheelwright informed the directors that Peacock had arrived 'having consumed nothing but Chile coal during the voyage'. In turn, the directors dispatched a reassuring extract to *The Times*.[46]

Writing to Peacock soon after, Scarlett highlighted the political importance of coal to Chile: '[t]he discovery of coal and its value as to quality are features in the history of Chile and Peru of extreme importance to them and of no small consequence to the steam navigation of all the commercial world'. Also in a

[45] PSNC Minutes (20 December 1843); Wardle (1940: 186).
[46] Peacock (1842: 413–15); *The Times*, 4 August 1841; Wardle (1940: 55).

letter to Peacock, Guayaquil's Governor and former Mexican ambassador in London, Don Vicente Rocafuerte, explained further the perceived link between coal and progress. 'It is a happy event [that you discovered] ... a mine of coal [and deposits of guano on the Island of Amortajado] whilst looking for the most eligible spot to erect a lighthouse', he wrote. These new discoveries contribute 'well to secure the future interest of the Company, and to promote the progress of science and commerce in general'. More widely, '[s]team Navigation being secured in the Pacific as it now is ... has brought Europe and America nearer together, and has placed the secret and marvellous treasures of nature – that lie hidden in this part of the new world – under the telescope and immediate investigation of the learned men of Europe'.[47]

Coal supply, however, continued to threaten the stability of PSNC's system. Early voyages, proprietors heard at the first GM, 'were interrupted for want of fuel, caused not only by the loss of two cargoes of coal [from Britain], and the long passages of other coal vessels', but also by the poor quality of a large shipment of Welsh coal sent out aboard the sailing vessel *Portsea.* As Talcahuano supplies came on stream, the *Portsea* and *Jasper* (both coal hulks) transported coal to Callao. This system meshed with the steamers' forty-day round voyages during which they now refuelled at Talcahuano's coal pier and from the stock at Callao. Estimated to cost $6 per ton, this local coal appeared to compare favourably with British coal shipped to the West Coast for $13 per ton. But at the second GM in 1844, the Company announced that experience had proved that the Talcahuano coal was *more* expensive than coal sent from Britain. Peacock and his associates in due course succeeded in making British coal even cheaper by offering sailing vessels the prospect of a valuable return cargo of guano for the home market.[48]

The Company's first GM heard the grim news that the Line was £111,630 in debt, that the two steamers had together shown a loss of over £13,500, that insufficient shares had been taken up and that substantial loans had been taken out from bankers Overend Gurney and the Royal Exchange Assurance Company. An astute proprietor, however, would have noted that the operating costs over the full year (1842) for each steamer were £1,000 under Wheelwright's projected figure (see above), though with coal at £600 and wages at £360 more than the original estimates. Indeed, overall expenses were a worrying £5,000 greater.[49]

[47] Scarlett to Peacock, 14 February 1842, and Rocafuerte to Peacock, 13 November 1841, in [Peacock] (1859: 62–5); Fifer (1998: 32, 35).

[48] *RM* (1843: 846), (1844: 1523); *The Times*, 2 June 1852 (including a letter from Peacock on disputed territorial claims on islands with guano deposits). See also Hollett (2008) (Peruvian guano trade).

[49] *RM* (1843: 846); Wardle (1940: 62–3) (finances).

The directors also outlined plans for a belated third steamer, but nothing came of the proposal for another two years (Chapter 16). Meanwhile, relations between Wheelwright and the directors deteriorated. Without explanation, the Board dismissed Wheelwright. Fighting back in a pamphlet aimed at the proprietors, Wheelwright indicted the Board for unwarranted expense over construction and operating costs. As a result of his appeal, he was reinstated as chief superintendent.[50]

The second GM at the end of 1844 heard that accumulated operating losses had risen to over £72,000. A month later the directors resigned. The Scarlett family played a major role in forming a new Board and by the spring of 1845 the Company had been reconstituted in Liverpool with most of the directors drawn from the local mercantile communities, including Boardman. Characteristic of the geographical proximity of some of the new Liverpool-based ship-owners, PSNC's office in Fenwick Street occupied the same building as that of the fledgling Lamport and Holt (Chapter 17).[51]

Wheelwright, appointed joint managing director with William Just from the end of 1845, negotiated a British mail contract worth £20,000 per annum for five years covering some twelve ports on the coast between Valparaiso and Panama. But when Just appeared before the SC on mail contracts in 1849 he accepted that for the first five years the Company's loss 'amounted to two-thirds of the paid up capital' and that over the period of nine years only two dividends, each of 5 per cent, 'have yet been paid to the proprietors'.[52]

Captain Peacock, having overseen the inauguration of the West Coast mail service between Valparaiso and Panama, returned to Britain at the end of 1846. Two years later he took up a new post, first as dock master and then from 1848 as superintendent, of Southampton docks, the home port for RMSP. There he 'had the satisfaction of aiding and assisting in the despatch of H.M. Mails, by night and by day, *to and from all parts of the world*', as well as ensuring the safe arrival or departure of Queen Victoria, other Royalty and members of the Cabinet to and from Her Majesty's residence at Osborne House on the Isle of Wight.[53]

'There was an abundance of fallacious promises and sanguine expectations', a sceptical shareholder told *RM* readers in the wake of the vexed 1843 GM, '[that were] diffused by means of circulars, and paragraphs in newspapers, before the failure and disasters contained in this report'. In contrast to the optimistic zeal

[50] PSNC Minutes (3 November 1843) (third steamer); (5 October 1843) (minute of dismissal); *RM* (1843: 845–7, 944, 1159–60) (first GM); *RM* (1844: 1523); *RM* (1845: 66–7) (second GM); Wardle (1940: 62–6) (Wheelwright dismissal and reinstatement).

[51] Wardle (1940: 66, 69).

[52] Wardle (1940: 69, 73); SC (1849), 206–07.

[53] [Peacock] (1859: 12).

of Peacock with respect to working practices on the Pacific coast, proprietors such as Peter Scarlett's brother Robert identified the Company's core problem as 'the general want of confidence in the Directors'. It was, he claimed, a state of affairs promoted by the barely quorate GM, the 'unhappy schism between Mr Wheelwright and the Directors', the absence of sufficient accounts other than those via Wheelwright, and the exclusion of Liverpool and Manchester proprietors from proxy voting.[54] In the eyes of contemporary proprietors, therefore, the prospects for the PSNC system in the early 1840s looked poor indeed.

[54] *RM* (1843), 944, 1159.

[illegible] with respect to [illegible] on the [illegible] property [illegible] Peter Scarlett's [illegible] the Company's [illegible] problem [illegible] management [illegible] of the Directors [illegible] CM, the [illegible] between [illegible] and the [illegible] the [illegible] and [illegible]

Part III

Eastward for India and China

12 'Built on a Large, Commodious and Powerful Scale'

Forging P&O's Eastern Mail Steamship System

I am more inclined to attribute the great delay [to a mail steamship service to the east] ... to its having never before fallen into *practical* hands, conversant with Steam Navigation on a large scale. It is now, however, fairly under way; and when the great benefits to Britain, as well as to India, which are likely to flow from it, are taken into consideration, and that the thousands of affluent individuals connected with our eastern empire, have now an opportunity of accelerating the march of its improvement, without risking any thing beyond the amount of the sum they may think proper to invest, and with every prospect of a profitable return for their money, I feel confident that the interests of 'Old India' will no longer be neglected.

Arthur Anderson promotes P&O's campaign to become the torchbearer of imperial communications between Britain and India (1840)[1]

Summary

On 23 April 1840 the Board Minutes of a new steamship enterprise recorded agreement among those present that the Company 'be called The Peninsular & Oriental Steam Navigation Company and that the capital be fixed at £1,000,000 – say 20,000 shares at £50 each'. The Board appointed three managing directors – Brodie McGhie Willcox, Arthur Anderson and Francis Carleton – who were henceforth 'entrusted with the general agency and management under control of the Board'.[2] At the same meeting, the abbreviation 'P+O' appeared for the first time on paper. The new Company represented the combined efforts and interests of a number of very different shipping men who had gradually come to work with one another at various points during the preceding quarter of a century. These projectors, first of Peninsular Steam and then of Peninsular and Oriental, relentlessly put into practical form a geographically vast steamship system. In less than a decade, the Company's lines linked Britain to the Iberian Peninsula, Gibraltar, Malta and Alexandria with an overland passage to Suez and onward to Ceylon, Madras and Calcutta. Indeed, by the mid-1840s, P&O was implementing plans for eastward extensions to China and contemplating future services to Australia.

[1] Anderson (1840: 21–2).

[2] P&O Minutes (23 April 1840). 'P+O' later became the more familiar 'P&O'.

12.1 'The Affairs of the Peninsula': Making Allies in Portugal, Spain and Ireland

Of the various interests that came together to form P&O in 1840, the partnership of ship-brokers Willcox and Anderson had existed formally since 1825. According to Lindsay's *History*, Willcox had been born in Ostend in 1785 of English and Scottish parentage and had 'spent his boyhood at Newcastle-on-Tyne, where he received the chief portion of his education'.[3] His maternal uncle, Brodie Augustus McGhie, was a ship-owner and shipbuilder in London. By 1815 Willcox was married to the daughter of a London-based Belgian merchant and had entered into a ship-broking partnership with an office in Lime Street in the City of London.[4]

Arthur Anderson's father, Robert, was a native of Unst, the northernmost of Shetland's three principal islands. A celebrated herring fishery brought annual armadas of drifters to its spacious natural harbour of Baltasound. At first dependent on sea fishing and fish-curing for a living, but skilled in both woodcarving and violin playing, Anderson senior had become factor (land agent) to one of Shetland's landowners, Sir Arthur Nicholson, after whom Anderson junior may have been named. Born in 1792 at Gremista on the shores of Lerwick's natural harbour, the young Arthur received an early education from his father, who then arranged for him to attend a private Shetland school run by the Revd John Turnbull, a friend of the celebrated Scottish romantic novelist Sir Walter Scott.[5]

By the age of about twelve, Arthur had left school to work at gutting, cleaning and drying fish on the island of Bressay on the eastern side of Lerwick harbour. His employer was Thomas Bolt, factor for Lord Dundas. Dundas' father, known as 'the Nabob of the North', had accrued immense profits as a contractor to the army. In the mid-1760s, he had purchased the lordship of Shetland for £63,000. Diverse family projects included Edinburgh's New Town, foundation of the Royal Bank of Scotland and construction of the Forth & Clyde Canal that would allow industrial products to be shipped from the basin at Port Dundas in Glasgow to the Firth of Forth and onward across the North Sea to continental markets.[6] In 1802 the younger Dundas promoted trials on the canal of William Symington's experimental steamboat, *Charlotte Dundas*. In one demonstration the steamboat towed two seventy-ton barges against gale-force winds and covered almost twenty miles in six hours.[7]

[3] Lindsay (1874–6: vol. 4, 378).
[4] Rabson and O'Donaghue (1988: 13); Harcourt (2006: 34).
[5] Jones (1935–8: vol. 1, 51–2). On Anderson's early life see also Nicolson (1914); Cable (1937: 9–13).
[6] 'Dundas family of Fingask and Kerse (per. 1728/9–1820)', *ODNB*.
[7] Lindsay (1874–6: vol. 4, 37–8).

On the Shetland outpost of Dundas' empire, Anderson advanced from fisher lad to boatman and then became clerk to Bolt. Such aristocratic patronage offered him protection from the raids of naval press gangs during the Napoleonic Wars. At the age of sixteen, however, Anderson volunteered for the Royal Navy. The same patronage secured his status as a midshipman rather than an ordinary rating. This gentlemanly route demanded a lifestyle far beyond the Shetlander's means and so he transferred to the humbler paymaster's branch as a captain's clerk. His own narrative now began to bear the hallmarks of Scott's romantic literature. After his ship paid off at Portsmouth in 1814–15, he recalled, he tramped to London and arrived with just enough resources for one twopence loaf and a pennyworth of cheese per day. Following Napoleon's escape from Elba, Anderson returned to Portsmouth on foot, 'in the last stages of exhaustion and starvation', to be taken on by one of his former officers, now promoted to captain of a naval brig.[8]

After Napoleon's defeat at Waterloo, Anderson finally left the service and returned to London, where this time he obtained the support of an uncle whose brother-in-law, Christopher Hill, was a Scarborough ship-owner. Strongly impressed with the narrative of Anderson's trustworthiness, Hill introduced him to Willcox, who employed him as a clerk. Anderson duly married Hill's daughter, Mary Ann, in 1822. Soon after, he replaced Willcox's previous business partner and from 1825 Willcox and Anderson began to build up ownership (as sixty-fourth shares) in about ten small sailing vessels trading to the Iberian Peninsula and even as far as the west coast of South America. Among these were the 174-ton wooden brig *Iberia* (1825) and the 226-ton wooden barque *Tyrian* (1828), both launched at Scarborough initially in the name of Anderson's father-in-law, but later registered to the ownership of Willcox and Anderson. By 1835, the partners operated from offices in St Mary Axe at the heart of the City of London's shipping district. In Lindsay's later judgement, he stated that the partners 'were plodding and industrious, and, consequently, soon insuring that success which industry, honesty, and economy must ever command'.[9]

From the late 1820s, civil war in Portugal threatened the stability of Willcox and Anderson's sailing ship trade with the Iberian Peninsula. From his domicile in Brazil, the late king's son, Dom Pedro, renounced the Portuguese throne in favour of his daughter Maria. His brother Dom Miguel, however, led an 'absolutist' faction against the 'consitutionalists' whose figurehead was now the young Queen. After taking ship to Portugal disguised as 'Mr Smith', Anderson returned to London with representatives of Dom Pedro. Meeting

[8] Jones (1835–8: vol. 1, 52–3).

[9] Rabson and O'Donoghue (1988: 13, 535–6) (listing the sailing ships in the period c. 1825–35); Lindsay (1874–6: vol. 4, 378–9).

at his home in Norwood, the Portuguese guests secured a loan to fight their cause. Anderson also arranged for the swashbuckling Scottish naval officer Sir Charles Napier to take command of Dom Pedro's fleet. Around 1832 the partners further negotiated the charter from British owners of five paddle-steamers, including the new Aberdeen-built 300-ton *Royal Tar* from Richard Bourne's Dublin and London Steam Marine Company. By 1834 the Portuguese Queen, securely on the throne, offered generous support in the form of trading facilities to her British backers.[10]

Hard on the heels of its neighbour's conflict, Spain's Queen Isabella received similar support from Britain to stabilize her own position. Willcox and Anderson arranged the charter of the *Royal Tar* to act as a warship. With the partners as agents, Bourne played an increasing role in establishing commercial steamer services to the Peninsula. Born into a well-to-do Protestant Anglo-Irish family in 1780, he had joined the Royal Navy at the age of seventeen and rose to command a twelve-gun schooner. Seriously wounded off the coast of Spain two years later, he retired from active service. On returning to Ireland he entered his brothers' stagecoach business which included the construction and operation of roads, coach building, the conveyance of mails and the ownership of hotels and inns on the coaching routes. Richard's brother William had founded the Dublin and London Steam Marine Company to operate passenger steamers linking Belfast, Dublin, Falmouth, Plymouth and London from 1827. Competition soon threatened the prosperity of the young Company. Taking over command, Richard effectively bought out the opposition by 1832.[11]

In late August 1834 the 'Peninsular Steam Navigation Company' issued a prospectus with Brodie Willcox named as chief agent and the Spanish minister in London, Juan Alvarez y Mendizabal, listed among ten proposed directors, most of whom were merchants with interests in the Peninsula. Also listed was Robert Cotesworth, who later played defensive roles on RMSP's Board (Chapters 7 and 8). Spanish merchants with London domicile included Pedro Juan de Zulueta (from a Basque family) who fled Cadiz for England in 1823 and founded a mercantile firm in London where he became an agent for the Spanish Government. Under the secured Spanish Crown, he was made the first Count de Torre Diaz. His son, Pedro Jose, later married a daughter of Willcox, served as a highly controversial P&O Director (Chapter 13) until 1866 and became second Count from 1855. Pedro Jose's first son, Brodie Manuel, became the third Count while the second of four sons served as Secretary to the Spanish embassy in London. One of the two daughters was mother to both

[10] Jones (1835–8: vol. 1, 53–4); Rabson and O'Donoghue (1988: 13, 24); 'Napier, Sir Charles (1786–1860)', *ODNB*.

[11] Harcourt (2006: 28–34). She shows that the Bourne family enjoyed wealth and status, deriving in part from the long-term lease of land in Ireland as well as from their later transport businesses.

a future Spanish Ambassador (Marquess Merry de Val) and Cardinal Merry de Val. The family also acted as agents for Peninsular Steam in Cadiz.[12]

'Political circumstances have hitherto rendered such an attempt hopeless'; Peninsular Steam's promoters confidently announced. '[T]here is now the strongest probability of the affairs of the Peninsula being permanently settled, and strong prospects are held out of the great increase in the trade and intercourse with this country'. They considered that 'the time is now come when the means of communication must be improved, and rendered more rapid and regular: this can only be effected by means of Steam Packets, built on a large, commodious and powerful scale'. Confidence in political stability therefore rendered credible the launch of an ambitious steam packet system linking Britain to the Iberian Peninsula, a route reaching well beyond existing short-sea steamer services.[13]

When the project stalled, Willcox and Anderson acted for a time as brokers for an extension to Iberian ports of the London & Edinburgh Steam Packet Company's North Sea coastal services (Chapter 3). In early 1835, however, they chartered Bourne's 206-ton *William Fawcett* (built in 1828 for Irish ship-owner Joseph Robinson Pim (Chapter 4) and Merseyside engine-builder William Fawcett) to operate a regular service under their own management. By late summer, Bourne had also purchased the 330-ton coastal paddle steamer *Liverpool* from G. & J. Burns (Chapter 1) for charter to the Peninsular service. Crucially, common usage in the press of the name 'Peninsular Steam Navigation Company' endowed the line of steamers with an independent identity in the absence of formal incorporation. Indeed, the Company operated not as a joint-stock company but, like the Cunard line of steamers, as a private partnership that relied heavily on a few investors. Unlike Cunard, however, the vessels themselves seem to have been chartered by Peninsular Steam while remaining in the ownership of individual partners.[14]

Anderson later traced the origins of the Peninsular Company to late 1835 'when a few private persons hired one or two steam vessels, and ran them occasionally to Lisbon and Gibraltar, in order to test the feasibility of establishing a steam communication with the [Iberian] Peninsula'. Each trip, Anderson claimed, lost them about £500 and the total loss sustained 'before it became remunerative' amounted to almost £30,000. But they persevered and resolved

12 Rabson and O'Donaghue (1988: 12) (facsimile of prospectus); *RM* (1843: 969) (Zulueta as son-in-law of Willcox); Cable (1937: 67–8) (Zulueta family).

13 Rabson and O'Donaghue (1988: 12) (prospectus).

14 Rabson and O'Donaghue (1988: 13–14); *RM* (1843: 6, 1074) (as private partnership). Haws (1978: 31, 34) gives rather less reliable data on these early steamers. Harcourt (2006: 3, 34) dismisses the initial prospectus of Willcox and Anderson as 'a flop'. The claim forms part of her desire to make Bourne and Williams (see below) P&O's 'two most important founders' (p. 1), *pace* traditional histories which confer that status on Willcox and Anderson.

to construct 'vessels of an improved description for establishing the communication'. Unlike the tonnage chartered from Bourne's cross-channel fleet, the projected new steamers would be purpose-built for the service.[15]

Under a new Government policy introduced in 1836–7, it was the Admiralty and no longer the Post Office that advertised for tenders and managed the mail contracts. For the Iberian route, it called for a weekly service with vessels of not less than 140-hp. Two companies submitted bids in mid-1837 (Peninsular Steam offering £32,860 with five steamers in the range 160–320-hp and Commercial Steam Packet £33,750 with five steamers of between 140 and 180-hp). Rejecting both bids, the Admiralty set a fresh deadline of mid-July. This time Peninsular offered £29,600 and Commercial £29,560. On the grounds that Peninsular now operated the better class of vessels, the Admiralty accepted the higher bid. Each vessel would depart from London, with forty-eight hours in summer and seventy-two hours in winter allowed for the passage to Falmouth where the mail would be taken on board. Calls would then be made at Vigo, Oporto, Lisbon, Cadiz and Gibraltar, from where Admiralty vessels would convey onward mails to Malta with further links to Corfu (for Greece) and Alexandria (for Egypt and India). In his capacity as chairman of Peninsular Steam and majority shareholder in three of the four new ships, Bourne signed the mail contract in late-August 1837. The Admiralty also approved, 'for political reasons', the Company's flying of 'a particular flag' that incorporated the colours of the royal houses of Portugal (blue and white) and Spain (red and gold).[16]

Reviving the name of their earlier brig, the 516-ton *Iberia* became the first major steamer ordered and owned by Willcox and Anderson. Built by Curling & Young at the high cost of £22,000, the *Iberia* extended the maiden voyage in October 1836 to carry passengers away from Europe's winter to the Portuguese island of Madeira. Three further vessels, owned by the much wealthier Bourne's, followed in 1837, including the 932-ton *Don Juan* and 688-ton *Braganza* built on the Thames and the 743-ton *Tagus* built by Scotts of Greenock. The *Don Juan* and *Tagus* were advertised on 1 September 1837 (the date on which the official mail contract took effect) as 'the largest and most powerful that have been yet put afloat'. Both the *Tagus* and the third steamer, *Braganza*, were engined by Scott, Sinclair of Greenock, close family associates of the shipbuilders. This early link with Greenock was soon to prove significant to P&O's face-to-face connections with Clyde iron shipbuilders and engineers (Chapter 14).[17]

[15] *The Times*, 3 November 1852 (reporting Anderson's speech verbatim).

[16] SC (1849: 59–60) (evidence of Thomas Crofton Croker, the Admiralty official handling the submissions); Daunton (1985: 154); Harcourt (2006: 3–4, 37).

[17] Rabson and O'Donoghue (1988: 14, 26); Cable (1937: 34–5) (advertisement).

In the first month of the Iberian service, the *Don Juan*, with Anderson and his wife aboard, was returning from Gibraltar when the ship ran aground in fog on Tarifa Point at the southern tip of the Peninsula. Anderson wrote a full account of the episode for his *Shetland Journal*. Resonating with Scott's romantic literature, it conveyed a number of messages. No lives, either of passengers, crew or rescuers had been lost and the mail had been given priority. Anderson himself had travelled by fishing boat to Gibraltar where he secured the assistance of HMS *Medea* under the command of Captain Horatio Austin (Chapter 11). The *Medea* arrived at the scene too late to salvage the steamer, but used naval authority to recover some $21,000 worth of specie from plunder. Anderson's romantic narrative thus told a heroic tale of shipwreck and survival. With losses limited to the ship, the line's reputation had been saved.[18]

The steamer had nevertheless cost £43,000 – some £15,000 more than the *Tagus* – and was apparently only partly insured. The unforeseen loss may have forced Bourne to sell his Dublin and London service to Charles Wye Williams' British and Irish Steam Packet Company, a subsidiary of his City of Dublin Company. Indeed, within a month of the loss, Bourne realized £40,000 from the sale to Williams of two such steamers.[19]

Williams had built a reputation as one of the most experienced, indeed wily, operators of cross-channel steamships. Trained originally as an Irish lawyer, his vocation lay in mechanical and chemical science, including the design of mill machinery, steam engines, paddlewheels and questions of fuel combustion and consumption. Associated with Liverpool designer and promoter of iron ships John Grantham (Chapter 14), Williams was also a keen advocate in the 1820s of iron hulls for river and canal traffic. He deployed Laird-built steamers on the network of waterways linking the River Shannon in the west to Dublin in the east, connecting with his cross-channel services. Like Grantham, Williams eagerly presented his findings to scientific audiences – by reading papers, for example, on combustion of peat to the ICE and on the value of watertight iron bulkheads to the Dublin meeting of the BAAS (1835).[20]

Well connected to the Anglo-Irish establishment in commerce and banking, Williams launched a steamer service between Dublin and Liverpool in 1824 that soon became known as the City of Dublin Company (later the CDSP), after successfully absorbing a rival concern owned by Birkenhead shipbuilder

[18] Cable (1937: 38–41); Rabson and O'Donoghue (1988: 14–15). I thank Stephen Courtney for insights into Anderson's 'heroic' representation here.

[19] Rabson and O'Donoghue (1988: 14, 26) (costs of *Don Juan* and *Tagus*); Harcourt (2006: 38) (sale of Dublin and London).

[20] Harcourt (2006: 22–8, 61n); Harcourt (1992: 141–62); 'Williams, Charles Wye (1779/80–1866)', *ODNB*. Harcourt's claim (p. 22) that Williams 'was one of the greatest pioneers, perhaps the greatest, in the first two decades of ocean steam navigation' is historiographically suspect given the arguments presented in the present study (esp. Chapter 13).

William Laird and Irish ship-owner Pim (Chapter 4). Since 1826 the Company had an understanding with the Post Office steam packet service that both operators would raise or lower fares simultaneously, but this agreement ended with the transfer of the packets to the Admiralty in 1837. Coexistence gave way to both financial and physical aggression. Williams reduced his Company's cabin fares by more than 25 per cent. Commander Chappell, then in charge of the packet service out of Liverpool, reported that Her Majesty's steam packet *Avon* had its course 'impeded by the unfair crossing of the *Athlone* City of Dublin Steam Company's private passage vessel'. The Admiralty responded with a warning that if such conduct were to continue, 'their Lordships would be obliged to take serious notice of it'. The competition eased when the CDSP accepted a mail contract for £9,000 per annum even though their initial bid had been for some £34,000.[21]

The *NM*'s Naval Chronicle section (1840), concerned with making the merchant steamer fleet 'effective and available for warlike purposes as it is now for those of peace', took a very positive view of Williams' activities. It reported on the strength of his fleet as comprising twenty-one seagoing steamers 'amounting to nearly 9,500 tons gross measurement, propelled by engines of 5,550 aggregate horse-power; they also have several river steam-vessels on the Upper and Lower Shannon [in Ireland]'. Proclaiming that Williams' 'observations carry with them the weight of long and extensive experience', the account further acknowledged that the 'science and safety of steam navigation owe much to Mr Williams'. Most notably, his Shannon steamer *Garryowen* (1834) had marked the introduction of the division of vessels into compartments by wrought-iron, watertight bulkheads. The report therefore assigned the highest credibility to Williams' values of strength over splendour in the design of commercial passenger steamships. His specification for the hull of a 600-ton steamer, it enthused,

> supplies some standard by which the public may be guided in its choice of steam-conveyance, rather than by the gildings, Gothic mouldings, mirrors, paintings, and other attractions lavished on the cabins of the advertised 'splendid and noble vessel' to the neglect of the manner in which the frame is put together and strengthened, and upon which the safety of the vessel depends ....[22]

Between 1830 and 1839, the CDSP prospered: sixteen new steamers joined the fleet and dividends totalling well over £200,000 were distributed to shareholders. In the same period, Williams promoted the Company's secretary, Francis Carleton, to the status of joint managing director. With the Company's primary base in Liverpool, Williams resided there from the 1830s. He was

[21] SC (1849: 43–5) (Croker's evidence).
[22] *NM* (1840: 850–2).

accordingly well placed to participate in the ambitious new projects for ocean steam navigation centred on the Mersey at the end of that decade (Chapters 3–6).[23] As we shall see, however, Williams' activities were not to receive universal acclamation (Chapter 13).

12.2 'The Economy of Space and Time': Projecting Passages to India

'In no quarters of the world could the application of the power [of steam] ... prove so advantageous to the commercial and the political interests of Great Britain as in the East Indies, in the West Indies, and in those places connected with these quarters' wrote James McQueen in the introduction to his *General Plan* (1838). While focusing primarily on the projected westward mail system (Chapter 7), the *Plan* also included a chapter exploring the feasibility of eastbound steam navigation that would link Britain both to Asia and to Australia. Adopting estimates of coal consumption and other operating costs similar to his West Indies scheme, he investigated the relative merits of the Red Sea and Cape of Good Hope routes. And although he calculated that the total annual costs of the latter track would exceed those of the former by almost £40,000 on the assumption of three 240-hp ships, he strongly supported the Cape option. 'The obstruction which the land barrier between Alexandria and Suez offers, and must always offer, even when unobstructed by hostile force', he argued, '... is a great drawback indeed'. Additional difficulties would arise both 'in the event of hostilities taking place between any of the great powers connected with the affairs of the Mediterranean' and from the shorter route affording easier competition from steamers of rival nations. He therefore concluded that 'the free communication' provided by the long-distance Cape passage more than compensated for its extra expense.[24]

McQueen's conclusions, however, were at variance with those of the earlier SC on Steam Navigation to India (1834). Macgregor Laird had submitted estimates of capital and running costs for two overland options by way of the Red Sea or the Persian Gulf. His projected annual operating costs for the former were some £20,000 higher than the latter, which would include Laird-built river boats as well as seagoing steamers. In its conclusions, the Committee recommended Parliamentary financial support for testing both options in practice.[25]

A pre-eminent witness before the Committee, EIC servant Thomas Love Peacock, reviewed its evidence on the rival merits of the three routes to India. In a persuasive essay contributed to the *Edinburgh Review* (1835), he concluded that the Cape option, considered physically and commercially, was best

[23] Harcourt (2006: 22–8).
[24] McQueen (1838b: 1–2, 64–81, esp. 78–80).
[25] Headrick (1981: 30).

Figure 12.1. Traditional East Indiaman

left to sailing vessels which, 'running with the trade winds and the monsoons, must always out-run a steam vessel in the entire passage' (Figure 12.1). For such ships there were no political obstacles: 'the ocean is our own highway, and we have no one's leave to ask, and no one to conciliate'. The Red Sea option, on the other hand, favoured steam navigation. Although it was 'one of the most dangerous navigations in the world for sailing vessels ...', he argued, 'steam vessels can keep the mid channel, and make the passage with little danger'. Similar to the Euphrates route, however, the overland component bristled with physical, political and commercial complexities.[26]

Peacock's essay delivered a forceful, even satirical, account of the pitfalls awaiting any inexperienced projector of steam navigation between Britain and the East. He cited the Bengal Government's charter of the 400-ton *Forbes*, 'a much-bepuffed steam-vessel ... to do not only all that the coolest heads had doubted of, or pronounced impossible, but much more than before had been dreamed of'. Leaving Calcutta, however, the *Forbes* soon ran into trouble, reportedly suffering a burst boiler and broken paddlewheels. Rather than admit their own folly, Peacock wittily observed, all its promoters agreed 'that the monsoon had nothing to do with the matter; [and] that there had been in the

[26] [Peacock] (1835: 445–54); Headrick (1981: 24–8).

boiler a flaw'. Of a second attempt to send the ship out against the monsoon, the 'intention, not the result, is all that is yet known ... and perhaps the patrons of the undertaking in England believe she is now in Egypt'.[27]

In Peacock's judgement, the Persian Gulf route offered 'the economy of space and time'. Immune from the difficulties presented by the monsoons, it was 'the natural line for expeditious communication'. The projected system would involve a steamship passage to a Syrian port, a journey overland to the Euphrates near Aleppo, an iron river-steamer to the Gulf and a 200-hp sea-going steamer to Bombay. At 5,800 miles, the distance compared favourably with 6,000 miles for a Red Sea service, and at least twice that distance for a Cape voyage. Peacock repeatedly supported the Committee's desire to see the Euphrates route 'brought to the test of a decisive experiment' which, if it should 'be attended with perfect success', would be 'a great triumph of art and enterprise'. He also pointed out that the delivery voyages from Britain of the seagoing steamers could act as a conclusive experimental test of the viability or otherwise of a Cape service. The Euphrates trial, however, suffered from a serious combination of delays, mishaps and harassment.[28]

From the late 1830s, the systems for the carriage of the Indian mails were twofold. First, Peninsular Steam conveyed the mails to Gibraltar for shipment by Admiralty steamers to Egypt, with an average transit time from London to Alexandria of about twenty-four days. Second, an overland route across France to Marseilles linked to naval steamers for the passage to Malta and Alexandria. In both cases, the mail crossed the desert before being loaded at Suez on to the EIC vessels for Bombay. The total transit time from London to Bombay varied between forty-five and sixty-four days. With Peninsular Steam delivering on promises with respect to the Iberian mails and with the Government concerned about increasing French dominance in the Mediterranean, the Admiralty approached the Company with a view to submitting proposals for a service to Alexandria.[29] Another major driving force had been Lord William Bentinck, India's reforming Governor General who had done much to introduce steam to the EIC mail service between Suez and Bombay (Chapter 2).[30]

Bentinck also encouraged the hitherto fractious communities of British merchants in India, along with their associates in Britain, to follow through on their aspirations to secure steam communication by the Red Sea to the

[27] [Peacock] (1835: 454–6); Bonsor (1975–80: vol. 1, 182–3) (*Forbes*). The *Forbes* was almost certainly owned by wealthy Boston merchant R. B. Forbes (Chapter 6) who had sent the vessel on an experimental 'steamship' voyage to Asia with a view to assessing the potential of steam navigation. I thank Ian Higginson for this insight.

[28] Bonsor (1975–80: vol. 1, 448–81); Headrick (1981: 30–1) (Euphrates trials).

[29] *RM* (1843: 1000–1); Harcourt (2006: 48–9).

[30] 'Bentinck, Lord William Henry Cavendish (1774–1839)', *ODNB*. The entry argues that the strong evangelicalism of Bentinck's wife Mary dovetailed with his notions of Britain's 'imperial trusteeship' of India.

three Presidencies (Bengal, Madras and Bombay). Bank of England Governor Timothy Abraham Curtis chaired public meetings around 1836–7 that agreed on a provisional directorate of City merchants and bankers. A later meeting in January 1839 resolved to form the East Indian Steam Navigation Company (EISNC), a title that suggested, misleadingly, associations with the EIC and, more accurately, links with the eastern side of the subcontinent. Some £134,000 (initially with 10 per cent paid up) was subscribed in India by 1840. The directors proposed to the EIC that it would convey the mail between England and the Presidencies for £100,000 per annum. The EIC's Court rejected the proposal, determined to protect its own steam line between Suez and Bombay.[31]

Closely involved in these moves, Captain Andrew Henderson later testified to the SC (1849) that EISNC soon fractured into three different parties 'each with separate objectives'. First, the Calcutta or 'Precursor' party aimed to confine its operations to the line between Suez and Calcutta. Second, the so-called 'Comprehensive' party, made up of residents in Bengal and Madras together with associates in Britain, 'wished to establish the communication throughout those Presidencies'. And third, there was the 'London' party 'whose object was to obtain the management and patronage of steam communications in the East under Government contracts'. The very general goals of this metropolitan group suggested ambitions to establish a vast eastern system that mirrored McQueen's original grand western system (Chapter 7). Divisions apart, by early 1840 negotiations were pending between the unstable EISNC and Parliament's Indian Board of Control 'for the whole line to India'.[32]

Meanwhile, the Peninsular Company had submitted a plan of its own to the Admiralty and received private assurances regarding the mail contract between Britain and Alexandria. Early in February 1840 the managers proposed a 'junction' with EISNC's 'Precursor' party, probably on the grounds that this party's plan (Suez to Calcutta), unlike the other parties, contained no territorial consequences for either the EIC (Suez to Bombay) or Peninsular Steam (Britain to Gibraltar and Alexandria). Anderson duly waited upon one of the EISNC directors, James MacKillop, and 'verbally acquainted him with what had been done'. He suggested that the Peninsular plan, as now conceived, 'would form a very important step towards carrying out the comprehensive plan, and that the parties I represented were willing to cooperate with his [EISNC] colleagues to

[31] SC (1849: 102–3) (testimony of Captain Andrew Henderson); Peacock (1835: 454–6) (earlier quarrels among the Presidencies). Henderson had been a Director of the EISNC and relentlessly lambasted P&O's aggressive strategy from its beginnings for the benefit of the SC. See also Harcourt (2006: 47–8, (EISNC), 59 (Henderson's hostility to P&O)).

[32] SC (1849: 103). Henderson's threefold characterization here differs somewhat from that of Harcourt (2006: 47, 52). She represents the 'Comprehensives' as the overarching group aiming to project a line from London, via the Red Sea, to all three presidencies. In this version, the 'Precursors' and another group of India-based merchants (including Henderson) split from the comprehensives in April 1840.

that effect'. In other words, Anderson's strategy of integrating the 'Precursors' with Peninsular Steam would deliver a 'comprehensive plan' of their own that would render the EISNC 'Comprehensives' obsolete.[33]

Accepting that MacKillop needed to consult his chairman (Curtis), Anderson duly furnished him with a detailed memorandum setting out the views 'of the [Peninsular Steam] gentlemen for whom I was acting'. Dated 14 February 1840, the document named a 'new' concern called the Peninsular and Oriental Steam Navigation Company. The title was a decisive statement of an eastward extension of the existing Peninsular Company's services rather than a wholly new project. The document explained that the Company 'is formed by the junction of a considerable portion of the proprietary and directors of the "City of Dublin Steam Packet Company" with the "Peninsular Steam Navigation Company"'. Of the projected 1 million pounds capital, some £300,000 was already invested both in 'the vessels and establishments of the Peninsular Company, and in the vessels intended to form a new line of communication between England and Alexandria'. Its objects were threefold: to carry on the existing steam packet line between Britain and the Iberian Peninsula; to 'establish at once an accelerated and otherwise improved conveyance for Mail and Passengers between England and Alexandria, *via* Gibraltar and Malta, by means of vessels of 1400 tons and 450 horses' power', and 'eventually to carry into effect the "Comprehensive Plan" of Steam Communication with India, by the establishment of a line of large and powerful vessels between Suez and the three Indian Presidencies, Ceylon &c'.[34]

Designed to preempt EISNC's as-yet incomplete prospectus to institute just such a 'comprehensive system', Anderson's memorandum also claimed that P&O's Alexandria proposal had been 'submitted to and adopted by her Majesty's Government' and that two vessels were 'in a great state of forwardness'. A gentlemanly readiness to co-operate, moreover, barely masked the powerful message that P&O alone possessed the credentials to implement in material and practical form '*that long contemplated enterprise to its full extent*'. Anderson therefore suggested two options. First, his Company would work the line as far as Alexandria while the EISNC would operate, in concert, the line from Suez to India. Second, subscribers to the 'Comprehensive Plan' should 'form a junction' with the Peninsular Company by becoming shareholders and by receiving appropriate representation on the Board of Directors and in its management.[35]

Anderson presented the case for the second option mainly in terms of Peninsular Steam's suitability for a 'mere extension of Capital' compared to

[33] Anderson (1840: 8).
[34] Anderson (1840: 8–11). See also Harcourt (2006: 49–50).
[35] Anderson (1840: 9–10). My italics.

the difficulties of raising capital by an entirely new concern such as EISNC. In doing so, he glossed over the inconvenient fact that Peninsular Steam was an established private partnership with sixty-fourth shares, while P&O, like EISNC, had not yet even been formed as a joint-stock company. Instead, he stressed the advantages of Peninsular Steam's existing mail contracts and profitability in giving a potential value to P&O's shares such 'as no new and untried enterprise could hope to command'. Were the junction option to be followed, he argued, steps could be at once taken to purchase or construct steamers suitable for the Indian Ocean project, to persuade the Government to award a third mail contract to the Company and to apply 'for a Charter of Incorporation ... limiting the liability of the shareholders'. The EISNC, however, held back for two months while it purported to consult subscribers in India.[36]

The memorandum gave no hint that Peninsular Steam intended to amalgamate with the struggling Transatlantic Company, owner of the underperforming *Liverpool* and prospective owner of the incomplete *United States* (Chapter 4). Transatlantic's managing director was CDSP's Williams and the principal shareholders were Richard Williams (his brother), James Hartley (a significant partner in Bourne's companies) and James Ferrier (of Williams' British and Irish subsidiary).[37] Some form of arrangement between the Transatlantic and Peninsular Companies would allow a rapid launch of a two-ship Alexandria service of the required power and speed.

The first formally minuted meeting between the Transatlantic and Peninsular companies took place at Peninsular Steam's offices, 51 St Mary Axe, in the City of London on 23 March 1840. At this early stage, the participants were overwhelmingly steam shipping men, in contrast to the City merchants that characterized RMSP (Chapter 7). Bourne, Anderson and James Hartley represented Peninsular Steam and Francis Carleton and James C. Ewart Transatlantic Steam. Any amalgamation of the two companies was still a 'proposed' one. Anderson and Carleton reported on communications with both the Admiralty and Treasury in which £30,000 had been named as the lower rate for the service between Southampton and Alexandria. Recognizing the value of promoting an image of the Company's managers as men of practical experience, Anderson also directed attention to practical matters such as coal stocks (nearly 12,000 tons) at Malta and the depth of water in the waterways between Alexandria and Cairo to suit small iron passenger steamers.[38]

McKillop finally replied to Anderson in mid-April. No junction between the two companies could now be made, he asserted, 'except in conformity with

[36] Anderson (1840: 10–11).
[37] Harcourt (2006: 44) (Transatlantic shareholders); Rabson and O'Donoghue (1988: 27) (*United States/Oriental*).
[38] P&O Minutes (23 March 1840); Harcourt (2006: 50).

the [EISNC] Prospectus'. Claiming that it dictated that the 'Comprehensive Company must have steamers for the whole line; in short, they must do *all* or *nothing*', he concluded, rather disingenuously, that the parties in London were more influenced 'by the desire of meeting the wishes of their friends in India than the expectation of pecuniary advantage'.[39]

In April 1840, the EISNC issued its prospectus 'under the influence of influential gentlemen' including Curtis (Chairman) and J. P. Larkins (Deputy-Chairman). Among the sixteen directors were at least three captains (Henderson, Lempriere and Nairne), one Member of Parliament (Frederic Hodgson) and an Alderman of the City of London (John Pirie). One of the auditors was Sir John Rae Reid, MP, of the bankers Reid, Irving (Chapters 3, 7–10).[40] There was no obvious hint of the party divisions: the project looked in every sense 'comprehensive'.

The EISNC prospectus proposed that 'the largest steam vessel procurable' would be dispatched for the Suez to Calcutta line in the first year. No actual vessel, however, was specified. At the heart of the project from the second year onward was the introduction of six new 2,000-ton steamships of 550-hp (and thus significantly larger than RMSP's steamers (Chapter 7)) with the ability to mount twenty-four large guns. With their size and power applicable to the Cape route if required, under the plan two of them would operate between Britain and Alexandria and the remainder in the Indian seas east of Suez. The estimated return from passengers and freight was £163,000 per annum.[41]

12.3 '[Nor] Under the Influence of any Particular Party': P&O's Political Ambition

Just five days after McKillop's reply, the Peninsular and Transatlantic representatives held their second meeting (with Bourne in the chair) at which they assigned the new title to the projected amalgamation, specified the capital and division of shares and appointed three managing directors (see above). The Peninsular Steam proprietors would receive £140,000 in shares, equivalent to the value of the five steamers owned 'and including their goodwill in the line and the heavy expenses incurred by them in opening the Peninsular Station'. Transatlantic proprietors would receive £80,000 'for the full value of the Liverpool Steam Ship with new machinery and other improvements as an equivalent for their interest in the present Transatlantic Steam Co'. The meeting also agreed that the timing of formal amalgamation of Peninsular and Transatlantic into Peninsular and Oriental would be decided when (and not if)

[39] McKillop to Anderson, 18 April 1840, in Anderson (1840: 12).
[40] SC (1849: 103) (testimony of Henderson).
[41] SC (1849: 103).

the mail contract 'shall have been secured'.[42] The confidence of the projectors rested on the knowledge that they themselves had practically defined the contract in terms not only of the availability of vessels large enough to launch the service, but above all of their horsepower.

The Admiralty had set a deadline of 19 May for the submission of tenders and 1 September for the launch of the service. It stipulated two steamers of not less than 400-hp and one of not less than 140-hp. As the SC (1849) Chairman (J. W. Henley, MP) inquired of his witness, this tight schedule 'was manifestly insufficient for any parties to build vessels for the service'? Thomas Crofton Croker, the Admiralty official most involved in the 1840 contract, agreed. 'Therefore the power of public competition was confined to those parties who had vessels of 400-horse power at the time?' Henley continued. Again, Croker replied in the affirmative. All other aspiring competitors were thus effectively excluded, unless they could purchase secondhand vessels from a very limited list.[43]

Committee member Captain Charles Mangles, MP (Chapter 9) probed the matter further. 'Were there such vessels existing in the market at that time, to be bought for money, do you suppose?' Croker's answer was decisive: 'there was the *President* and the *British Queen*; both were in the market at that time I believe'. The first, he explained, was 560-hp and the second 475-hp (Chapters 4–6). Indeed, just prior to the *President*'s disappearance in 1841, there were press reports, duly denied, that they had been sold to the Belgian Government for £140,000. But their performances would have required expensive remedies if they were to meet the mail schedule. Only P&O had privileged access to two large steamers of exactly the requisite horsepower.[44]

For the Alexandria contract, the Admiralty had received three other bids: Pim (£51,000), Macgregor Laird (£49,000) (Chapters 3 and 4) and the unknown G. M. Jackson (£37,950). Laird may have planned to use the *British Queen* and *President* himself, leaving EISNC without the requisite tonnage and power. A timely June meeting took place between P&O managers and Whig Government patron Edward Parry. Unsurprisingly, P&O secured the five-year award for a monthly service from Britain to Alexandria to begin in September 1840. Croker told the SC (1849) that the contract was worth £37,000 in the first year, decreasing annually to £32,000 in the fifth year.[45]

As early as spring 1840, P&O initiated planning the construction of a 1,600-ton, 500-hp steamer for a Suez to Calcutta service. In the autumn Anderson addressed head-on the EISNC threat. In twenty-two intensely argued printed

[42] P&O Minutes (23 April 1840).

[43] SC (1849: 86). Henley was a Conservative MP who later became President of the Board of Trade.

[44] SC (1849: 86); *RM* (1841: 298) (sale of Smith's steamers).

[45] P&O Minutes (6 June 1840); Lambert (1999: 37) (Parry's politics); Harcourt (2006: 51, 54) (naming the bidders); SC (1849: 68–9) (Croker). Harcourt cites figures of £1,000 more in each year than Croker (p. 51).

pages, he centred a literary strategy on the word 'practical' or its opposite no less than sixteen times. By labelling his quarry as 'men of high commercial *status* and extensive practical experience', he sought both to flatter and then disabuse them of the unfulfilled promises of EISNC. In the seventeen years since steam communication with India had first been mooted, he argued, private enterprise had, prior to P&O's recent involvement, 'not practically advanced the establishment of the "Comprehensive Scheme", so far as to effect the formation of the plank of a vessel or the pin of an engine'. His primary aim was thus to demonstrate that P&O, and not EISNC, possessed the practical experience to implement a 'comprehensive scheme'.[46]

Anderson also announced that the Company was now incorporated by Royal Charter (formally awarded at the end of December 1840) 'for the express purpose of establishing a steam communication to all the Indian Presidencies'. Within twelve months, Anderson predicted, his correspondents would see a monthly line established with the most important parts of British India, bringing Calcutta within six weeks of London. 'Such is the progress made in a few months by individuals unconnected with India', he argued, 'in an undertaking which has apparently baffled for many years, the whole enterprise of the wealthy and influential merchants and others connected with that important Empire'.[47]

The November pamphlet further claimed that at least three quarters of the East Indian subscribers, mainly impatient Calcutta investors under the banner of the 'Precursors', were now in alliance with P&O. Thus he represented the project as a non-partisan and 'enlightened' one, not committed to narrow agendas or specific political, commercial or religious dogmas but instead offering 'a wide and effective channel for the co-operation of all who may desire to promote the welfare of British India'. It was, in short, a project 'not formed on any exclusive principles, nor under the influence of any particular party'.[48]

Anderson explicitly aligned this vision with those of leading political and religious figures connected with India. The late Lord William Bentinck, he claimed, was 'the most enlightened, the most patriotic and popular Governor that India ever possessed' and in his endeavour to get steam communication established had declared that 'it would be cheaply purchased at any price'. The Indian Board President, Sir John Cam Hobhouse, had declared in Parliament that steam navigation was calculated to benefit India 'beyond the power of the most ardent imagination to conceive'. And the Bishop of Calcutta, 'that enlightened prelate', had spoken of the moral effects of steam communication, stating that 'it would open the floodgates of measureless blessings to mankind'.[49]

46 Anderson (1840: 3–5, 17).
47 Anderson (1840: 5–7); Cable (1937: 43) (Royal Charter).
48 Anderson (1840: 12–18).
49 Anderson (1840: 20–1. See also Harcourt (2006: 40).

The persuasion worked. The EISNC passed a resolution in late November 1840 'to the effect that such a union was expedient' between the two companies. Curtis and his deputy chairman Larkins undertook to negotiate the terms. Curtis' merchant house failed, however, and he was unable to take part in the future proceedings of the EISNC/P&O union. Three EISNC Directors (Alderman John Pirie, Captain Alexander Nairne and Robert Thurburn) transferred their allegiance to P&O and duly took their seats on the Board by the spring of 1841. George Larpent, much associated with the 'Precursors', was elected as P&O's first chairman in May. Originally planned to upstage P&O on the Suez to Calcutta line, the impressive 1,751-ton, 500-hp wooden steamship *Precursor*, built by Hedderwick and Rankine at Govan and engined by Robert Napier for a total price of £65,000 in 1841–2, eventually came under full P&O ownership in 1844 after languishing without work in London's East India Docks.[50]

Negotiations also opened between the EIC and P&O. The latter made clear its resolve to commence in 1842 'a communication between Suez, Bombay, Ceylon & Calcutta' with two 1,600-ton, 520-hp steamers (*Hindostan* and *Bentinck*) contracted with Wilson (hull) and Fawcett (engines) of Liverpool. It further affirmed that it was ready to order two more steamers for the same service. This show of practical determination produced rapid results. By the end of March 1841, Melvill announced that the EIC proposed 'to make a grant of twenty thousand pounds [per annum] to the Company for a period of five years'. The EIC, however, resisted P&O's attempts to appropriate its Suez to Bombay territory.[51]

With its rhetoric of practicality and its claim always to have the right mail steamers in readiness, Anderson's political strategy drove P&O's geographical system building. For its conquest of the Mediterranean, the Company played two winning cards: first, arranging the 'junction' with the failed Transatlantic Company that delivered two large vessels into P&O's hands; and second, persuading the Admiralty that vessels of at least 400-hp, of which none were currently available to rival bidders, were necessary for the service to Alexandria. The Company continued to deploy similar tactics east of Suez, placing orders for the *Hindostan* in advance of a full mail contract, but with the annual grant secured from the EIC for the Calcutta line. Combined with Whig patronage from Parry and others, the strategy looked unbeatable. But the clever negotiating manoeuvres, together with an aura of secrecy surrounding the funding of the steamers, generated a plethora of critics alleging morally and financially questionable practices (Chapter 13).

[50] *RM* (1843: 1072–5) (correspondence between EISNC and P&O respecting the union); Harcourt (2006: 55–9); Rabson and O'Donoghue (1988: 31); Napier (1904: 148–9).

[51] P&O Minutes (10 February and 24 March 1841); Harcourt (2006: 80–2) (EIC).

13 'So Great a Cloud of Obloquy and Mistrust'

Locking and Unlocking the Secrets of a Maritime Empire

> The [P&O] Directors for some reason appear very unwilling that we should know but as little as possible of their secrets. Why this is so, we cannot tell, as we have lately made the most of their secrets, and as far as in us lies, have taken care that they shall be troubled with any such unpleasant things no longer than we can help; simply because, in a Public Company we do think that all matters, we mean all honest matters, gain by being widely known.
>
> *The Railway Magazine challenges the perceived culture of secrecy surrounding P&O's management (1843)*[1]

Summary

The control of P&O over its nascent and vulnerable Victorian maritime empire depended heavily upon the management of recalcitrant external elements. In our period, for example, the system relied on the goodwill and cooperation of Egypt's ruler for the orderly transit of passengers and mail between Alexandria and Suez. Just as challenging, however, were the criticisms and complaints appearing in the British press, notably those flowing from disaffected proprietors leaking confidential and controversial information to Herapath's *Railway Magazine*. Contesting the Company's crafted narratives of onward and upward progress, this counter-narrative threatened the carefully choreographed performance of P&O and thereby threatened to undermine public confidence in both the monetary and moral values of the Company.

13.1 'Puff! – Puff! – Puff!' Challenging the CDSP's and P&O's Credibility

'Mr EDITOR, – A paragraph is going the round of the papers', wrote the pseudonymous 'Veritas' of Water Street in Liverpool in a well-informed letter published in the *RM*, 'stating that the *Hindostan* Steamer saved 210 tons of coal on her voyage from England to Calcutta, by using the patent apparatus of C. Wye Williams, Esq., for consuming smoke'. The claim, 'Veritas' observed, was spurious. Since the steamer had never made a voyage without the apparatus, 'we should like to know how the amount of saving was ascertained'. Carrying the

[1] *RM* (1843: 687). Herapath's style, as here, is often of an ironic character.

subheading 'Puff! – Puff! – Puff!' his letter contained a further sharp criticism of P&O's first purpose-built steamer for the Suez to Calcutta mail service. 'From the unwieldy size of the *Hindostan*', he reported, 'she took three days to get from Calcutta to the Land's Head' at the mouth of the Hooghly River in the Bay of Bengal. Thus 'people begin to doubt after all, whether these fine vessels are suitable for the trade, as there is no [dry]dock in India that can take them ... if they touch the ground'.[2]

Frequently attributed to pseudonymous correspondents and often merging with contemporaneous criticisms of RMSP's practices (Chapter 8), the intensity of the *RM*'s sustained attacks on P&O reached a crescendo during 1843. In many cases the letters were probably written by disaffected proprietors, but they contained more than expressions of personal bitterness or a broader resentment over the power of joint-stock companies, especially those protected by Royal Charter. Rather, they often conveyed to the reader a level of detail and an insider's knowledge that gave the correspondents and their claims a dangerous credibility, especially when those claims were subsequently woven into the Editor's satirical broadsides.

'Veritas's' reference to the savings claimed for Williams' patent apparatus preceded further sarcasm. Were such savings made, he asserted, it would be 'very disgraceful that not one of the City of Dublin's splendid fleet of Steam Vessels, of which Mr Williams is managing director, has the apparatus attached to her boilers'. Either this was due to the modesty of the inventor, 'Veritas' suggested or it was 'another illustration of the old proverb, namely, that a prophet has no honour in his own country'. The latter, in plain language, meant that 'where the inventions of C. Wye Williams, Esq., are best known, they are least thought of'. Even after Williams himself insisted that some thirteen of his City of Dublin steamers had indeed been thus fitted, 'Veritas' retorted that 'to the naked eye, the quantity, and quality, of the smoke from their funnels, seem much about the same now as ever'![3]

Nor was 'Veritas' a lone voice. 'A Constant Reader and Subscriber', writing from the CDSP's base at Liverpool's Clarence Dock, told of witnessing the apparatus being taken out of David MacIver's paddle-steamer *Admiral* as 'they could not get sufficient steam, and they made as much smoke as ever'. The correspondent accepted that some of the CDSP's 'fleet of old rattle-trap steam-boats' had got Williams' 'cold-air patent in them' and were 'the greatest nuisance we have about the dock'. On the other hand, he believed that Joseph Williams' 'hot-air patent', where fitted to the St George Company steamers, allowed 'plenty of steam, no smoke, and a great saving of fuel'. As a result, the

[2] *RM* (1843: 393).
[3] *RM* (1843: 393, 571).

respective inventors of the two systems were known, with a touch of sarcasm, as 'cold air Williams' and 'hot air Williams'.[4]

The *Railway Magazine* correspondents had other, broader questions to ask concerning the trustworthiness of Williams and his associates. Early in January 1843 'Argus' (from Liverpool) explained that some five companies (P&O, City of Dublin, British and Irish, Dublin and Glasgow, and Liverpool and Dublin Steam Shipbuilding) 'are connected in a peculiar manner by some [three] of the Directors serving on all the Boards'. Some five-sixths of the large capital (estimated at £1.3 to £1.4 million), he claimed, was Irish. Indeed, all three directors (Carleton, Williams and Hartley) common to the six companies were Dubliners. When, for example, the St George Company of Cork attempted to break into the Dublin to Liverpool route, the Dublin group made common cause against the interloper by using its British and Irish member to oppose the St George Company on its London service.[5]

Most striking was the purpose of the 'Ship Building Company' which was not, as its title suggested, to construct ships but rather to raise capital to build ships, especially when other constituent companies had no capital to spare or had legal constraints on the amount of capital they could raise. According to 'Argus', shareholders in the Ship Building Company received interest of 6 per cent per annum from the vessels running under the colours of the other companies, but these ships were 'in reality the property of, or pledged to, this Company'. 'Argus' identified this 'ingenious plan for increasing the number of vessels nominally belonging to a Company that has exhausted its capital' as the 'invention of a retired barrister [Williams], who for the last twenty years has taken a great interest in steam navigation'. The system, he observed, had its advantages and its perils. Thus the CDSP had an iron steamer 'of immense power' under construction in Liverpool. To be placed nominally under its colours, the shipping line would 'get the credit with the public of increasing their fleet, and of being in a prosperous condition'. But such was no more than puff. Behind the public facade, the CDSP was 'far from being in a condition to lay out £35,000 in a first-class iron steamer of 360 horse-power'. Indeed, it was known to have had to mortgage its property in the Liverpool docks. Drawing attention to a six-month slide in the CDSP shares from £105 to £82, 'Argus' warned that P&O shares (currently at £50), and closely connected with the other companies under Williams', Hartley's and Carleton's control, could come to be offered at a discount should the Dublin company continue to lose ground.[6]

[4] *RM* (1843: 762–3); Duckworth and Langmuir (1977: 189) (Admiral).

[5] *RM* (1843: 6).

[6] *RM*(1843: 6). The iron steamer under construction in Liverpool was most likely the 629-ton paddle-steamer *Iron Duke* built by Thomas Wilson. See McRonald (2005–07: vol. 1, 195).

A vigorous attack on the CDSP's character appeared in the *RM* under the name 'Solon'. Taking as his text 'by their fruits ye shall know them', 'Solon' offered a contrast between three Glasgow steam packet companies and the Dublin one, all of whom ran into Liverpool. The former three were all private partnerships. None were 'chartered, incorporated, have government contracts, or any exclusive privilege; every partner is liable for the whole debts of the concern'. None of their vessels, two iron and four wooden-hulled, were more than four years old. No accident to life or property had ever occurred to them. As a result of the economy exercised by the companies, he claimed, freight rates were half the Dublin rates even though the distance steamed from Glasgow was nearly double that from Dublin. By replacing vessels over three or four years of age with vessels 'of the most improved form and power, they have paid and do pay better than any steamers out of Liverpool, and no such vessels in appearance, safety, speed and comfort, are to be found in the coasting trade of any part of the kingdom'.

In contrast, the CDSP 'reigns supreme' with its monopoly in the Dublin trade for some twenty years. Incorporated by Act of Parliament, it had £450,000 paid up capital of which £200,000 had limited liability. With the remaining £250,000 thoroughly mixed up with the £200,000, shareholders 'generally imagine that they have the non-liability clause for the whole'. All this property was backed up by the Liverpool and Dublin Steam Ship Building Company with capital of £150,000. Yet, he stated, none of the seagoing steamers was of iron, their average age was twelve years and, instead of selling off the older vessels, the Company 'have gone on laying out their money in patching up their old craft, lengthening some by the bow, others amidships, putting new engines into old vessels, building half a new vessel round an old engine, and boasting that they were keeping them in efficient order'. Moreover, shareholders would eventually find that the Company had been paying dividends out of the principal.[7]

13.2 'Like Gold on a Lawyer's Desk': P&O's Practices Under Scrutiny

Given these insights into the character of the City of Dublin firm, it is scarcely surprising that critics asked simultaneous questions about the stability of P&O, itself closely intertwined with the Williams system of finance. The *Railway Magazine* correspondents increasingly probed the financial foundations of P&O, particularly those arrangements between Peninsular Steam and Transatlantic companies. The usually positive 'Fair Play', for example, expressed surprise at the premiums paid 'for liberty to share in such property as the *Great Liverpool*', Transatlantic's former 1,050-ton *Liverpool*

[7] *RM* (1843: 156).

(Figure 13.1). This steamer had been purchased from Sir John Tobin for £44,000 in 1838 at a time 'when the world went mad about transatlantic steam navigation' (Chapter 4). Now transformed with 'false sides put to her to increase her beam from 31 to 37 feet ... [she] came out as the *Great Liverpool* of 1540 tons' and was valued at some £65,000! 'Fair Play' thus voiced, with a hint of irony, 'the highest opinion of the ability displayed in getting the Transatlantic Company out of the serious scrape they were in, not only without loss, but with a bonus of 30% on their property'. In less measured tones, 'A. B.' responded by highlighting the profit accrued to Transatlantic's directors for a vessel 'which never was, and never will be, worth half the money paid'. On the other hand, 'A Shareholder' in Transatlantic claimed that the directors received 1,000 paid-up P&O shares worth £50 each for the *Liverpool* in 1840. He also alleged that the same directors never called a shareholder meeting and kept the money (in the form of shares) to themselves.[8]

'A. B.' had already called upon readers to judge 'if the Company can ever pay a dividend of 7%' and followed up his challenge with a detailed analysis of estimated annual running costs over the three areas of current P&O operations: the Peninsula, the Mediterranean and the Nile. As with McQueen's and Wheelwright's projections of RMSP and PSNC, 'A. B.' focused on coal consumption, wages and a range of lesser costs including port, pilotage and agency

Figure 13.1. P&O's troublesome *Great Liverpool*, formerly *Liverpool* (1838), labouring heavily in dark stormy seas

[8] *RM* (1843: 6–7, 42).

charges. Coal accounted for half the total costs in the case of the Peninsula service and almost three-quarters in the case of the two Alexandria steamers. He had, he admitted, been induced to undertake this inquiry because 'the more I look into the affairs of the Company, the more I am convinced all is not as it should be'. His aim, he insisted, was to hope that shareholders would 'not in future allow themselves to be laughed at for being so credulous'. And he left it to readers to point out any unfairness in the assumptions he introduced.[9]

'A shareholder' did soon complain to the Editor that 'A. B.' had greatly overestimated both the coal consumption and price in his calculations of the line's viability. He cited various steamers that had undertaken relatively long-distance voyages (HMS *Cyclops*, *Royal William*, *Great Western* and even *Great Liverpool*) at a consumption of between six and seven pounds of coal per horse-power per hour. The shareholder demanded that 'A. B.' 'show some authority for his different statements', especially as he was writing anonymously. "A. B"', he suggested, should 'show that he is neither a discarded servant of the Company he maligns, nor a disappointed candidate for their employment; and that he is not an intended purchaser of shares, when his criticisms shall have succeeded in bringing down the price'. In contrast, the correspondent held up the P&O directors, 'known to be honourable men', for their 'plain and straightforward statement as to the unencumbered [debt-free] status of P&O property'.[10]

Correspondents now began to redirect their attention to ownership of Transatlantic's former *United States*, eventually completed as P&O's Alexandria mail steamer *Oriental*. A correspondent calling himself 'Steam', noted that previous contributors had stated that £74,800 had been paid in scrip by P&O to the Transatlantic for the vessel. Yet other correspondents had claimed that the Ship Building Company, having raised the cash to build the steamer, had been holding the *Oriental* as security. 'Steam' therefore wanted to know whether the Building Company still had a claim on the ship. He surmised that either 'the Building Company must still have a lien on the vessel' or it must have accepted the amount of scrip paid to Transatlantic. All this mattered, he argued, because P&O proprietors had just paid a very large sum for the steamer over which they presumed ownership. Were the Building Company still to have a lien then it would also receive a large sum for interest, drawn from P&O's earnings, on the capital thus employed. Shareholders would accordingly have to expect a lower dividend.[11]

[9] *RM* (1843: 42, 120–1).
[10] *RM* (1843: 187).
[11] *RM* (1843: 42). Harcourt (2006: 66n) notes that in 1842 P&O mortgaged the *Oriental* to the Royal Exchange Assurance Company for three years, but this was not recorded on the Certificate of Registry.

'Argus's' investigations at the custom house soon revealed that the *Oriental* was indeed registered in traditional sixty-fourth shares in the ownership of Williams (twenty-two sixty-fourths), Ewart (twenty-one sixty-fourths) and Carleton (twenty-one sixty-fourths). Williams had there declared that he and the two others were the sole owners of the vessel. The Register made no mention whatever of the name of P&O. The Secretary, James Allan, responded to the growing criticism with a brief letter stating that the 'Directors do not intend to enter into any controversy with anonymous writers'. He also conveyed the Board's view that the claims 'are composed chiefly of either exaggerations and distortions of facts, or of assertions entirely devoid of truth'. In his response, 'Argus' was scathing about the lack of P&O's openness. 'The Secretary ... is quite right in declining to enter into any controversy', he observed sardonically. 'It only improves a good *bona fide* concern to have public attention drawn to it; anything of a questionable nature is much better without it'. He denounced the directors' assertive denials with the comment that if the claims were 'worth noticing at all, they were worth answering' and he further questioned the Secretary's offer to proprietors (and not representatives of wider interested public such as the press) to call on P&O's managing committee: '[i]f the concern is a good one, the more that is known of it the more the shares will rise; if the contrary, confidence is shaken, and shares fall'.[12]

In April 1843 the *RM* reported facetiously that on the first of that month P&O shareholders had been requested by a circular bearing the Dublin postmark to meet the directors at the Custom House. There they were 'to witness the transfer of the *Oriental* to the rightful owner'. Nineteen gentlemen, 'including seven half-pay officers of H.E.I.C., five retired civil servants, and seven independent men with more money than wit ... attended punctually'. After waiting an hour and a half in vain, they adjourned to P&O's London office where the managing directors called their attention to the day of the month but assured them that 'trifling objections' would be 'speedily settled by application to the Board of Trade, the Admiralty, and the Treasury, all which departments had shown a decided preference to their interests over that of the public'. Whether 'their interests' referred to shareholders or to the managing directors was left to the readers to decide.[13]

A couple of months after his previous letter, 'Argus' reported to the *RM* that still no reply had been forthcoming from the Secretary on the specific matter of the *Oriental*'s true ownership. 'I take it for granted that this splendid steam-ship, that figures in the advertisements of the Company as their own', he inferred, 'is in reality the property of the gentlemen above named [Williams, Ewart and Carleton], who may sell her to the Grand Turk or Mehemed Ali [sic]

[12] *RM* (1843: 58, 147, 308–9).
[13] *RM* (1843: 400–1).

tomorrow, if they get a tempting offer'. He also noted that questions had been asked about payment for the engines of the new *Hindostan* which, he suggested, may resemble the case of the *Oriental*: 'Like gold on a lawyers' desk, guineas set for show, like nest eggs to make clients lay, and will turn out ... to be held by 'FRIENDS' of the concern'.[14]

The Editor himself delivered a stinging critique of P&O's business practices on 3 June 1843. Reminding *RM* readers that early the previous year several shareholders, 'severely affected by the depressed price of the shares, had communicated their alarm at public rumours as to the stability of the Company'. The *Magazine*'s reporter had been declined admission to the half-yearly meeting in May 1842, but the Company had supplied a copy of the directors' report which was duly published in the paper. Prior to P&O's second annual meeting in November 1842, the *RM*'s leading article set out a number of the rumours in circulation in order to give the directors the opportunity either to contradict or explain the allegations. The article 'showed that at that period there was a want of confidence in the management or stability of the concern, or both, very evident from the shares being at a discount, while the concern is paying 7 per cent'. The high dividend, he inferred, pointed to a price of £70 rather than the current £47 on shares of £50. With an *RM* reporter present at the meeting, a proprietor (Mr Potts) read out the leading article and the chairman (Sir John Campbell) replied to the rumours with 'the fullest information'.[15]

One hour before the start of P&O's half-yearly meeting in 1843 however, a letter from the P&O Secretary communicated the directors' wish to exclude the *RM*'s reporter. 'We believe there is no Railway Company with a capital of five millions which would dare so far to brave their distant Shareholders, and jeopardise public confidence in their property, as to exclude any member of the press from their meetings', the Editor wrote in his 3 June edition, 'neither do we believe that the Peninsular and Oriental Company can do so with impunity'. What the directors had achieved, he continued, was to teach proprietors that there was indeed 'something rotten at the core which will doubtless lead the shareholders to suspect ... the goodness of their property, or the honesty of their Board'. The directors, in short, had exposed themselves and their concern to 'so great a cloud of obloquy [public disgrace] and distrust'.[16]

The following week the Editor resumed his critique with renewed ardour. He accused the Company of attempting, by their action of excluding the paper, to secure 'that secrecy which they knew, by experience, they could neither bribe nor menace us to'. He asserted that P&O did not see that 'they are in

14 *RM* (1843: 547).

15 *RM* (1842: 534 (half-yearly report), 1215 (leading article), 1229 (second annual report), 1231 (replies to rumours)); *RM* (1843: 566–7) (Editor's commentary).

16 *RM* (1843: 567).

part public property, and that the public – we mean the general public – have a right, therefore to know how they are going on, and what security there is for the performance of that duty they are paid for doing'. Some £63,000 per annum of public money went to the Company 'for executing certain duties', that is, carrying the mail. Consequently, he argued, 'they who pay them thus liberally have a right to know, and ought to be informed, not only whether they are physically able to do their work ... but whether by mismanagement, bad conduct, or neglect, they have jeopardised their competence in their relations with other parties'.

As a prime example of incompetence combined with concealment, he cited as a matter of serious interest to Britain and to its Indian territory the claim that P&O had 'lost the power of traversing the isthmus of Suez'. Now, he asserted, 'the transit across that indispensable link in the chain of communication is in the entire possession, and at the mercy, of a foreign potentate'. While P&O's directors disingenuously insisted to the proprietors that they 'are on very friendly terms with the Pasha', Company servants had recently been turned 'neck and heels out of the Isthmus'. The Editor thus concluded that although the Pasha was 'a man of whose honour every one speaks well', Egypt's ruler might at will 'sever the link he holds in twain, and leave us, through the bad illiberal policy of the Peninsular and Oriental Company, with the useless head and tail of that communication'.

Such concealment, the Editor insisted, 'is treason against the public, and unjust to their own shareholders'. But it was only one instance of 'that uncandid system' similarly practised in the matter of the Company's dividends and vessel ownership. As shown by able letters of 'A. B.' and others, the directors had been 'cramming their Shareholders with a dividend unwarranted by the state of their finances, but misleading them with the idea of their being possessed with a property which does not belong to them'. Proprietors and public alike would eventually discover that the game had been kept up both by vessels had 'on tick', and by new capital 'every now and then raised' so that dividends continued to be paid at unrealistic levels. Thus, for the moment 'the Company present[ed] to the public the face of a flourishing concern, while in reality they may be in a state of Insolvency'.[17]

By mid-June 1843 the Editor raised the stakes still higher in the *RM*'s confrontation with P&O. '[I]t may be as well just to glance at this very pretty Whig job, and the manner in which the Government enabled a few clever fellows to bamboozle the public', he suggested with an echo of the patronage theme that ran through the period of mail contracts awarded by the Melbourne Government to Cunard, RMSP and P&O. At the time, he noted, two or three individuals (Willcox, Anderson and Bourne) 'possessed four patriarchal tubs

[17] *RM* (1843: 590–1).

[making up the Peninsular fleet] ... on which they set a price, not what they would have sold for in the market, but what they were pleased to say they were worth, namely £108,000'. To make the deal easier to swallow, they reduced the value to £81,000 'which, then, perhaps, was nearly, or quite, what others would call their full value'. To this property they added the two former Transatlantic vessels at £139,800 along with old machinery at Greenock valued at £17,200.

The property, the *RM* stated, now stood at £238,000 (worth not more than £150–160,000 in the market). With a mail contract and a charter granting limited liability, they added another £66,000, making a grand total of £304,000. Since these 'worthy sellers of old ships and old machinery' chose to take shares, the effect was thus to raise the value from, say, £160,000 to £304,000, which put '£144,000 into the pockets of these dextrous old ship and old marine store Proprietors'. In the Editor's view, a 'more gross job or barefaced delusion was, perhaps, never perpetrated on the public'. Worst of all, he insisted, was 'the trick upon the public of limiting the liability by the charter', for correspondents had already cast doubt on its validity were it to be shown that the capital was not *bona fide*, subscribed, paid for and unencumbered.[18]

In sardonic vein, 'Looker on' purported to reveal 'the great secret of dividend paying Steam Navigation Companies, which is simply to be constantly calling up fresh capital, or borrowing money, or getting credit if there is not'. Thus the CDSP was regularly paying dividends 'while fresh capital was [still] paying in by the poor devils, who had the last issue of shares, and how quickly they stopped [the dividends] when they had nothing further to call upon or property to pledge'. Worse still was the insolvent St George Company which 'regularly paid their 8 per cent per annum, while their Proprietors were paying up, and their credit was good; and mark how it turns out now that the dividends have all been paid up out of capital from the first'.

He alleged that P&O were no different in simultaneously calling up capital and paying dividends. And the Steam Ship Building Company, following closely on payment of a dividend, were now calling on their proprietors for a payment of £5 per share on those issued only two months previously. 'Looker on' concluded by recommending that RMSP managers, with 40 per cent still to call up from proprietors, follow the example set by these Irish directors. The RMSP would then be able to declare the much-anticipated but elusive dividend. He facetiously accepted that such 'old-fashioned men as Messrs Baring, Irving, and Colvile may object, but that would only prove what is shrewdly suspected that they are not fit to manage Steam Navigation Companies' (Chapter 8).[19]

[18] *RM* (1843: 615, 639–40, 687) (Editor); *RM* (1843: 6–7) ('Fair Play's' initial doubts on the validity of the Royal Charter).
[19] *RM* (1843: 763).

In late July 1843 the *RM* published a one page satirical report on P&O's half-yearly shareholders' meeting. It purported to reconstruct the meeting 'as it ought to have been'. Putting words into the mouths of real life P&O directors and proprietors, the fictitious report even included RMSP's Captain Chappell, Andrew Colvile and John Irving among the speakers in order to enhance the ironic power of the piece. At its heart, the attack was directed against P&O's persistent exclusion of the *RM*'s reporter from such meetings and the concomitant lack of openness concerning the Company's financial and material condition.

The satire began with a tongue-in-cheek statement that the following was 'a sketch of what we were informed took place during our exclusion'. A proprietor (Mr Potts) rose to ask if the press were invited to the meeting. '[A]ll the daily papers had been invited', the Chairman (Sir John Campbell, KCH) responded evasively. 'Is there any reporter from *Herapath's Journal* (hear, hear)?' asked the proprietor. 'Of course not', Sir John retorted. The proprietor expressed incomprehension since on a previous occasion 'that journal was the only one which gave a full and accurate account of the proceedings'. He stated his intention to move that the paper be admitted.[20]

In this fictitious but targeted account, the P&O Board now closed ranks to kill off Potts' move. Acting true to form, one unnamed director urged resistance to the motion on the grounds that 'it would show a want of confidence in the Directors – (laughter)'. Mr J. C. Ewart, also a director, assured the meeting that the Company's affairs 'were in the most satisfactory condition possible'. Ignoring a proprietor's intervention asking about the current price of the shares, Ewart continued with unconscious irony concerning the establishment press. 'The Directors had shown every disposition to court publicity'; he proclaimed, '[and] they had invited the attendance of a reporter from the *Times*, and they knew that the Directors could not shut the mouths of any one belonging to that establishment'. Francis Carleton, director and one of the managers, accused the *Railway Magazine* of more widely 'creating broils between Directors and Shareholders, and in creating a feeling of want of confidence' over an eight-year period. Indeed, he concluded, 'It was a nuisance, and every Director in the country ought to combine to put it down'. A more recent director, Sir John Pirie from the EISNC stable, suggested that 'for all he knew to the contrary, the paper might be a very reputable paper', thereby inadvertently confessing to his sheltered existence. He strenuously supported 'publicity on all occasions' on account of his 'benevolent motives' – a point that reflected EISNC's disingenuous claim not to be motivated by profit (Chapter 12).[21]

[20] *RM* (1843: 754).
[21] *RM* (1843).

Inserted into the meeting were the comments of RMSP's leading lights. Captain Chappell (Secretary), seemingly oblivious to his own reputation for jobbing, agreed with a proprietor that the journal had 'done much to expose jobbing in Steam Navigation Companies'. He also unwittingly referred the reader to another of his reputed characteristics when he declared that the publication had 'done more than any other paper to repress the insolence of officials'. Colvile (Deputy Chairman) also appeared to support the admission of the *RM's* reporter. 'What had they to complain of?' he asked P&O's Board, doubtless thinking that, by comparison with RMSP, they had no reason to feel aggrieved at the *RM*'s criticisms. John Irving (Chairman) was even more explicit. It was his fate, he acknowledged with sadness, 'to represent a concern [RMSP] that was not in a sounder condition than their own [P&O]'. He weakly chastised the journal for 'perhaps [having] been too strong in their language and too harsh in the construction they had put upon his own conduct and that of his brother Directors'. But he felt bound to give the paper 'full credit for the extreme accuracy and impartiality of their reports'. The reporters had 'done their duty fearlessly, manfully, and candidly'. And they had opened their 'columns to any defence of the Directors, as well as any attacks upon them'. His words here simultaneously deepened the satire upon RMSP *and* used it as a foil to vindicate the truth of the *RM*'s reports over the sort of 'puffing' that characterized the establishment press.

As the meeting drew to a close, Mr Potts pressed the motion to readmit the journal's reporter. Every member of the Board opposed the motion, while all the proprietors voted in favour. Unconvinced, the Chairman made the parties 'hold up their hands ... nine times' before accepting the 'democratic' verdict. He then read the half-yearly report 'in a very low and depressed tone of voice, so as to be scarcely audible'. A proprietor, who confessed that he had never before attended a meeting, was not a man of business and did not understand figures, rose to propose the adoption of the report. Praising the 'very satisfactory and lucid report', he expressed his 'unbounded confidence in the Directors' while also stating that 'thanks ought to be given to them for the care they had taken of themselves – (a laugh)'. A fictitious seconder, 'Mr Suckeggs', similarly professed his 'great confidence in the Directors' on the grounds that he 'had a brother in their employ [who] had been treated very liberally'. He further urged that 'the salaries of all the Directors and officials should be doubled'. The meeting concluded after Mr Potts read extracts from a fictitious edition of *Herapath's Journal*. The faces of certain directors were observed to present a very curious appearance by changing all manner of colours. The Secretary appeared several times 'to wipe foam from the mouth of the Chairman'. When the Board expressed themselves 'unable and unwilling to give any satisfactory account of the state of the Company, a vote of want of confidence was passed by acclamation, and the Directors were called upon immediately to resign'.[22]

This satirical piece reflected most of the critical questions that correspondents, as well as editorial items, had levelled at the new P&O Company over the preceding months. In doing so, these questions attempted, with varying degrees of success, to lay bare the secrets of P&O's rapid, seemingly unstoppable, rise to imperial prominence. But the *RM* was about to release a far more explosive set of allegations that, if allowed to set off a chain reaction, threatened to shake the moral and financial credibility of the Company to its very foundations.

13.3 '[M]any False and Scandalous Statements': The Slavery Case

'[W]e must not forget', the *RM* urged its readers in mid-September 1843 concerning news that a member of the P&O Board was to appear in court on a serious but as yet unnamed criminal charge, 'that Mr {Pedro Jose] Zulueta, who is to be tried for felony, is the son-in-law of Mr Willcox, one of the three managing Directors'. The journal considered it almost certain that the accused would 'get off, either through a flaw in the indictment, or witnesses not coming forward, or some defect in the evidence'. Had the case been that of an Englishman residing in Spain accused of such an outrage against the laws of that country, however, the paper was certain that he 'would, upon much worse evidence, have been convicted and shot for it in little more time than we have been writing this'.[23]

Just three days after the publication of this report, the *Railway Magazine* received a letter from P&O's attorneys, De Mole and Browning. 'Your weekly publications for some time having teemed with unfounded imputations and anonymous writings, conveying to the public many false and scandalous statements, regarding the affairs and management of the Peninsular and Oriental Steam Navigation Company, and a most slanderous attack, wholly devoid of truth, upon its solvency and credit'. Their letter, however, made no mention whatever of the Zulueta case. Instead, it concerned a claim by 'Veritas' published in a very short letter early in July. 'Veritas' had simply stated as a matter of information for readers and for 'those interested in the Company' that P&O had paid for the *Hindostan*'s engines by three bills, the first of which when due had been 'returned and renewed by the drawer'. The P&O minutes show that the Company construed this as a libel against it, most likely because the subtext of 'Veritas's' letter suggested that P&O lacked creditworthiness. Now the attorneys demanded that the journal communicate to them the identity of

[22] *RM* (1843).
[23] *RM* (1843: 969).

the author and publish a contradiction to the allegation. They threatened that, unless the *RM* complied, 'proceedings would be taken against Mr Herapath'.[24]

The Editor duly issued a carefully worded statement that barely acknowledged the attorneys' demand for 'correction of the error'. He instead emphasized, with a hint of irony, his wish that 'our correspondents would be more careful in their communications, and not state things the truth of which may be questioned'. And he furthermore asserted the right of the press to watch closely and comment freely upon the proceedings of a public company 'receiving large sums annually from the public purse'. He also took steps to identify 'Veritas'.[25]

What became clear in the exchanges was Herapath's conviction that the desire to identify 'Veritas' was merely P&O's pretext to gag the journal over an intensifying series of criticisms that culminated in the Zulueta crisis. He thus explained that the Company's 'bile has been excited, not by that harmless letter [of 'Veritas'], but by our notice of Zulueta's case, and by our compelling them to register the *Oriental* and restore that ship to the Shareholders'. The P&O minutes show that the Company was planning the legal challenge as early as July when the 'Veritas' letter first appeared, that is, some time before the *Magazine* published its first report of the Zulueta case. Yet the despatch of the attorneys' first letter just three days after the journal had reported the connection between Zulueta and Willcox looked scarcely coincidental. But the case against the journal over 'Veritas' was undoubtedly an easier matter for Company lawyers to handle than the potentially incendiary matter of a P&O director heading for trial at the Central Criminal Court of the Old Bailey on the charge of 'being concerned in the infamously inhuman traffic of slave dealing'.[26]

The *RM* now unleashed a full onslaught on the consequences of slavery for Britain. Twenty million pounds, the Editor claimed, had been paid to the West Indies as compensation for the abolition of slavery and more than 22 million for its prevention on the African coast to the end of 1838. That total of £45 million towards suppressing 'this diabolical traffic' represented around one-fifteenth of the national debt. Were the charges against Zulueta proven, he asked, 'could we then quietly sit down under such a cost, and see our laws openly violated by men who are living in the heart of our capital, and in almost daily communication with the heads of our Government, and by one, if not more, of the Directors of a public company, carrying a very great portion of the correspondence of our country, and now straining every nerve to be entrusted

[24] *RM* (1843: 1074) (P&O's solicitors); *RM* (1843: 1069–70) (the legal challenge); *RM* (1843: 691) (engine payment); P&O Minutes (22 August and 12 September 1843).

[25] *RM* (1843: 996) ('correction' of 'Veritas').

[26] *RM* (1843: 1069–70) (Editor); P&O Minutes (11 July, 22 August and 12 September 1843). See also [Anon.] (1844) (Zulueta trial). The defendant had been arrested on 23 August and was tried between the 27th and 30th of October 1843.

with more – the whole of our Indian correspondence? We think not'. But such a verdict, he concluded, was not likely. 'Governments too often wink at offences committed by the powerful and wealthy', he argued. 'In their cases the forms of justice are, unfortunately, scarcely any thing but forms'. Thus the trade had been carried on for years 'and large sums made by it'. According to common reports, Zulueta bought vessels through others who fitted them out in return for a commission. The ships were then sent out under an English crew and commander to a known individual at Cadiz. Under a Spanish crew, they then headed for Africa.[27]

After a three-day trial, Zulueta was, as Herapath predicted, acquitted by the jury in the face of evidence that, to the 'most able and impartial' judge, seemed to point to a guilty verdict. First, 'he found money for the purchase of a certain slaver; that he fitted her out under the name of "*Augusta*", as agent to one Martinez and Co., slave-dealers, and despatched her to the Gallinas, a place on the coast of Africa, proved to deal in nothing but slaves'. Second, 'she had the forbidden articles, shackles, certain water tanks, and shifting decks'. Much of this evidence had apparently been taken from Zulueta's own statements before a House of Commons Committee. In sole defence 'was the character of the prisoner for honour, straightforwardness, humanity, and the high respectability of his connexions, and the improbability therefore that he would engage in such a trade'. Herapath's own verdict arising from the case was simply that if 'rich men are in felonious cases to substitute the testimonies of their friends to general character, in lieu of giving direct evidence of their innocence ... why, then, there is an end to all justice as far as wealthy men are concerned. ... They may carry on the nefarious traffic of slave-dealing – which none but rich men do – laugh at our Acts of Parliament with impunity, and set our attempts to defy them at defiance'.[28]

In November 1843 Herapath issued a follow-up report on the Zulueta case based on a recent article in the *Patriot*. The defendant had protested his innocence to the House Committee on the grounds that the vessel had had to put into Cadiz owing to the stress of weather and that to have done so under other than such perils of the sea would have invalidated the Lloyds insurance policy that covered only a direct voyage from Liverpool to the Gallinas. He had therefore claimed that the Lloyds agent (who was also the British consul) at Cadiz was well aware of the circumstances and had to sanction the claim for weather damage upon the underwriters which they duly paid. All this was tendered towards proof that the Cadiz call was 'previously unknown to him (Zulueta), and therefore to be taken as presumptive evidence of his innocence of the infamous object of the voyage'. Now, however, the insurance broker

[27] *RM* (1843: 1070) (Editor).
[28] *RM* (1843:1144).

claimed before the witness to the prosecution lawyer that 'the LOSS WAS NEVER PAID NOR CLAIMED'!! and consequently that Zulueta's statement to the Committee was 'a falsehood deliberately and artfully imposed on the Committee'. To add to P&O's woes, Herapath referred to a highly contentious report in the *Morning Chronicle* alleging that Messrs Anderson and Willcox had been connected with the sale of another vessel 'to a well-known slave dealer'.[29]

A mere three days after publication of this piece, Herapath's solicitors received a legal declaration of proceedings against the *RM* in which the P&O shareholders were made the plaintiffs. The journal was declared 'to have injured them in their character, credit, and trade'. Herapath responded by publishing an open letter to the P&O proprietors in which he not only repudiated the charge of injuring their interests, but sought to demonstrate in detail that the *RM*'s whole aim had been 'to open your eyes and to do you good'. Recommending that the shareholders appoint a committee of inquiry to examine the truth or falsity of Herapath's claims, he urged that such a committee investigate some eleven matters, all of which had been raised by correspondents over the previous year or so. These included: how the *Oriental* had been financed; whether the transactions thereof had not vitiated the charter; whether the dividends had been paid out of profits or capital; the reasons behind the low share price and high dividends; the discrepancy between original capital price and actual value of the assets; the method by which the *Hindostan*'s engines were paid for; the earnings of the three managing directors and their retirement pensions; and the alleged jobbing whereby Company servants 'are related or connected by some sort of tie' to directors.[30]

Herapath further published a personal letter in the *RM* arguing that, contrary to the charges laid against the journal by Francis Carleton at the recent P&O general meeting, there was no quarrel between the Company and the paper. While declaring that 'the object of the prosecution ... [is] a vindictive determination to destroy this journal', he maintained a relatively conciliatory argument. He thus expressed his personal belief that 'the whole of this [legal action] is more or less the work of others, whose interests are not what may be for the solid benefit of the Company'. Although he did not name these 'others', he almost certainly adhered to his conviction that the Zulueta case was the real driving force behind the whole prosecution. He thus explicitly exempted Carleton, Ewart and some other unnamed colleagues who 'should not so easily have suffered themselves to be led into a contest' with, whatever the result, consequences likely to be detrimental to the Company and shareholders. The

[29] *RM* (1843:1195). See also *RM* (1843:1275–6) (strong support for Herapath's moral stand against the slave trade).
[30] *RM* (1843: 1222–3, 1273–4) (Editor).

P&O Board minutes three days later referred to a 'proposition made on behalf of Mr Herapath about coming to some amicable understanding with a view to stop the action brought against Mr Herapath by the Company'. The minute, however, was erased after a week![31] The war, it seemed, was still on.

Closure of the case finally came in July 1844. A brief notice in the *RM* simply stated that it had been terminated by Mr Herapath 'submitting to a verdict against him, in consequence of the gross and cruel manner in which he was at first, and has been throughout, misled and deceived in this affair'.[32] The significant practical consequence was that criticisms of P&O from the end of 1843 were entirely, or almost entirely, absent from the correspondence columns of the journal, with reporting of P&O confined to largely factual matters and to verbatim accounts of the half-yearly and annual meetings. The Company had, it seemed, successfully gagged the *Magazine* and its editor, thus imposing a large, if heavy-handed control over the freedom of the press to question corporate activities.

'[W]ho will speak out when he sees such flagrant acts of iniquity perpetrated by Directors who, drest in a little "brief authority", would ... assist the poor man, because his skin is black, to be torn from his "native land", regardless of the sacred ties of consanguinity, and made to endure the tortures of

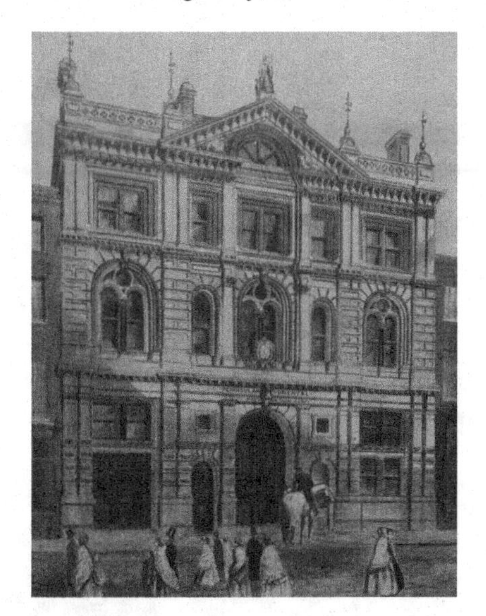

Figure 13.2. P&O's Leadenhall Street Head Office in the City of London in 1859

[31] *RM* (1843:1273–4); P&O minutes (12, 19 December 1843).
[32] *RM* (1844: 792).

a slave-ship', wrote the *RM*'s correspondent 'Pro bono publico' in praise of Herapath's stance. If 'Pro bono publico' was at all representative of the early Victorian public, John Herapath had touched an emotional, highly moral nerve with his unlocking of an unpalatable secret of the Peninsular & Oriental Steam Navigation Company. His exposé of the Zulueta affair had every prospect of spreading moral contamination to the very heart of the P&O project that aimed to enlist Government, Admiralty, City and public in one of the 'great and enlightened undertakings' of the age.[33]

Having aggressively appropriated from the hands of prospective rivals vast tracts of geographical territory for its aspiring maritime empire of steam navigation, P&O fought to regain command of a moral environment that, unchecked, threatened to undermine and even destroy the Company's still-fragile reputation, credit and trade. It thus not only faced the challenge of building, sustaining and remoulding this eastern mail steamship system in physical form, but also increasingly realized that its position as a servant of the British and British Empire's public carried with it severe moral penalties if it failed in its capacity to live up to the standards demanded by those who ultimately made possible its existence and its prosperity (Figure 13.2).

[33] *RM* (1843: 1275–6).

14 'A More Desirable Result in the Performance of the Vessel'

P&O's Mail Steamers in Action

> In consequence of its appearing that the speed of the [new iron paddle-steamers] *Haddington*, *Pottinger*, and *Ripon* not being equal to what was anticipated or being at least a knot slower than the [wooden paddle-steamer] *Oriental*, the Directors were desirous of ascertaining what could be done to secure a more desirable result in the performance of the vessel now building by Messrs Caird of Greenock. The under signed [managing director Francis Carleton and assistant manager James Allan] accordingly went to Greenock to see Messrs Caird & Co upon the subject ...
>
> *P&O's managers act to address the under-performance of the Company's newest steamers (1847)*[1]

Summary

In the 1840s, the wooden-hulled paddle-steamer *Oriental* set a standard of seagoing performance that P&O's newer, iron-hulled designs struggled to match. Conspicuously less subservient to the Admiralty than RMSP, the Company displayed a policy of design diversity contrasting with their peers' preference for uniformity in steamship systems. Embracing an experimental venture with the small iron-hulled *Pacha*, P&O chose iron for a quartet of mail steamers to serve both Mediterranean and Indian Ocean routes. Ordered from Thames and Mersey shipbuilders and engine-makers, these steamers often performed below expectations in terms of speed, stability and sea kindliness. In seeking solutions, P&O cultivated face-to-face connections with Clyde contractors, most notably Tod & MacGregor in Glasgow and Caird & Company in Greenock, both with growing reputations for reliable steam engines and strong iron-hulled steamers. The former delivered P&O's first large iron-screw steamers in the early 1850s, epitomized by the *Bengal* (1852). The performance of this class of vessel, however, also had its drawbacks, especially in the confined waters of the Hooghly River.

14.1 'She is the Best on the Water': P&O's *Oriental* Sets a Standard

After P&O's *Great Liverpool* failed an Admiralty survey, the newly commissioned *Oriental* inaugurated the Alexandria service in September

[1] P&O Minutes (16 March 1847).

1840. Constructed by Merseyside's Thomas Wilson, closely associated with Williams, the single-funnelled paddle-steamer was driven by side-lever engines built by Fawcett, Preston. While fitting out in Liverpool as the *United States* in the spring, the *Oriental*'s hull-form had drawn local admiration (Chapter 6). And the *RM* noted two years later that '[she] beat the [RMSP] *Avon*, in a run down the Channel, we believe about 160 miles, [by] no less than 9 hours'.[2]

The *Oriental* earned the praise of RMSP's Captain Mangles when he visited the ship at Southampton while preparing to depart with fifty passengers in 1844. Mangles' aim was that of 'witnessing her departure & of comparing the systems adopted by the two [mail] companies'. As a result, the RMSP's Secretary concluded that it was 'certainly no more than justice to the P&O Company to state that their proceedings seemed carried on in an extremely business like way & that their vessel left the Port in a very clean & orderly state' – a standard not always met by the West India steamers (Chapters 8 and 9). Passenger opinions of the *Oriental*, however, were sometimes rather less laudatory with complaints of disagreeable smells, inadequate access to cabins and overcrowded accommodation.[3]

After extensive rebuilding, *Great Liverpool* finally entered P&O service at the close of 1840, but remained burdened by the previous poor form. Writing in the *RM*, correspondent 'R. C.' characterized the ship as a vessel 'unfit for the purpose intended' and 'at variance with all proportion in naval architecture'. Three years later the steamer struck a reef south of Cape Finisterre while homeward bound and broke up a few days later on a lee shore as P&O's first significant loss.[4]

The contrasting performances of the *Great Liverpool* and the *Oriental* entered Merseyside religious culture early in 1845. At the reopening of Liverpool Collegiate Institute's lecture hall, the evangelical Church of Scotland minister John Todd Brown used his status as guest speaker to make 'some remarks upon the deluge and the history of the wondrous preservation of Noah and his family in the Ark'. In the chair was one of Burns' Tory friends, Ulster-born Revd Hugh McNeile. 'He is a truly militant divine', Liverpool's satirical magazine *The Porcupine* later observed of the Church of England evangelical.[5]

McNeile seized the occasion to press home a Biblical message. Thomas Wilson, 'one of our chief ship-builders', had been 'impressed with the fact of *mistake after mistake* having been made in the construction of steam-vessels – that

2 P&O Minutes (22 August 1840) (Admiralty rejection); Dawson (2009: 35–41); McRonald (2009: 27–32) (Fawcett, Preston); *RM* (1843: 996) (performance).

3 RMSP DM (2 April 1844); Artmonsky and Cox (2012: 103–4) (complaints); Howarth and Howarth (1986: 27) (accommodation plan); Rabson and O'Donoghue (1988: 16–17, 27).

4 *RM* (1843: 645–6); Rabson and O'Donoghue (1988: 28) (loss of *Great Liverpool*).

5 *Liverpool Journal*, 18 January 1845; *The Porcupine* 6 (1864–5: 230); Hodder (1890: 180); Secord (2000: 216, 220); Smith et al. (2003a: 459n).

some had been made too broad for adequate rapidity, and some too narrow for sufficient safety, and that divers other mistakes had been made'. Wilson, he claimed, turned to the scriptures for instruction and accordingly 'took his Bible [in]to his workshop ... and ordered the construction of a steam-ship on the exact proportions [6:1 length:beam ratio] of the ark'. The result was the *Oriental*, a vessel, McNeile contended, 'that will bear comparison with any now in operation, – perhaps, I might say, that she is the best on the water (cheers)'.[6]

If, therefore, it could be shown McNeile inferred, that 'the very best directions for shipbuilding which the most improved scientific artist, in this time of science, can adopt, are those given to Noah by his Divine Instructor, we have another argument for the wisdom of that volume [the Bible] baffling the inventions of men, baffling all the ingenuity of infidelity to gainsay (cheers)'. What he neglected to mention was that, apart from the *Liverpool*, most transatlantic steamers out of the Mersey, including *Great Western*, *British Queen*, *President* and the original Cunard quartet, also conformed to the Ark's purported length-to-beam ratio.[7]

Liverpool's liberal *Mercury* contested McNeile's claims. 'What a pity it is that the Reverend Gentleman had not consulted Mr Wilson himself ... or looked into his Bible', it declared. Had he done the former, the shipbuilder would 'probably have told him that he was misinformed'. Had he considered the latter, 'he would see that the instructions given by God to Noah (Genesis vi: 14, 15, 16) ... for the building of the [rudderless and mastless] ark could have been of no assistance to Mr Wilson in the building of a modern steamship'. The paper therefore warned against McNeile's conclusion that *if* the story were true, it supported the notion of the Divine inspiration of the Bible: 'Aye, but "*if*" the anecdote be *untrue* ... the infidel might turn Mr McNeile's own array of "ifs" against him, and that too, in a very effective manner'.[8]

The prestigious new eastern mail steamers of P&O followed the same 6:1 ratio. Wilson's *Hindostan* (1842) and the *Bentinck* (1843) were slightly longer and beamier than the *Oriental*. The side-lever engines for each new vessel delivered 520-hp. At around 2,000 tons, the ships were about 200 tons larger. With their two funnels (entailing boilers both forward and abaft the paddle-crank) and three masts, they conveyed the stately if lumbering appearance, as well as the solid quality, of a traditional East Indiaman. At £88–89,000 each, the owners spared little expense to make them agreeable to their passengers and patrons. Completed in advance of the eastern mail contract, P&O intended

[6] *Liverpool Journal*, 18 January 1845. My italics.
[7] *Liverpool Journal*, 18 January 1845; Bonsor (1975–80: vol. 1, 59, 66, 140–1) (dimensions).
[8] *Liverpool Mercury*, 31 January 1845; Secord (2000: 216n).

Figure 14.1. P&O's *Hindostan* (1842) built in Liverpool to serve the Suez to Calcutta line

these two ships to project the Company's power both at home and across the eastern seas (Figure 14.1).[9]

The Company eagerly displayed these steamers to its patrons. In the summer of 1843, the Board planned to promote the new *Bentinck*, named in honour of the late Lord William, with 'an entertainment' at Southampton for 'Her Majesty's Ministers, Directors of the East India Co and such other influential parties'. Ordering the steamer instead to the more accessible Blackwall, the Company hoped to 'invite the Duke of Buccleuch to the dejeuner ... and also that his Grace be requested to enquire if H.R. Highness Prince Albert will honor the Directors with his company'.[10]

The new *Illustrated London News* (*ILN*) provided both a textual and a visual account of the *Bentinck*. As with *Great Liverpool* and *Oriental*, the steamer was fitted with the iron watertight bulkheads long promoted by Wye Williams. A powerful force-pump apparatus could 'instantaneously' extinguish fire in any part of the vessel. The ship had patent paddle-box lifeboats and Williams' 'smoke-consumers, which prevent the issue of smoke from the funnels'. The 'elegant, commodious and complete' interiors offered cabin accommodation for one-hundred and two passengers comprising twenty single, twenty-two double and

[9] Rabson and O'Donoghue (1988: 17, 30–1).
[10] P&O Minutes (18, 25 July and 1 August 1843); Harcourt (2006: 57).

twelve family cabins, each with its 'marble-covered basin-stand, mirrors, drawers and writing apparatus'. Every effort had been made to ensure a 'constant circulation of wholesome and refreshing drafts of air'. Cold, hot and shower baths were in close proximity. The spar deck offered a 'magnificent walk, the full length of the ship', while the main deck below was also 'comparatively open and airy' but also formed 'a spacious and well-lighted arcade, which may be resorted to in showery or boisterous weather'. Under the quarterdeck aft, the saloon (about thirty-feet square) ran the full width of the vessel with full views of the sea through large stern windows and spacious ports on each side. Unlike the first Cunarder's and *Great Britain*, there were no cabins around the sides of the saloon. The machinery followed the pattern of those constructed by the same engine builder for the *President*, with Gothic style framing.[11]

Ever alert for opportunities to puncture corporate pretensions, the *RM*'s Editor in July 1843 reported that 'rumour tells us that the *Hindostan* is performing very indifferently her portion of the [P&O] farce'. Navigators experienced in the ways of the Hooghly say 'that she is much too long, and draws too much water ... They affirm, that it is only at very particular states of the tide that she can navigate the Hooghley [sic]'. 'A. B.' in the same number noticed that P&O's office had stated that the *Oriental* had recently arrived at Southampton with 140 passengers. He asked the managing directors to state publicly 'if, out of the 140 passengers named, 80 were not discharged seamen ... belonging to the Company's steamer *Hindostan* ... sent home as totally unfit for the service[?]'. The *Railway Magazine* mischievously speculated that it was 'through the unfitness of the vessel [rather] than that of the [unfitness of the] 80 [sea] men (we beg pardon, passengers) who came home in the *Oriental*'.[12]

The *Hindostan*'s sister ship fared no better. 'The *Bentinck* we hear is a failure', the *RM*'s Editor reported with regard to a trial trip across the Irish Sea for the benefit of Irish and English shareholders. 'She was beaten out and out the other day between Liverpool and Dublin by one of the Government mail boats built several years since'. In contrast, the Company's autumn report recorded that upon arrival in London 'the highest authorities ... pronounced that she is one of the finest vessels of her class that was ever constructed'.[13]

Complementing these attacks, another anonymous correspondent attempted to deflate P&O's attempts to puff the steamers by providing evidence that the EIC's new but smaller and less powerful 1,200-ton *Akbar* could outperform them. He likened the P&O steamers in their 'great accommodation for passengers' and 'great speed' only in 'moderate weather' to the *President*,

[11] *ILN* (12 August 1843: 107); Harcourt (2006: 57); Howarth and Howarth (1986: 31) (accommodation plan). See also Burgess (2016: 227–8).

[12] *RM* (1843: 687, 691).

[13] *RM* (1843: 759–60, 1234).

British Queen and RMSP vessels. Five years later, even the Company minutes recorded the embarrassing episode when the *Bentinck*, while on a passage from Madras to Calcutta, exhausted the coal supply and resorted to burning spars and a hawser to enable the vessel to reach the mouth of the Hooghly, where the ship obtained ten tons of coal from a tug (Chapter 9).[14]

With strong resistance from the EIC to P&O's territorial ambitions, the Government prevaricated over the award of an Indian mail contract to P&O. The Company, however, seized on the end of hostilities with China to argue the case for additional steam communication to Hong Kong. This larger geographical ambition persuaded Government departments to recognize the national importance of P&O's projected system. Early in 1845 the MDs reported to the Board that after 'sixteen months continued negotiation ... they have now the pleasure to lay before the Board the Contract duly executed for the conveyance of Her Majesty's mails to India China &c granting this Company £160,000 a year for the services'. This news injected fresh momentum into fleet expansion plans for the remainder of the decade.[15]

14.2 'No Wood Ships can Compete with Iron Ships': Building Trust into P&O's Steamers

Early in October 1840, the new P&O Company received a valuation from 'Messrs Sinclair, Grantham and Tod' for steam engines and boilers originally intended for Peninsular Steam's *Royal Tar* but lying for some time at the Scott, Sinclair engine-building works in Greenock.[16] Grantham and Tod were each consequential in P&O's gradual introduction of iron-hulled steamers into its growing fleet over the coming decade.

Partners in Glasgow marine engineers and iron shipbuilders Tod & MacGregor, David Tod and James MacGregor both acquired their marine engineering skills with David Napier (Chapter 2) at his Glasgow sites. They also spent intervals as seagoing engineers in steamers engined by Napier. Both partners had intimate connections with the Barony Church where Revd Burns (Chapter 1) officiated at David's marriage to Jean Walker 1823 and baptized several of James' and his first wife's children. The Revd Macleod (see Introduction) conducted James' second marriage at the Barony Church in 1851.[17]

[14] *RM* (1843: 1314–42); P&O Minutes (5 January 1849).

[15] P&O Minutes (7 January 1845); Harcourt (2006: 5–6, 79–83) explores negotiations between P&O, various departments of the Government (especially the Admiralty, the Treasury and the India Board) and the EIC (1841–5).

[16] P&O Minutes (4 August and 7 October 1840).

[17] Tod & MacGregor's personal and shipbuilding history is well presented in www.gregormacgregor.com/Tod&Macgregor/tod_and_macgregor_main.htm [accessed 9 November 2011]. I thank Trish Hatton for researching the family histories.

When David Napier moved to the Thames in the mid-1830s, he offered the Lancefield works to his former artisans but, anxious not to overreach themselves, 'Tod & MacGregor, Engineers' set up their workshops in Carrick Street which led down to the Clyde just west of the Broomielaw. The engine-building business expanded with the establishment of their Clyde Foundry at Lancefield. Around 1838 the partners began building iron vessels on the south bank of the river next to the works of Thomas Wingate, builders of the *Sirius'* machinery (Chapters 2 and 4). Tod & MacGregor now founded their shipbuilding reputation on the construction of small iron paddle-steamers for Clyde, coastal and cross-channel services.[18]

Tod & MacGregor's business flourished in part because of a well-publicized, hands-on approach to marine engineering. In 1844, for example, the *RM* reported that their twin 600-ton paddle-steamers, designed for an Ardrossan to Fleetwood service, underwent a trial trip down the Clyde 'which satisfactorily tested their sailing [that is, steaming] qualities, the average speed being upwards of fifteen miles an hour'. Identical in 'model and construction', each vessel was fitted with a tubular boiler and direct-acting engines operating at no more than 6 psi. The hulls were divided into five watertight compartments. But they differed for the purpose of the trial insofar as one was fitted with solid paddle-floats and the other with divided floats. The builders wished 'to test the capabilities of the different paddles, Mr Tod acting as chief engineer on board the [*Royal*] *Consort* [and] Mr MacGregor filling the same situation on board *Her Majesty*'. They found the solid float to be superior to the divided one as regards to speed. Reporting the trial, the *RM* and *Glasgow Herald* proclaimed the addition of 'another triumph to the ship-building capabilities of the Clyde'.[19]

Whereas Tod acted as an independent consultant in the valuation of P&O's engine at Greenock, John Grantham informally represented the ship-owner. Grantham's close association with Wye Williams extended back to the 1820s when Grantham was working on Ireland's waterway systems, both in relation to civil engineering projects and to the introduction of inland steam navigation. Mentored by Williams in Liverpool, Grantham set up his own practice there as naval architect and consulting engineer. One of the most enthusiastic advocates of iron-screw steamers, he interacted regularly with Mersey and Clyde builders of iron vessels. A founder member of the Liverpool Polytechnic Society, he used this platform to advance his iron agenda, beginning with his presidential address on 'Iron as a material for ship-building' in March 1842. 'No wood ships can compete with iron ships in profit or accommodation', he concluded

[18] Bell (1912: 120–1); Duckworth and Langmuir (1977: 129–98) (fleet lists); Shields (1949: 53–4).

[19] *RM* (1844: 1126) (drawing the report from the *Glasgow Herald*); Duckworth and Langmuir (1977: 117–18) (Fleetwood and Ardrossan Steam Packet Company).

the lengthy pamphlet version. 'This must be the case even with the "Great Western", "British Queen", and "President" ships; they will hardly be established on their stations before they must be removed'.[20]

In early autumn 1841 Tod & MacGregor dispatched a letter to P&O with 'their terms for building an iron vessel for the engines of the [P&O] Company' now lying at Greenock. The P&O Board were aiming to lose no time in building a vessel to receive the stored engines and had already identified Wilson as the likely builder of the hull. But Wilson was occupied with the *Hindostan* and *Bentinck*, for which the spare set of engines was too small. Now Tod & MacGregor offered to tailor the engines to a new hull of their own. The Board accepted the terms, and less than a year later were agreeing further terms for the completion of the cabins of the new 548-ton iron paddle-steamer *Pacha*, named in honour of the Egyptian ruler.[21]

Following delayed delivery, the *Pacha* entered the Iberian service in April 1843. When the steamer grounded in Cadiz Bay a year into service, the absence of damage enhanced the credibility of the iron hull in P&O's eyes. Moreover, in mid-summer the Company's superintendent at Southampton dispatched a letter to the Board 'giving a very satisfactory account of the "Pacha's" bottom as being almost entirely free from corrosion'.[22] Anticipating a successful outcome to the negotiations over the extended mail contract, P&O now considered ordering iron steamers with a much larger tonnage.

At the close of 1843 the Board requested their managing directors to submit 'accurate & detailed information with plans specifications and estimates for the construction of an additional vessel *wood or iron* with engines of not less than 500 horse power for the India Seas'. The steamship would have to be 'of first rate speed, power and capacity, combining all recent improvements for the India Seas, to meet the extension of the passengers' intercourse to and from India'. Company discussions over the choice of material continued during the first half of 1844. By June, the Board had decided on a new vessel, with iron hull, for the Alexandria service and urged the managing directors that 'Mr Williams from his practical skill & knowledge be associated with them in the undertaking'.[23]

Tenders were invited by P&O from five contractors in Liverpool, Glasgow and London. The Company also investigated the purchase of both the *Great Western* and *British Queen* in the same period, but surveys showed that neither wooden-hulled steamer was fit for service.[24] Of four projected iron steamers of

[20] Grantham (1842: 80, 1875: 270–2); Dawson (2012: 22–7).
[21] P&O Minutes (28 September 1841, 23 August and 20 September 1842).
[22] P&O Minutes (30 April and 23 July 1844); Rabson and O'Donoghue (1988: 30).
[23] P&O Minutes (12 December 1843, 9 April and 18 June 1844). My italics.
[24] P&O Minutes (5 March, 21 May, 18 June, 16 July, 13 August, 17 September and 19 November 1844; 9 and 16 March 1847).

1,100–1,300 tons, the owners allocated two orders to Wigram (Blackwall) and one each to Vernon (Liverpool) and Fairbairn (Millwall). Miller & Ravenhill obtained the engine contracts for the three Thames-built hulls and Bury, Curtis those for the Mersey vessel. All four sets of engines were of a controversial oscillating (or vibrating) type in which the whole cylinder oscillated about a fixed axis in such a way that the piston rod drove the crank directly, rather than by connecting rods. Although the oscillating system appeared to expend less power in overcoming internal friction, concerns focused on the possible snapping of the piston rod in a heavy sea acting on the paddlewheels.[25]

Scottish-born William Fairbairn, whose wealth and reputation derived from his Manchester works for the construction of mill machinery, boilers and stationary steam-engines, established a shipbuilding yard at Millwall, opposite Greenwich, in 1835. From the 1820s he had built, and experimented with, small iron-hulled steamers for canal use. Ambitious to expand in an area close to the seats of power, he borrowed heavily to develop a large Thames site for the iron shipbuilding 'factory'. Soon the yard had orders for twelve vessels from the EIC for navigating the Ganges, and four others for different parts of mainland Europe.[26]

In order to provide local management, Fairbairn installed one of his former pupils, Andrew Murray, as yard manager. The owner later admitted that '[w]e made many blunders as to prices &c. in a business which we had yet to learn'.[27] During the summer of 1844, Murray's brother Robert was hard at work in the drawing office. A prize of £100 had been offered by P&O for the best plans for an iron steamer boat of 1,200 tons and Murray was trying for the award. Alongside him was a twenty-one-year-old aspiring apprentice engineer from Glasgow, James Thomson. Corresponding with his brother William (Chapter 16), James explained that '[s]everal times of late I have been working in the drawing office from six in the morning till eight in the evening'. His time was wholly devoted to copying Murray's drawings for the P&O vessel, as Murray himself did not have the time to meet the deadline. Evidently the submission paid off and Fairbairn won the contract for the steamer two months later.[28]

The Company planned to make this steamer occupy a pre-eminent role in their projected China mail service. In mid-June 1845 the Board arranged an entertainment at the Albion Hotel, London, for Sir Henry Pottinger, who

[25] P&O Minutes (26 September and 1–23 October 1844); Banbury (1971: 114–29, 181–6, 289–91 (Wigram), 206–10 (Miller, Ravenhill)). Bury, Curtis and Vernon probably had links with EISNC chairman T. A. Curtis (Chapter 12). For a contemporary discussion of oscillating engines see Smith (2013b: 516–17) (Thomson brothers). RMSP also adopted the engine c. 1847 (Chapter 9).

[26] Smith (2013b: 154–5).

[27] Smith (2013b: 154–5); Banbury (1971: 171–4).

[28] James to William Thomson, 4 August 1844, Kelvin Collection (KC), University Library Cambridge (ULC).

had returned to Britain from China the previous year. Appointed by Foreign Secretary Palmerston as envoy and plenipotentiary, as well as Superintendent of British Trade, Sir Henry had taken the leading role in forcing China to agree to the treaty, signed aboard a British warship at Nanking in 1842, to cede Hong Kong to Britain and open five ports (Canton, Amoy, Foochow, Ningpo and Shanghai) to British trade. In 1843–4 Sir Henry served as Hong Kong's first governor. The P&O celebration was staged to mark the 'occasion of opening the monthly steam communication with China'. Cabinet ministers, EIC directors and the heads of several departments of Government made up the list of guests. The patronage ritual was completed by P&O assigning the name *Pottinger* to the Fairbairn steamer.[29]

Even before the Pottinger celebration, however, all was not well in the shipyards and engine works on the Thames. In early December 1844 the Board received from P&O's superintendent of construction (George Bayley) the first hint of problems with some of the contractors over the quality of materials and workmanship, as well as schedules. Fairbairn, for instance, blamed delays to their 1,200-ton vessel on Miller, Ravenhill's failure to furnish plans of the engine space.[30]

The Company expected the *Pottinger* to be the first into service of the four large iron steamers when launched at Millwall in March 1846. Leaving the builders for Southampton in early September, however, *Pottinger*'s troubles were only just beginning. By November, it seemed the new ship was still far from satisfactory. In one recent incident the ship had been on shore near Cowes, Isle of Wight, and had required the assistance of an Admiralty tug to refloat the vessel. But the problems were less to do with the trustworthiness of the navigating officer and more to do with the competence of the designers and builders. The P&O Board appointed a Nautical Committee, 'consisting of Captn Bourne, Captn Thornton and Captn Nairne, for the purpose of enquiring as to whether there was any departure from the original specification'. The principal concern was stability. The Board wanted the Committee, aided by any 'competent persons as they may consider necessary', to determine 'why a necessity existed for the large quantity of kentledge [ballast] introduced into that vessel – whether the top weight & masts & rigging have been judiciously arranged, and generally to investigate & report to the Board fully in detail upon the whole question connected with the original plans for & completion of that vessel & machinery for sea'.[31]

The *Pottinger*'s early performance on the intended service was apparently so poor that P&O soon ordered the steamer to return to England. The ship later

[29] 'Pottinger, Sir Henry, first baronet (1789–1856)', *ODNB*; P&O Minutes (13 June 1845).

[30] P&O Minutes (13 August, 17 September and 9 December 1844; 8 July 1845).

[31] P&O Minutes (24 March and 24 November 1846).

underwent re-engining and lengthening by Wigram at Blackwall.[32] Even after all these changes, however, the vessel still attracted poor publicity. 'The Calcutta agents of the company ... have sent us in the *Pottinger*', a group of more than forty disgruntled passengers out of seventy-six who boarded at Calcutta for Suez in 1852 wrote to *The Times*, 'a vessel confessedly unfitted to carry any considerable number of passengers in a tropical climate' on account of poor ventilation below decks and inadequate accommodation throughout the ship.[33] The letter also made clear that the passengers deemed, in contrast, the traditional *Hindostan* just such a 'fit' vessel, doubtless in part because of the superior ventilation combined with heavy timbers and planks that rendered *Hindostan*'s cabins relatively immune from the tendency of the iron-hulled *Pottinger*'s accommodation to become like an oven in the tropical heat. The experienced Captain Henderson also had very unkind comments to make to the SC (1849). East of Singapore, he explained, '[t]he typhoons are so extraordinary that a paddle steamer is not so seaworthy' as a screw vessel and that the *Pottinger* (and similar P&O mail steamers for the China Seas) were overlarge paddle-steamers 'built for a floating tavern'. He also proclaimed the *Pottinger* as a vessel that 'out Herod[s] Herod', that is, exceeding King Herod in violence and extravagance.[34]

At the end of 1847, the Company faced a crisis when Allan suddenly tendered his resignation as secretary. Seeking to avoid the loss of such a trusted servant (see Introduction), the MDs persuaded Allan to return as assistant manager, with an understanding that upon departure of one of the three MDs Allan would be promoted to fill the vacancy. Within weeks of his new appointment, he and Carleton travelled to Greenock to consult with James Caird on the question of the poor performance, especially as to speed, of the new iron steamers (see Epigraph). The episode well illustrates Allan's low-key, pragmatic and altogether critical roles in restoring or maintaining the Company's relative stability through its many troubled periods. Not for nothing had Lindsay characterized his friend as someone of uncompromising integrity (see Introduction).[35]

Having recently completed the 764-ton *Tiber* for P&O, Caird's were now working on a 1,200-ton iron steamer of very similar lines to the earlier P&O quartet, but some twelve feet shorter amidships. The meeting with Allan and Carleton aimed to agree on design changes in order to achieve three goals: increased speed, a diminished draught at the waterline, and rendering the Caird ship easier as a passenger ship in a sea.[36]

[32] Rabson and O'Donoghue (1988: 33).

[33] *The Times*, 25 June 1852.

[34] SC (1849), 114–16; *Hamlet* Act 3, Sc. 2: 'It out Herods Herod'.

[35] P&O Minutes (2 January 1844; 18, 22 December 1846); Harcourt (2006: 153–4) (a conspiratorial spin on Allan's resignation). P&O Minutes (4 December 1844 and 25 March 1845) suggest the stress of work.

[36] P&O Minutes (16 March 1847).

Figure 14.2. P&O's iron-hulled paddle-steamer *Ripon* (1846) after lengthening

Armed with Caird's practical advice, the managers' report to P&O ruled out time-consuming and costly lengthening aft (suggested in London). Instead, the key to improved performance lay in the reduction of 'the top weight as much as practicable'. The managers therefore recommended 'cutting off the spar deck and building a [deck]house abaft similar to those on board the North American steam vessels [of Cunard]'. As a consequence, they reported, there would be a positive reduction in dead weight aloft of at least 140 tons, the height of the vessel would be reduced by at least five feet, the masts and yards could also be reduced and the draught would decrease by at least a foot. Moreover, the deck-house would accommodate as many passengers as that on Wigram's *Ripon*, but would offer much better ventilation, while by a rearrangement of the cabins on the main deck the passenger accommodation would be nearly as great as that on the *Pottinger*.

Taken together, the managers concluded that 'the vessel would be rendered a stable ship – a good carrier – and a favorite with passengers; whilst it is confidently believed the speed realized would be at all events equal if not superior to that of the Oriental'. Completed as *Euxine* in late 1847, the lower profile distinguished the ship from the Thames-built steamers with their spar decks. Fairbairn's *Pottinger* and Wigram's *Ripon* and *Indus* (1847) were all lengthened between 1848 and 1852 in an effort to improve their performance with regards to speed and sea-keeping (Figure 14.2).[37]

[37] P&O Minutes (16 March 1847); Rabson and O'Donoghue (1988: 34–8) (ship data). The similar *Malta* came from the same yard in 1848.

These discussions offer a rare glimpse into the ways in which ship-owners and shipbuilders engaged in building confidence into the iron steamship. The RMSP's practice of vesting authority in its superintendent engineer (Chapter 10) stood in stark contrast to P&O's very different practice under Allan of trusting in the practical knowledge and skill of specific shipbuilders to arrive at a consensus on how to address matters of underperformance. The fruits of this policy were borne out by the first of many contracts handled by Tod & MacGregor.

From about 1843, the Glasgow partners relocated their shipbuilding yard to a much larger site at Meadowside at the junction of the Clyde and Kelvin rivers. The firm had already delivered CDSP's 350-ton iron steamer *Prince of Wales* and had the much larger *Trafalgar* on the order books for the Dublin Company. For a mere £38,500, P&O purchased the 1,100-ton *Trafalgar* (with side-lever engines from Caird & Co.) in mid-1846 and renamed the ship *Sultan*. Completed at Meadowside within a month of the launch in June 1847, the *Sultan* received high praise from Nautical Committee captains Thornton and Nairne for a 'very satisfactory' trial trip from Glasgow to Southampton. The ship's master, Captain Brooks, then reported his arrival at Malta by mid-August after a ten-day passage from Southampton, including a sixteen-hour stay at Gibraltar. Even in 'a heavy gale and head sea', he enthused, his vessel had managed to steam at seven knots, only three knots below its service speed. Arriving at Constantinople in late August, the *Sultan* won still more praise from Captain Ford, the general superintendent there, who stated that the steamer 'had been much admired since her arrival, and that he considered her in every way suitable to the trade on that line'.[38]

Ordered by P&O, a second Tod & MacGegor iron paddle-steamer, *Pekin*, was marginally smaller than the original four large iron steamers, but cost only £46,000 compared to £66,000 for the Fairbairn-built *Pottinger*. Completed by early 1847, P&O expressed the view that 'the Contractors had behaved very liberally in the construction of the ship, especially in adding 40 horse power to her machinery without extra charge'. These remarks were indicative of a strong bond of trust developing between the owners and contractors. Within three years, P&O's experience with both vessels led them to award a contract for two 1,100-ton steamers (priced at £69,500 for the pair) to the Meadowside yard. The Board dispatched its resident engineer Andrew Lamb and marine surveyor Mr Ronald to conduct face-to-face negotiations with the shipbuilders over the precise specifications. The *Sultan* acted as a standard: 'ten feet longer, eight inches deeper than the *Sultan* ... 3 inches more diameter of cylinder and

[38] P&O Minutes (6, 20 May and 9 December 1845; 21 July 1846, 27 June, 25 August and 14 September 1847); Bell (1912: 120–1); Shields (1949: 53–4); McRonald (2005–07: vol. 1, 196–7); Rabson and O'Donoghue (1988: 37).

6 inches more length of stroke'. In thus scaling up the successful model, the engineers from both firms 'estimated that the vessels now proposed will realize a rate of speed exceeding that of the Sultan by about one knot per hour ... The extra cost for each ship will amount to £3500'.[39]

The Company agreed 'that a formal contract be dispensed with and that a letter ... drawn up and signed' by both the Company Secretary and Tod & MacGregor would be sufficient. Their trust proved well founded. After he had inspected the new ships, Lamb submitted 'a very satisfactory report of their progress and of the quality of the materials and workmanship'. Progress was indeed astonishingly rapid compared to Thames builders. Allan attended the launch of the first vessel, *Singapore*, in late September. By the end of November, the builders reported 'a satisfactory trial of the [oscillating] engines'. And on 20 December 1850, Lamb wrote to the Board from Southampton 'giving a very satisfactory account of the ship and machinery on the voyage' from the Clyde.[40]

That same year, however, an Admiralty order threatened P&O's commitment to iron steamers. '[N]o vessel commenced after the date of this letter will be approved of, under the terms of the contract', came the directive to the Company in June, 'if built of iron or of any material offering so ineffectual a resistance to the striking of shot'. A similar missive went to Cunard, Burns & MacIver whose line kept clear of iron hulls for its mail steamers. In a vigorous appeal in January 1851 against the ruling, P&O claimed that the clause in the original contract (1840) specifying wooden hulls had been inserted in deference to the wishes of the then Chancellor of the Exchequer (Baring), now First Lord of the Admiralty. It had not been intended 'as a prevention to the Company's adopting such improvements as the advanced state of science might suggest, either in giving a greater acceleration and safety to the transmission of mails, or in economizing, in combination with that improvement, the cost of performing the service'. Iron vessels, it proceeded to explain, demonstrated in practice just those advantages. In reply four days later, the Lords Commissioners saw 'no reason to alter their decision'.[41] But, as we have seen, even before the *Amazon* catastrophe in January 1852, the order lapsed (Chapter 10).

14.3 'One of the Finest Screw Steamers in the World': The Iron Screw Steamer in Practice

In June 1844, P&O's paddle-steamer *Tagus* (1837) and the Admiralty's experimental screw-steamer *Rattler* (1843) took part in a trial of speed on the Thames.

[39] P&O Minutes (27 June 1847; 11 January and 4 February 1850); Rabson and O'Donoghue (1988: 33–5) (costs of *Pottinger* and *Pekin*).

[40] P&O Minutes (4, 19 February, 7 May, 20 September, 29 November and 20 December 1850).

[41] HCO (1851: 1–3) (published letters).

It was HMS *Rattler*'s twentieth experimental trip to test screw propellers of different designs. As Andrew Lambert has shown, Brunel played a key role in the *Rattler* project with the aim of obtaining reliable data for the design of the *Great Britain*'s screw. This trial now brought together two steamers of similar tonnage and power. The *RM* reported that 'the two vessels had the same speed while in company, the average of 12 trials of the *Rattler* on that day giving a speed of 9.579 knots'. From the evidence, the *Railway Magazine* envisaged future mail steamers of smaller tonnage deploying a screw rather than paddles. In spring 1845 P&O's Board requested that the MDs 'collect every information in their power both from the Admiralty & from private individuals regarding the advantages or disadvantages attending the adoption of the Archimedean screw propeller'.[42]

Thanks largely to Grantham and his contributions to the Polytechnic Society, Merseyside celebrated screw propulsion earlier than most of its commercial rivals. In 1843 the *RM* carried a report stating that the experimental tug *Liverpool Screw* 'continues to excite great curiosity and surprise on the river, twisting and turning ... as she does amongst the numerous craft off St George's Pier, with a perfect snake-like power and facility'.[43]

Under Grantham's supervision, in 1846 Liverpool shipbuilders James Hodgson and engine-builders Bury, Curtis delivered the 1,300-ton iron-screw steamer *Sarah Sands* to Sands & Co., 'a company of Liverpool gentlemen'. Thomas Sands, to whom Grantham had dedicated the published version of his 1842 Polytechnic lecture, was chairman of the Liverpool Dock Committee and had recently served as Mayor of Liverpool. Ostensibly built to initiate an ambitious line to Australia, the ship remained a one-off, comparable in size to the wooden Cunarder's and the *Great Western*. The *Sarah Sands* was immediately placed under the agency of Robert Kermit – operator of the Red Star Line of New York to Liverpool sailing packets – with an experienced sailing packet master, William C. Thompson, in command. In almost three years of transatlantic service, the ship completed twelve and a half round voyages with passengers in first, second and steerage classes. The steamer's performance provided independent Liverpool ship-owners with a credible exemplar of the deep-sea iron-screw steamer barely three years after the *Rattler* experiments and almost contemporaneously with the *Great Britain*'s debut (Figure 14.3).[44]

Grantham was again called upon in mid-1847 to design an auxiliary steam yacht for Spencer Wynn, Lord Newborough of Glynllifon in North Wales, who was a neighbour of Assheton Smith (Chapter 2). It was Grantham who

[42] *RM* (1844: 994–5); Lambert (1999: 27–52, esp. 41–9); P&O Minutes (29 April 1845).
[43] *RM* (1843: 499). See also Grantham (1843–4: 60–4).
[44] Grantham (1842: 1); Dawson (2012: 24); Albion (1938: 30–1); Tyler (1939: 185); Bonsor 1975–80: vol. 1, 184–5).

Figure 14.3. First Mersey-built seagoing iron-screw steamer *Sarah Sands* (1846)

recommended accepting Tod & MacGregor's tender of £7,800 and who also undertook to supervise construction at a charge of 2.5 per cent of the total cost. Fitted with an oscillating steam-engine driving a two-bladed screw, the schooner-rigged *Vesta* commenced trials in August 1848, some six months after the order had been confirmed.[45]

As P&O's *Singapore* and *Ganges* took shape on the Meadowside slipways in 1850, Tod & MacGregor were fitting out a larger vessel on their own account for North Atlantic service. But the *City of Glasgow* was not a paddle-steamer. This latest version of screw propulsion drew for inspiration on Tod & MacGregor's old master, David Napier, whose steeple engine had found favour with owners of small, fast river and estuary paddle-steamers. In the *City of Glasgow* Tod & MacGregor placed the cylinder to one side of the propeller shaft that ran aft along the centre line. The piston acted vertically upward on an iron beam running transversely above and across the shaft. The connecting rod at the other end of the beam drove a massive geared wheel located parallel to, and engaging with, a smaller wheel mounted on the shaft. The propeller shaft thus turned more rapidly than the large gear wheel.[46]

[45] Grantham (1866); Dawson (2012: 24–6). Fox (2003: 176–7) wrongly asserts that the *Vesta* was built for William Inman.

[46] See Banbury (1971: 214–18) (David Napier); Tyler (1939: 196) (engine drawing); Fox (2003: 172–8).

In April 1851 Lamb informed P&O that Tod & MacGregor had a large iron-screw vessel for sale. The Company considered the purchase of the steamer either for the Constantinople or Alexandria line, but 'altogether independent of a mail contract' since the screw steamer would be unlikely to have the power and speed to maintain the schedule. While the builders quickly sold the steamer to the Inman Line (by this time also owners of the *City of Glasgow*), Tod & MacGregor quoted terms for the construction of two screw steamers (1,600–1,800 tons) with delivery in twelve months from the time of the order. Allan held a meeting with the firm in Glasgow. After he had reported that 'all the other shipbuilding firms at that port [are] very full of work', P&O resolved to accept Tod & MacGregor's tender.[47] The *Madras* and *Bombay* were the first large screw steamers for P&O. Between 1843 and 1858 the Clyde yard constructed some seventeen iron steamers, eleven of them screw, for the Line.

On the afternoon of Saturday 30 October 1852 the largest iron steamer ever constructed up to that time on the banks of the Clyde stood poised for launching from Tod & MacGregor's yard. Ten feet longer than the *Great Britain*, the 330-foot *Bengal* measured 2,300 tons. As the men systematically knocked away the shores around the stern, 'the immense mass', in the words of *The Times*, 'was suddenly seen to move'. Gathering speed, the ship tore the supporting cradle to pieces. Workmen leapt for their lives. With 'the force of an avalanche' and minutes ahead of schedule, the *Bengal* plunged, without injury to anyone, into the river.[48]

From early 1853, the *Bengal* served initially on the Company's mail service from Southampton to Alexandria. In an article extolling the 'great merits of the screw for ocean steaming', *The Times* claimed that the 'success of this vessel has been extraordinary in the point of speed, and it is considered by the most competent judges that that vessel has achieved a complete triumph over the more costly paddle-wheel steamers'. With about two-thirds the power of the similar-sized RMSP steamers such as the *Orinoco* (Chapter 10), the *Bengal* made greater speed and consumed only half the tonnage of coal per day. The Company 'announced its determination gradually to discard the paddle-wheel in the whole of its extensive service'.[49]

In late summer, P&O ordered the *Bengal* to Calcutta. Allan issued strict instructions to Captain John Bowen concerning the passage via the Cape. 'You are to proceed under half steam with two boilers when necessary, and by making the best possible use of your sails upon all occasions where practicable, the Managing Directors look forward with confidence to the performance of the voyage in a shorter period than it has hitherto been accomplished', Allan told

[47] P&O Minutes (1, 7, 25 April and 9, 16 May 1851).
[48] *The Times*, 3 November 1852.
[49] *The Times*, 20 June 1853.

the captain. 'You will understand that you are not under any circumstances to exceed a consumption of 25 tons of coals per diem and the Managing Directors fully expect that in the ordinary way even this quantity will not be used'.[50]

Allan then turned to the handling characteristics of large screw steamers. 'You have now been entrusted with the command of one of the finest screw steamers in the world', he reminded Bowen. But '[e]xperience has shown that from the extreme length of this class of vessels as compared with the ordinary steamers or sailing ships, they do not under certain circumstances answer the helm so quickly as is usual, and it is therefore desirable to stop the engines preparatory to heaving the helm hard over either to Port or to Starboard'. Such a practice was, he stressed, indispensable when meeting other vessels at close quarters in fog or darkness and when under way in a confined harbour, roadstead or river. Thus it was 'essential to get the ship's head pointed in a right direction before giving her speed with the engines'. Allan made clear that these warnings had special force when navigating the Hooghly 'where the tides run strong and uncertain in their action upon the bow of a long vessel'. It would be 'more expedient and prudent', he insisted, 'to take ample time in going up and down the Hooghly and to employ a powerful tug to convey the passengers and mails, rather than to run the slightest risk with these long vessels in so dangerous a navigation'.[51]

Furthermore, in keeping with standard practice aboard all P&O vessels, the master would receive on board a comprehensive set of forms which he must ensure were filled up and transmitted to headquarters at the termination of the voyage. These included log books of the ship, the officers and the engineers, as well as reports by the engineers, the carpenter, the boatswain and the surgeon. There were also reports on officers and crew, the muster roll and the coal account. With that, Allan wished Captain Bowen 'a speedy and prosperous voyage'.[52]

In late December 1853, Company officials received preliminary reports, based on a letter from Captain Bowen dispatched from Mauritius some two months previously, of the *Bengal*'s less-than-perfect passage. Captain Engledue, P&O's marine superintendent for a long time at Southampton, extracted some worrying news, including the inaccurate state of the compasses, the poor handling of the steamer under sail with damage to the masts and the loss of three lifeboats. In the superintendent's opinion, the compasses had been rigorously adjusted at Southampton, the loss of the boats was inexcusable and the master's claim that his vessel was not 'seaworthy under canvas' was 'a perfect absurdity'. Captain Robert Guthrie, P&O's superintendent of

[50] Allan to Bowen, 18 August 1853, in P&O/3/3, NMM; [Anon.] (1846: 214–15).
[51] Allan to Bowen, 18 August 1853, in P&O/3/3, NMM.
[52] Allan to Bowen, 18 August 1853, in P&O/3/3, NMM.

navigation, further reminded the managers that Bowen had 'not only expressed himself satisfied before leaving England with the [compass] corrections given to him, but was supplied with every means of discovering any error wherever it existed'. He also suspected that the reported loss of the steamer's fore topmast and damage to the main topmast were due either to 'the rigging being slack' or 'from want of proper attention in not shortening sail in time [for the squalls]'. In Guthrie's opinion, there appears 'no good ground for Captain Bowen's complaints'.[53]

In a carefully worded letter Bowen defended himself. His eyewitness account laid bare matters of considerable concern to the Company. Thus he asserted that damage to the masts had occurred in relatively light winds. Instead, 'what occasioned the accident of losing the mast was the sudden and excessive rolling, [even though] the yards and booms were properly secured'.[54] By implication, therefore, the *Bengal*'s very design as a large screw steamer, lacking the stabilizing influence of paddlewheels, was called into question.

Again, he accepted that the compass corrections given to him were quite correct in the Northern Hemisphere and that he had been supplied with an azimuth compass for making astronomical checks on the steering compass at intervals during the ocean voyage. But, he asked, what use was this 'when I have been many days without a chance of seeing the sun, and of taking amplitude or azimuth observations'. Referring to the researches of Dr Fisher of Greenwich Hospital, Bowen claimed that the masts 'have to me an unaccountable effect on the compasses – they being above have a great and overuling [sic] power'. He emphasized that he was not the only reliable witness. His officers had participated in a regular comparison of the three compasses: '[s]urely I could not as well as the Officers who would be looking at two of the compasses while I attended to the third, in comparing them daily [or every few hours] ... be constantly making mistakes'.[55]

Finally, Captain Bowen challenged Engledue's curt remarks as 'likely to be prejudicial to my future prospects in the Company'. Setting aside fears of 'being thought vain or presumptious', Bowen asserted 'that as a Commander in your service I am equal to any and inferior to none, and also that I am capable of commanding the largest ship the Company owns, or the smallest in their possession'.[56]

Steamship performance and public trust went hand-in-hand. In order to attain and maintain that trust, P&O insisted on an extraordinary level of attention to detail in matters of shipbuilding, engineering and navigational practice.

[53] Guthrie to the MDs, 2 January 1854 (copy), in P&O/3/3, NMM.
[54] Bowen to the MDs, 22 April 1854 (copy), in P&O/3/3, NMM.
[55] Bowen to the MDs, 22 April 1854 (copy), in P&O/3/3, NMM; Roberts (2009: 57–72) (Fisher).
[56] Bowen to the MDs, 22 April 1854 (copy), in P&O/3/3, NMM.

As the first of the ocean mail companies to order iron-paddle and then iron-screw steamers, the Company faced the additional challenges of the new systems. Heavily dependent on sail for their vaunted superior fuel economy, iron-screw steamers exhibited tendencies to excessive and potentially damaging rolling, to serious compass deviation due not simply to the hull and engines, but to the massive iron masts and to limited manoeuvrability in confined waters. Thus, the oceangoing iron-screw steamer possessed no guaranteed track to dominance over other systems of motive power.

Part IV

Engineering an Oceanic Economy

15 'She Would Be Perfectly Stable and Strong'

Rival Systems of Engineering Economy

> Lines of swift and magnificent ocean steamers now traverse the seas, east, west, north and south carrying with them the seeds of civilisation and progress. Whenever peace is restored, these mighty vessels will resume their natural mission and become the heralds of good will to every quarter of the globe.
>
> *William Schaw Lindsay, in time of war, reflects upon the beneficent character of ocean steam navigation (1854)*[1]

Summary

British steam coal provided a quality fuel, but its source was often found a very long way from the steamers that consumed it. Through an acquaintance with James Allan, former sailing ship master Captain Lindsay quickly recognized how the regular working of P&O's mail system depended on the supply of coal to both sides of the great Egyptian divide. Lindsay therefore constructed a ship-broking system that offered P&O (and others) the means to secure the availability of that coal wherever it was required. In contrast to this stood the contemporaneous ambitions of Russell and Brunel to build the 'Great Ship' with the goal of carrying on board enough coal for a round trip voyage to Australia or India. Beset by engineering and financial crises following her problematic and protracted launch in 1858, the *Great Eastern* struggled against public mistrust to complete a series of Atlantic voyages. Cunard's trusted line of steamers, meanwhile, had further consolidated its oceanic proprietorship following the tragedies that beset the much-acclaimed Collins Line.

15.1 'Work! Work! Work'! Lindsay and the Steam Coal Trade

The north-east ports of Sunderland and Hartlepool provided the sites for Lindsay's transition from deep-sea sailing shipmaster, through coal agent, to ship-broker and charterer in the early 1840s. Contesting current colliery practices of financially bribing shipmasters to load coal from a particular pit, he introduced methods of persuasion that benefitted shipmasters, ship-owners and merchants by providing masters with a small initial quantity of coal with which

[1] William Schaw Lindsay (WSL) Journals (LND35/3) (57–9).

to stabilize their sailing vessels at the moment when, light ship, they were most vulnerable to capsize. As a result, Lindsay's employers prospered. Moving to London on their behalf around 1845, he targeted 'the general consumers of steam coal' such as P&O, PSNC and RMSP, as well as 'those mercantile houses who shipped steam coal to foreign parts'. Despite scepticism, he felt sure 'that there was a wide field open for steam coals, which was daily expanding to an extent of which few could form any conception'.[2]

After many fruitless calls, Lindsay finally met face-to-face with P&O's James Allan (Introduction, Chapters 12, 14 and 18). 'Our conversation led to Glasgow coals, then Glasgow people', Lindsay recorded of his first meeting. He quickly established that Allan knew his brother-in-law Robert Stewart, ironmaster. A letter of introduction from Stewart provided Lindsay with the formal confirmation he needed. 'I called and presented it', he wrote of the second meeting with Allan. 'I found him at the office and had a long chat with him. From that hour the tide turned. ... a kind Providence gave me understanding to see my way distinctly'. Indeed, he noted in his *Merchant Shipping* that, after his visit to London, Allan became 'one of my earliest friends, and our friendship remained unbroken until his death in September 1874'.[3]

Allan advised Lindsay that P&O 'never bought coals at a price delivered on board at a port of shipment' but only 'at a price delivered at the port where they were consumed'. He provided Lindsay with a list of P&O coaling stations and told him 'the price they were ready to pay for steam coals delivered at Southampton, at Gibraltar, at Malta, at Alexandria, at Aden, Bombay, Singapore and Calcutta and Hong Kong'. Allan offered to begin by taking one or two cargoes 'to test their quality' at any of these stations and at the price indicated.

Knowing no ship-owners except his old employers Greenwell & Sacker in Sunderland, Lindsay's challenge was to charter ships to do the work. Taking P&O's price of thirty-one shillings per ton delivered at Alexandria, he calculated that the price of coals loaded at Hartlepool was seven shillings per ton and that sixpence per ton would cover insurance, leaving twenty-three shillings and sixpence for freight. Searching the London docks for a ship prepared to go to Alexandria, he negotiated a freight of twenty-two shillings and sixpence which yielded one shilling per ton for contingencies – or profit. He then sent a letter to P&O advising them that he would deliver a cargo to Alexandria at thirty-one shillings.[4]

[2] WSL Journals (LND 35/2)(7–8). This section relies on Lindsay's Journals written in the mid-1850s. Although his account there of his initial involvement in the steam coal trade lacks specific dates, contextual evidence points (for example, the foundation of Lamport & Holt in 1845) to the period 1845–9.

[3] WSL Journals (LND 35/2); Lindsay (1874–6: vol. 4, 379n).

[4] WSL Journals (LND 35/2) (9–10); Clark (2012: 46).

Table 15.1. *Lindsay-brokered P&O coal shipments (1851–2)*

Date	Tonnage	For shipment to	Price per ton (shillings/pence)
6 June 1851	7,000	Aden	41/7d
28 Oct 1851	4,800	Aden	41/3d
27 Jan 1852	4,800	Singapore	30/10d
30 Aug 1852	4,800	Hong Kong	43/3d
30 Aug 1852	4,800	Singapore	34/9d

Lindsay developed the practice of conducting daily rounds of the docks 'making it known to all whom it might concern, that I was open to take ships at certain rates for all or any of the places Mr Allan had enumerated'. Offers were at first slow but at last 'struggling Ship Brokers began to come to [my office at] 35 Abchurch Lane in the City to offer me ships for charter'. The Hartlepool coals were found to be satisfactory by P&O, and they 'took all these cargoes from me at the prices named, with orders for more'.[5]

The P&O minutes reveal the importance to Lindsay of the P&O connection (see Table 15.1). These sailing ships had now become an integral part of the mail steamship system, ensuring a continual supply of best-quality British steam coal across P&O's vast geographical system. Supposing each vessel to carry about 300 tons, the smaller shipments would require at least sixteen ships with profits per vessel of around £240 to £350 for the outward passage. Establishing close ties with Sunderland and Liverpool ship-owners in the mid-1840s, Lindsay and his associates generated considerable capital which they directed towards the construction of independent fleets, both sail and steam, from the late 1840s.

William James Lamport provided Lindsay with access to members of a small but very prosperous dissenting denomination that flourished in nineteenth-century mercantile Liverpool. Son of a Unitarian minister from Lancashire and brother of Charles, shipbuilder in the Cumberland port of Workington, Lamport had served as a clerk in the House of Gibbs, Bright and Company, the well-respected mercantile firm based in both Bristol and Liverpool whose interests had included the GWR, the reconstructed *Great Britain* and the Eagle Line of sailing packets to Australia. In 1845 Lamport began a ship-owning partnership with George Holt, eldest son of a prosperous Liverpool cotton broker.[6]

New England writer Nathaniel Hawthorne summed up the character of Liverpool's Unitarian communities as possessed both of 'a kind of social and

[5] WSL Journals (LND 35/2) (10).

[6] Heaton (1986: 15); Corlett (1980:140–1); Smith et al. (2003b: 398–405).

Figure 15.1. One of Lindsay's auxiliary steamers of the 1850s

family aspect' and of 'great wealth and respectability'. The chapels were the meeting houses for tightly knit networks of ship-owners and merchants who frequently formed alliances by marriage, met socially, invested in one another's ventures, shared or exchanged practical skills, embarked on philanthropic (especially educational) schemes and engaged fully in the liberal politics of free trade and social reform. They were marked by high levels of mutual trust and a public reputation for great integrity that Lindsay found conducive to his own values and practices.[7]

Lindsay's steam coal trade transformed Lamport & Holt into a flourishing sailing ship enterprise. Alongside this prosperity, Lindsay, the Holts and their friends invested in iron-screw cargo steamers. Unlike the large mail steamers, these vessels were effectively auxiliary steamers in the mould of the *Sarah Sands*: modest, economical, independent of state contracts and often reliant on sail for their longer voyages (Figure 15.1). It was in this context that George Holt's younger brother Alfred first entered steamship-owning (Chapters 17 and 18).

15.2 'And That is all Scott Russell or Brunel Care About': The *Great Eastern* Project

In September 1854, as Lindsay was making ready two purpose-built auxiliary steamers for the Australian and Indian trades, the BAAS held its annual

[7] Stewart (1962: 50); Smith et al. (2003a: 444–53); Smith et al. (2003b: 383–8).

meeting in Liverpool's St George's Hall close to the Lime Street terminus of the L&MR. Indeed, this Parthenon-like building had only been inaugurated a few days ahead of the BAAS meeting in the presence of civic leaders, councillors and merchants, including George Holt (Sr) and his sons Philip, Alfred and Robert. Allocated the north entrance hall, the Mechanical Science Section welcomed Russell as President and William Fairbairn as a Vice-President. Given Liverpool's powerful commitment to iron-hulled sailing vessels and steamers, the theme of iron in shipbuilding was a pre-eminent one. Appropriately, Russell's presidential address promoted a project to build by far the largest iron ship ever contemplated up to that time.[8]

Following his move from Greenock to London in 1844, Russell had taken over Fairbairn's Millwall yard at the end of the 1840s. Among the contracts was a 100-ton schooner yacht *Titania,* built according to his wave-line principle, for Robert Stephenson in 1850. Two years later Russell delivered a pair of auxiliary steamships for the new Australian Royal Mail Steam Navigation Company whose consulting engineer was Brunel. These 1,350-ton iron vessels consumed, according to Russell's performance data, thirty-seven tons of coal per day at 11 knots. Under sail only, they could make speeds of up to 8.5 knots. The aim was to provide extensive passenger accommodation combined with as much space for high-value freight as the coal would allow. Lindsay, however, judged the company to have failed to design economical auxiliary steamers for long-distance voyages east.[9]

In the same year, Brunel proposed to the Eastern Steam Navigation Company a vessel of some 680 feet along the waterline for a service between London and Calcutta via the Cape. This new company had recently lost to P&O in a bid for the overland mail contract between Britain, India and China with a branch to Australia. Backed by influential Australian and Indian mercantile interests with £1.2 million in capital, the Eastern Company now proposed outflanking its rival by an unsubsidized Cape route to Calcutta with a cut in the passage time of up to a week. The decision to propel the vessel by paddlewheels *and* screw propeller resulted from Brunel's consultations with Scott Russell, Joshua Field, and the firm of James Watt in spring 1852. The arrangement may also have had its origins in John Bourne's Treatise (1852).[10]

The 'Great Ship' was laid down in spring 1854.[11] At the request of the BAAS's managers, Russell's address began by linking the project to the Association's

[8] GHD (18 September 1854); Marsden and Smith (2005: 101–02) (BAAS meeting).

[9] Banbury (1971: 242–6); WSL Journals (LND/35/2) (72–3).

[10] Bonsor (1975–80: vol. 2, 579); Lindsay (1874–6: vol. 4, 486–91); Bourne (1852: 227–35); Emmerson (1977: 65).

[11] On the *Great Eastern*'s tortured construction, see Emmerson (1977: 65–157); Griffiths et al. (1999: 136–53); Buchanan (2002: 113–33); Banbury (1971: 248) (yard fire). See also Dugan (1953) (popular history of the *Great Eastern*).

experimental work on hull forms (Chapter 7). He explained that 'since the opening of the Australian trade that line had been fixed upon as an alternative' to the Indian route. He also announced that the vessel 'would supersede smaller ships, in being able to carry sufficient coal to serve for the voyage to and from Australia', with immense savings in coal. The system of coal shipments by sail would thus be redundant. As for the earning capacity, there would be space for 6,000 tons of goods as well as 600 first-class and 1,000 second-class passengers.[12]

The BAAS gentlemen rallied to the cause. Fairbairn, for example, admitted that he now had not 'the slightest hesitation in saying that she would be perfectly stable and strong, and fully able to carry out her objects'. No one knew better than Fairbairn about Robert Stephenson's rigorous experimental testing at the Millwall yard of scaled down box-section girders with a top-and-bottom cellular structure designed to permit heavy railway traffic to pass safely over the proposed Britannia Bridge across the Menai Straits in North Wales. The 'Great Ship' would follow a similar structure with a cellular double-bottom extending up to the waterline.[13]

A dissenting voice was that of Captain Henderson (Chapters 12 and 14). The *Liverpool Mercury* praised his message as one 'based on long experience and practical acquaintance with those elements whose violence and fury so often overwhelm theories and wreck mechanical science'. His arguments were illustrated by comparisons of the relative speed, tonnage, length and breadth of the 'most celebrated clipper mail steam-ships which now sail the waters'. He cited P&O's new Thames-built *Himalaya* as the 'largest and longest vessel yet constructed'. He then explained that 'Holy Writ records that Noah's Ark' would equate to 11,905 tons with a 6:1 length:beam ratio compared to 7.41:1 for the *Himalaya.* Such a publicly expressed biblical literalism was rarely glimpsed at BAAS meetings, since the ruling coterie of broad Church of England gentlemen of science regarded it as undesirably divisive. Judged by literal Mosaic standards, the *Himalaya* appeared to have no great future (Chapter 17), while the prospects for Russell's steamer looked even worse.[14]

Henderson thus predicted that the modern Ark's 'huge size would be an effectual bar to her combining all the requisite qualities of an ocean steamer'. First, it was almost double size of the biblical Ark with an 8.19:1 ratio. As Edward Gillin has shown, the *Great Eastern* later became a text upon which preachers could inscribe a variety of culturally shaped and biblically inspired meanings.[15] Second, Henderson insisted that insufficient consideration had been 'given to *its security in heavy* SEAS, its capabilities for navigation,

[12] *Liverpool Mercury*, 26 September 1854.
[13] *Liverpool Mercury*, 26 September 1854; Pole (1877: 197–213) (Britannia Bridge).
[14] *Liverpool Mercury*, 26 September 1854; Henderson (1854: 152–65); Morrell and Thackray (1981: 224–45) (BAAS values of liberal Anglicanism).
[15] *Liverpool Mercury*, 26, 29 September 1854; Henderson (1854: 152–65); Gillin (2015: 928–56).

and the effect of an ocean wave of ... extraordinary length'. Invoking direct personal experience, he concluded that such wave conditions would make it impossible for a vessel of this size 'to be kept under command ... should any accident happen to her machinery'.[16]

Like Henderson, Lindsay had much personal seagoing experience. Around 1855 he visited Russell's yard. Judging that when finished 'she will cost not far short of a million sterling', he declared that 'as a commercial speculation a more monstrous failure cannot well be conceived'. The ship's size was itself the source of multiple problems. There was as yet no dry dock anywhere to facilitate repairs or bottom cleaning. Passengers and especially freight would 'not easily find their way to such places' as Milford Haven in south-west Wales, safe anchorage that it undoubtedly was. Shippers of goods, Lindsay reckoned, would 'much prefer sending them in a smaller vessel' to ensure a more speedy delivery. With the possibility of up to 10,000 passengers aboard, a contagious disease breaking out would have fearful consequences.[17]

Lindsay also questioned the projected speed, the strength of the hull and the scale and reliability of the machinery. 'She will be one of the wonders of the age', he concluded. 'That is all. And that is all Scott Russell or Brunel care about; for if silly capitalists and a gullible public would only find them the money these two gentlemen would build them a ship even double her size'. Alfred Holt later echoed Lindsay's harsh verdict when he told the Institute of Civil Engineers (ICE) in 1877 that 'he thought size was her chief peculiarity, and that there was not so much to learn from her as other vessels [such as the *Great Britain*]'. Indeed, he remarked sardonically, '[c]onsidering Mr Brunel's genius and the flow of capital his designs attracted, his [Holt's] only wonder was that she was so small'.[18]

As the 'Monster Ship' or 'Leviathan' neared the launching stage in the summer of 1857, Lindsay returned to the site. Accompanied throughout by both Stephenson and Brunel, Lindsay replied to Brunel's inquiry as to what he thought of the vessel with the judgement that 'she is the strongest and best built ship I ever saw and she is really a marvellous piece of mechanism'. Brunel's response was revealing. 'I did not want your opinion about her build', he responded. '... *How will she pay?* If she belonged to you in what trade would you place her?' Lindsay's reply, at which Stephenson laughed and for which Brunel never forgave him, was both facetious and serious:

> Turn her into a show ... something attractive to the masses ... She will never pay as a ship. Send her to Brighton, dig out a hole in the beach and bed her stern in it, and if well set she would make a substantial *pier* and her deck a splendid promenade; her

[16] Henderson (1854: 152–65); Winter (1994: 69–98) (Scoresby on ocean waves).

[17] WSL Journals (LND/35/3) (165–6).

[18] WSL Journals (LND/35/3) (165–6); Holt (1877–8: 70–1); Smith et al. (2003b: 419).

hold would make magnificent salt-water baths and her 'tween decks a grand hotel, with restaurant, smoking and dancing saloons, and I know not what all. She would be a marvellous attraction for the cockneys, who would flock to her in thousands.[19]

Addressing the Fredericton Athenaeum, New Brunswick, in 1857 on the past, present and future of Atlantic steam navigation, civil engineer Vernon Smith represented the *Great Eastern* project as one means of superseding the Cunard Company's 'merest apology for a connection' between Britain and Halifax.[20] In his 'paean for the vessel', he spoke of the ship's remarkable engineering features. Hull safety, for example, was 'almost perfect'. With fifty feet torn from its sides 'she would be comparatively unscathed; cut in two, neither end would necessarily sink; and with two or three of her [watertight] compartments filled with water, she would be scarcely inconvenienced'.[21]

The total bunker capacity, Smith claimed, amounted to 12,000 tons. He stated that the average price of RMSP's coal in 1851 was £3 per ton at the western coaling stations compared to fifteen shillings (seventy-five pence) per ton at home, while P&O employed 400 sailing vessels to deliver coal from Britain to the Company's stations as far as Hong Kong (at forty-two shillings per ton). Avoiding such freight costs, the *Great Eastern* would, on a voyage to Australia and back, save £9,000 on coal alone.[22]

Safety, economy and speed were also enhanced by the employment of three 'different systems of propulsion in different parts of the vessel', namely, paddlewheels, screw and sails. Serving 'a beautifully modelled clipper' hull, these systems of motive power formed interrelated components of a larger system, the ship. In so doing, they exhibited certain key characteristics. First, the four paddle and four screw engines were in separate compartments such that 'an accident occurring to one set of engines cannot therefore affect the other'. Second, the hull's design to achieve 'steadiness in the water will assist the efficiency of her [two] paddle-wheels'. Each revolution of these fifty-six-foot diameter wheels would advance the vessel some fifty yards, and at 10 rpm would enable an Atlantic crossing in six and a half days. Third, 'her six masts, spreading whole acres of canvas, and her four powerful screw engines [driving a single shaft], will be her main dependence'. Under screw alone at some 15 knots, the crossing would still be only eight days. Smith estimated the maximum speed under paddles and screw (but not sail) to be some 20 knots, rendering the ship also 'a powerful engine of war' that could run down any ship, demolish an enemy fleet

[19] Lindsay (1874–6: vol. 4, 513n); Marsden and Smith (2005: 106–07); Burgess (2016: 88) (erroneously describing Lindsay as Brunel's 'friend and fellow engineer' and claiming that they were joined by George (rather than Robert) Stephenson).
[20] Smith (1857: 3, 23).
[21] Smith (1857: 23–4).
[22] Smith (1857: 24).

and yet not be caught. Finally, passenger comfort would not only be increased by the promised reductions in pitching and rolling, but by 'apartments second in size, convenience and refinement to nothing we are accustomed to on land'.[23]

Although Smith's vision for the *Great Eastern* was unambiguously optimistic, he acknowledged two related shortcomings. First, the steamer's deep draft severely restricted the number of accessible ports. Second, the enormous size limited the range of accessible markets for passengers and cargo. He suggested a new, integrated geographical system centred on a 2,400-mile, seven-day crossing between Milford Haven and Quebec. For passengers from Cleveland and Chicago, as well as from Canadian towns, railroads and steamboats would easily converge on Quebec with a full complement. But it was not passengers alone that would supply the *Great Eastern*. Iron, especially iron rails, from South Wales would be loaded at Milford Haven bound for America, while Canadian grain would reach Quebec by canal, river and rail for shipment eastward on the 'Great Ship'. Small wonder traditional ship-owners did not want to believe in the venture.[24]

Smith concluded his lecture with a still more stirring vision of '"the good days coming" when we shall be able to eat our Christmas dinner at home, and meet our friends here on New Year's day or very soon after'. He reminded his audience that over the twenty-year history of Atlantic Steam navigation, there had been no great increase in speed from Britain to New York. The *Great Western* had generally taken fourteen days, the Collins Line vessels had averaged eleven and a half days and Cunard's *Persia* probably just under eleven days. The *Great Eastern*, he asserted, would sadly disappoint if the ship failed to make New York in nine days. He therefore urged 'our Steamboat builders ... [to] use better models, lighter engines, and high-pressure steam; let them see that it does not require a whole coal mine to carry a Steamer across the Atlantic ... and they will still give us an opportunity of dining one Sunday here, and going to Church the next Sunday with our friends in England'.[25]

In January 1859, almost five years after the *Great Eastern* had been laid down, *The Engineer* raised the question of the future of the vessel. Its editorial recalled that the projectors had estimated that a daily coal consumption of 182 tons would achieve a speed of 18 knots. Cunard's *Persia* (1856), 'the fastest ocean steamer in the world', burned 150 tons per day at 14 knots. At 18 knots, however, the consumption doubled. Scaling up by a factor of three, the journal estimated that for the same rate (equivalent to 6 lb per hp per hour) the *Great Eastern* would burn 900 tons a day. In this case, the ship would therefore exhaust the bunkers in thirteen days, making problematic even a round voyage to New York and back without refuelling! Allowing for the *Great Eastern*'s

[23] Smith (1857: 24–6).
[24] Smith (1857: 26–30).
[25] Smith (1857: 30–1).

Figure 15.2. The *Great Eastern* oddly dressed in Cunard funnel colours

engines to work at higher pressures and at a greater degree of expansion, it reckoned that consumption would be 500 tons per day.[26]

The *Great Eastern* finally began the maiden voyage in June 1860. Over the next three years, the ship made a total of nine round voyages to North America. Passage times, typically ten or eleven days to New York, were no better than the regular mail steamers. The best times were nine and a half days from Milford Haven to New York and eight days from Liverpool to Quebec. Even in this short career, the ship suffered serious damage: first, disabled by an Atlantic storm in 1861 and second, holed by an uncharted rock in Long Island Sound. Beset by poor returns and mounting debts, the owners went into liquidation in late 1863. The *Great Eastern*'s passenger-carrying days were over (Figure 15.2).[27]

15.3 'Most Providentially Protected': The Cunard Line of Steamers and its Rivals

On one January night in 1856, a journalist from the *Liverpool Courier* travelled by railway from Liverpool to Glasgow. As the night train approached its destination in the small hours, iron furnaces on both sides of the track 'lighted

[26] *The Engineer*, vol. 7 (1859: 67).
[27] Bonsor (1975–80: vol. 2, 579–85).

up the heavens with a supernatural grandeur' and 'threw an intense light and heat upon the carriages'. Upon arrival, he joined a special train conveying a large party, some 250 strong, to Greenock from where Robert Napier's steamer *Vulcan* conveyed the guests to the new Napier-built Cunard steamship *Persia*, largest and most powerful steamer afloat, lying at the anchorage known as the Tail of the Bank. As the 'Glasgow branch of the Cunard line', Messrs Burns wanted 'to give their neighbours an opportunity of witnessing the performance of the vessel' during two trial trips. As the *Vulcan* neared the Cunarder, the band of the Renfrewshire Militia played 'Rule, Britannia' and Captain Judkins, Commodore of the Cunard fleet and friend of the *Courier* journalist, received the party on board his latest command.[28]

Among the guests were the Revd McNeile (Chapter 14), the Revd Norman Macleod (current minister of Glasgow's Barony Church) (see Introduction), Dr John Strang (Chapter 1), William Brown (Liverpool banker and major Cunard investor), William Connal (Chapter 5), Anthony Trollope (representing the postal service), Lord Provost Orr of Glasgow (who served as chairman of the Clyde River Trust), three local Members of Parliament, one Admiral and two lieutenant colonels and their officers (many of whom had seen recent service in the Crimea). Also represented were Glasgow's Episcopal churches and its Free Kirks.[29]

After the *Persia* had completed a measured run between the Cloch and Cumbrae Lighthouses, the engines slowed to half speed to allow the company to sit down to a banquet in both fore and aft saloons. McNeile, as Dean of Ripon, said grace before, and Macleod gave a similar blessing after, dinner. The public speeches then commenced in earnest with George Burns in the chair. He toasted an absent Samuel Cunard 'as a friend of nearly twenty years standing' and acknowledged the Line's debt to Sir Charles Wood, Secretary to the Admiralty at the time of the original mail contract negotiations around 1838 and now First Lord, for being the first in the country who 'lent a helping hand to this undertaking'. The assistance, Burns stressed, of both Wood and Sir Francis Baring, had 'enabled Mr Cunard and the projectors of the company to link the old world and the new'. Sir Charles, he added, had privately told him that he looked upon 'the Peninsular and Oriental Company and your company as my two children'.[30]

In an avowedly diplomatic move, the Chairman now proposed 'success to their rival line, commonly called the Collins' – the steamers belonging to the American government, and running between Liverpool and New York

[28] *Liverpool Courier*, 16 January 1856; Smith (2014: 77–8).
[29] Smith (2014: 77–8). See also Napier (1904: 194–7) (*Persia* project). Hyde (1975: 13) shows Brown initially holding 116 shares (£11,600), equal to that of Connal.
[30] *Liverpool Courier*, 16 January 1856.

alternately with his company's'. As Burns' shrewd remarks indicated, the Line that had commenced its service in 1850 with four large paddle-steamers (see below) owed its existence and continuation to massive, if precarious financial support from the Federal Government – funds that made the British Admiralty's mail contracts with Cunard appear distinctly modest. In part because Liverpool's banker William Brown (and family associates) had a substantial interest in both lines, the rivalry was much attenuated by co-operation through, for example, agreed sailing days to New York on Wednesdays (Collins) and Saturdays (Cunard). Thus, as Burns told his guests, when the British Government called up many of the Cunard steamers as Crimean war transports in 1854, 'Mr [James] Brown [William's brother] and the other gentlemen who managed that concern [Collins' Line] ... at once handsomely entered into an arrangement to take up the blank Saturday [from Liverpool], on the understanding that when his company should resume, they [Collins] would return to their stated Wednesdays'. Moreover, the United States Government had 'most heartily' sanctioned the agreement. They could therefore 'live in competition, doing all they possibly could to promote the interests of their respective companies, and yet at the same time live in the most perfect friendship'.[31]

Turning to religious matters, Burns stated that he had always felt it his duty 'to acknowledge the favour of divine providence as the spring of all the success which had attended their undertakings'. Although the company now enjoyed prosperity, Burns emphasized, 'he could not forget its past and unseen struggles, which, humanly speaking, could not have been overcome, and which never could have been surmounted without the aid of a divine power'. From his audience arose sounds of 'Hear, hear'. He acknowledged too 'the influence of the clergy in elevating the character and stimulating the enterprise of a community' and accordingly proposed the health of three of the clergy present from the Anglican/Episcopalian, Church of Scotland and Free Kirk denominations.[32]

Disembarking many of the guests at the end of the first trip, the *Persia*'s second trial run took it the 175 miles from the Cloch Light to the Mersey Bell Buoy in ten and three-quarter hours, some five hours 'under the ordinary time' for a coastal steamer. Fuelled with 'Scotch' coal rather than the preferred 'Cardiff coal', 'great volumes of smoke and unconsumed coal were drawn into the funnels, and there ignited'. As the *Liverpool Courier* observed, 'This fact must have caused no little surprise to those who saw the vessel steaming down the [North] channel, with jets of flame shooting from her chimneys from time to time, and with her great speed'. Far from signalling danger, however, the

[31] *Liverpool Courier*, 16 January 1856; Hyde (1975: 37–45) (details of mutual arrangements).
[32] *Liverpool Courier*, 16 January 1856.

presence of unmelted tallow and white lead on the bright work of the engines throughout the passage indicated no overheating of any of the bearings.[33]

Berthing in Liverpool's new Huskisson Dock, the steamer created a sensation on the Exchange, many of the gentlemen there forsaking business in order to see the vessel. Other vessels were dressed overall with flags to mark the ship's arrival. Three days before the maiden departure, invited diners witnessed through the saloon windows the Collins Line steamer *Pacific* passing majestically down the river, outward bound for New York on a regular Wednesday departure. Aboard the *Pacific* were some 286 passengers and crew. On that late January day one of the *Persia*'s party proposed a toast, enthusiastically endorsed, to '[t]he health of Captain Eldridge and the good ship *Pacific*'. To the wish that 'she might have a speedy and prosperous voyage', the proposer expressed the hope 'that his friend Captain Judkins would outstrip her in speed'.[34] Informally at least, the travelling public expected the newest and largest of the Cunard line of steamers to vie with the Collins line for the Atlantic speed record. And so on the Saturday following, the *Persia* too departed for New York.

Son of a shipmaster, New Englander Edward Knight Collins had founded the Dramatic Line of New York sailing packets just before the financial panic of the mid-1830s. Building up to a fleet of five large packets, the Line competed on the Liverpool run against three well-established lines: the Black Ball, Red Star and Blue Swallowtail. Each of the well-named vessels – *Shakespeare, Garrick, Sheridan, Siddons* and *Roscius* – came from the prestigious yard of Brown and Bell in New York. With prior experience of running flat-bottomed coastal packets to New Orleans, Collins chose to model his ocean vessels on these fast craft. The Line quickly established a reputation for the ablest captains, the fastest passages, the best accommodation and the finest food on the Atlantic service. It also maintained an excellent safety record with no losses under Collins' ownership.[35]

By the early 1840s Collins was lobbying the Federal Government for a mail subsidy to operate four 2,500-ton Atlantic steamships. He found an ally in *Herald* proprietor Bennett (Chapter 4), keen to secure European news ahead of his competitors. In March 1847 Collins won a subsidy of $385,000 to operate a fortnightly service for eight months of the year and a monthly service during the remainder. The scheme projected five wooden paddle-steamers of around 2,000 tons, all to be available as ships of war should the need arise. The capital exceeded $1 million with Collins and James Brown each taking about 20 per cent. No expense was

[33] *Liverpool Courier*, 16 January 1856.

[34] *The Porcupine* vol. 9 (1867–8: 399); Albion (1938: 170).

[35] Albion (1938: 43–4 (founding), 84 (Brown and Bell), 90 ('flat-floors'), 123–4 (Dramatic Line), 138 (personal wealth amassed)); Fox (2003: 5, 116–17) (early career).

spared in construction. As costs spiralled, Brown had to raise loans of some $2 million. The number of projected vessels was reduced to four. Instalments paid to the builders faltered until the Federal Government agreed to advance $25,000 per month on each vessel from the time it was launched.[36]

New York shipbuilder William H. Brown, supported in laying down the lines by his skilled loftsman George Steers, shaped the design of the *Atlantic* and *Arctic*. The order for the *Pacific* and *Baltic* went to the unrelated firm of Brown and Bell. On behalf of Collins, two chief engineers of the United States Navy, William Sewell and John Faron, drew up the engine specifications, with Faron joining the Line as engineering superintendent in 1848. Inspecting marine engines in Britain, he returned aboard Cunard's *Niagara* and surreptitiously ascertained that the steamer's operating pressure was 13 psi. Until then, Collins' engineers had been assuming a working pressure of less than 10 psi for Cunard ships. As a result, engine specifications were altered to perform at the higher pressure. Unlike David Elder's unpretentious engines, however, three different designs of patent cut-off valves – cutting off the ingress to allow the steam to continue expanding under its own pressure – appeared in the four ships. The boilers, again designed by Faron, were also unique to the Collins Line with various features to economize on fuel consumption. Contracts for these side-lever engines went to the Novelty Works for the Brown pair and to the Allaire Works for the Brown and Bell duo. Delays with construction and working of the complex steam plants, however, meant that the four vessels entered service at least a year behind schedule.[37]

The first of the quartet, the *Atlantic* attracted wide press attention, not least on account of the quality of the accommodation. 'Between the panels connecting with the staterooms are the arms of the different States of the Confederacy painted in the highest style of the art, and framed with bronze-work', the *Illustrated London News* reported. Lindsay, who had visited the steamers, later recalled that even the *Pacific*'s engines had ornament. One part 'supported a beautiful turret used as an air reservoir, which rather resembles an old castle of the Middle Ages than a steam engine of the nineteenth century, and gave an imposing appearance to the whole structure'.[38]

Not all British visitors liked what they saw of the *Atlantic*. Too beamy to enter Liverpool's enclosed docks, the ship remained anchored in the Mersey. 'She is undoubtedly clumsy'; one observer commented in the *Edinburgh Journal*, 'the three masts are low, the funnel is short and dumpy, [and] there is no bowsprit'.

[36] [Pray] (1855: 375–6) (Bennett and his competitors); Tyler (1838: 150–1, 209); Bonsor (1975–80: vol. 1, 201); Fox (2003: 117–18).

[37] Abbott (1851: 722–34) (Novelty Works); Ridgely-Nevitt (1981: 149–54) (hull and engine variations); Lindsay (1874–6: vol. 4, 205–8); Bonsor (1975–80: vol. 1, 206–7); Fox (2003: 118–21).

[38] *Illustrated London News*, 25 May 1850 (quoted in Bonsor 1975–80: vol. 1, 202–03); Lindsay (1874–6: vol. 4, 209–10).

The ship's ratio, he noted without approval, maintained 'the proportion, as old as Noah's ark, of six feet of length to one of breadth'. A figurehead at the bow was 'of colossal dimensions, intended, say some, for Neptune; others say that it is the "old Triton blowing his wreathed horn", so lovingly described by Wordsworth; and some wags assert that it is the proprietor of the ship blowing his own trumpet'. Amidships, the thirty-six-foot diameter paddlewheels required floats fifteen feet in length, but the Atlantic Ocean had left many of these 'split and broken ... lying about in the water'. The writer acknowledged that despite these problems and a burst condenser, the new ship had made the late spring crossing in twelve days. But, he concluded, 'One voyage is no test, nor even a series of voyages during the summer months: she must cross and recross at least for a year before any just comparison can be made'.[39]

Such comparison centred on crossing times. It was next to impossible to convey to British readers, a *NM* correspondent reflected, 'an adequate idea of the interest which the contests between the English and American mails excite in Boston, New York, and Philadelphia. Each run is carefully noted and compared, fears are excited, hopes raised by every voyage, and half a dozen hours in the length of a trip of three thousand miles is thought a considerable variation'. At the moment, the 'struggle for mastery' lay between Cunard's *Asia* (1850) – which had recently made the fastest eastbound passage – and Collins' *Atlantic* – which had completed the same run in just four hours more. Although minimum passage times remained governed by the mail contracts, these small differences mattered: 'The fleetest vessels must carry out letters, orders, news, government despatches, and, having the prestige of scientific excellence and success, will generally command the choice of passenger traffic'.[40]

By 1851 Collins' steamers appeared to have gained an average advantage of about twelve hours and, accordingly, soon carried over one-third more passengers per annum on the New York run than Cunard. For their internal elegance the ships won the confidence of well-to-do passengers. 'We have Jews and Gentiles, Catholics and Protestants, on board [the *Arctic*], and all tongues are spoken', observed the United States clergyman John Abbott in *Harper's Magazine*. 'Most of them appear to be clerks or younger partners in mercantile houses going out to make purchases. There is, however, an amazing fondness for champagne and tobacco'. Even the disapproving Abbott, however, felt able to pass pleasant evenings shielded from a hostile ocean: 'The dark sea, the storm, the night, were all forgotten, as in that beautiful saloon'.[41]

The Cunard partners had ordered a pair of new 2,200-ton wooden paddle-steamers, *Asia* and *Africa*, in anticipation of competition from Collins' quartet

[39] [Anon.] (1850: 411–14).
[40] [Anon.] (1851: 99–100).
[41] Abbott (1852: 63); Bonsor (1975–80: vol. 1, 78) (passenger numbers).

on the New York line. These Cunard and Collins steamers all entered service in 1850, offering European and American passengers a clear opportunity to see for themselves the rival practices. Crossing on the *Asia* that year under Commodore Judkins, Revd Henry Ward Beecher, pastor of Plymouth Congregational Church in Brooklyn, evangelical preacher, and brother of the celebrated author of *Uncle Tom's Cabin,* duly provoked press controversy over the Line's strict practices regarding divine services at sea. Keen to avoid denominational disputes on board ship, the partners had adopted a Church of England form of service which was conducted by the captain. Furthermore, they only permitted Church of England (Episcopalian) or established Church of Scotland (Presbyterian) clergy to deliver the address 'no matter how many clergymen of other persuasions were on board'. Given the absence of a state-established church in the United States, Beecher strongly objected to the practice. He thus appealed to travellers to avoid 'the disagreeableness of [Established] Church and Steamboat' and asserted that '[w]e would rather wait a month for an American steamer rather than pass the ocean again in one of Mr Cunard's line'. American steamers, he stressed, were 'in speed and safety ... fully equal to the English, and in comfort, politeness and convenience immeasurably superior to them'.[42]

Beecher was by no means alone in finding fault with Cunard practices and in favour of Collins. In 1853, *Harper's Magazine* supplied its readers with authoritative testimony as to the superiority of Collins' system, both in relation to sea-keeping and to passenger–officer relations. Captain Mackinnon, RN, travelled outward to New York on Cunard's *America* (1848) and returned to Liverpool on Collins' *Baltic* (1850). While lauding the unrivalled 'smoothness and beauty with which the [former's] engines performed their work', he identified, ironically given the frequent complaints against the superior airs of RN officers appointed to command mail steamers, 'a degree of pompous mystery in the arrangements of the vessel, very much in contrast with the Yankee steamers'. A lack of information on the ship's daily run, sullen replies from officers to passengers' questions, and a denial of part of the upper deck to all passengers but those special friends of the officers were particular manifestations of this 'mystery'.[43]

In contrast, the *Baltic*'s officers, doubtless with an eye to favourable publicity, conferred 'universal and cordial civility' upon their British guest. Invited by 'the experienced commander', Captain Comstock, into his cabin to discuss 'the now clear indications of an approaching storm', Mackinnon represented his own role as that of an equal authority on those 'infallible tokens [of an

[42] Henry Ward Beecher, 'Church and Steamboat – Cunard's Line' (September 1850). Copy retained among CP, LUL.

[43] MacKinnon (1853: 205).

imminent gale] known only to experienced seamen'. Permitted on the bridge to witness for himself the steamer's performance, he conveyed to his readers his fascination with 'the wonderful triumph of the ship's course over the madly vexed waters, ... [we] remained in our exposed position, spell-bound, at her easy performance over such rough and formidable obstacles'.[44]

Such printed displays of confidence were just what Collins wanted. Rumours of mechanics working round the clock between voyages to repair damage could be discounted. But unwelcome news of major machinery problems at sea was difficult to suppress. Just a few months into service, the *Atlantic*'s main shaft broke after nine days of heavy gales some 900 miles off Halifax while bound for New York. Amid the January weather, the disabled steamer lay to under sail for thirty hours before the floats could be removed from the paddlewheels. Concerns about the non-arrival began to mount. Unable to make westward under sail, the *Atlantic* ran the 1,400 miles to Queenstown (Cork Harbour) from where two Cunarder's took over the passengers and freight. Eventually the steamer arrived under tow at Liverpool where repairs lasted six months. Ridgeley-Nevitt has shown that over the first four years of the Collins Line, one or more of the steamers was frequently absent from service for extended periods. In 1852, for example, the *Baltic* made only five round voyages and remained out of action completely between July and October. To the initiated, the evidence pointed to a story of heavy wear and tear linked to demands for rapid crossings. Even the *New York Times* reluctantly admitted in January 1852 the superiority of the Cunarders in terms of power, durability and strength.[45]

On 27 September 1854, seven days out of Liverpool, the *Arctic* found itself blanketed in a dense and blinding fog. At noon, Captain James Luce 'announced the splendid run of the twenty-four hours previous'. Passengers began seating themselves for lunch. Suddenly there came a crash, an alarm and a stopping of engines. The small French iron-screw steamer *Vesta* – probably the only other vessel in a hundred-mile radius – had rammed into the bows of the big Atlantic liner 'as a swordfish might wound a whale'. As the *NAR* expressed it, in 'the mysterious ordering of Divine Providence, the *Vesta* had struck the blow at the instant when the *Arctic*, just rising on the swell, exposed herself to the peril where she was most vulnerable'.[46]

Mortally wounded below the waterline, the *Arctic* began to settle by the head. As the order came to abandon ship, discipline broke down as 'some of the officers and most of the crew, throwing off all restraint, sought only to

[44] MacKinnon (1853: 206–7).
[45] Ridgely-Nevitt (1981: 156–7); Bonsor (1975–80: vol. 1, 203–4); Fox (2003: 125) (quoting the *New York Times*, 17 January 1852).
[46] [Anon.] (1864: 496).

save themselves'. Witnesses later reported 'the shameful sight of boats half filled with strong men pushing away from the vessel on whose deck a hundred women and children were standing helpless'. All the woman and children perished – including Collins' wife, son and daughter. James Brown lost two daughters, his son and daughter-in-law and two grandchildren. The exact number of deaths was never clear but the *NAR* gave the total as 322. Under one-third of the survivors were passengers.[47]

Critics who had previously hailed the Line for its progressive spirit now attributed to it every kind of extravagance and recklessness. Heading the list was speed. Echoing a pervasive blame, the *NAR*'s mantras were stark: '[t]oo impatient and too fast' and '[t]his expensive and reckless navigation'. The high value attached to speed was underlined too by the publication of league tables (often contested on account of the absence of agreement on the parameters) of the fastest passages.[48] Alongside 'speed' went recognition of other ways in which Providence had been tempted. There had not been sufficient lifeboats even if all the places had been taken up. The steamers had no watertight compartments. The crew, appointed afresh each voyage, seemed to lack loyalty and discipline. Captain Luce had undertaken an ill-judged attempt to steam the damaged *Arctic* for the distant coast. While many sermons invoked Jeremiah's words that 'there is sorrow on the sea', the Revd Beecher, ignoring his earlier recommendation to entrust their lives to American rather than Cunard steamers, now pronounced that 'God is striking thundering strokes at the wealth of the whole community'.[49]

Five days out of Liverpool on the maiden crossing in January 1856, the iron-hulled *Persia* was steaming at 11 knots when it 'struck heavily on a field of ice'. According to near-contemporary reports, the collision resulted in 'a large hole through the plates of her iron bow, tore rivets asunder for sixteen feet on her starboard side, and bent and twisted the rims of her paddle-wheels ... The first compartment instantly filled, but the water-tight bulkhead saved her, and though laden down heavily by the head, she was enabled slowly to keep on her course, and reached New York in safety, though much behind the anticipated time'. But the *Pacific* had not arrived. Ominously, the steamer *Edinburgh* on a passage from Glasgow to New York picked up cabin furniture and a lady's workbox in the same area in early February. Later that month, several transatlantic steamers (including the *Atlantic*) reported seeing an island of ice. By March, George Holt recorded in the family diary: 'No accounts of the steam

[47] [Anon.] (1864: 497); Fox (2003: 128–35). Brown (1961: 168–76) gives passenger and crew lists that produce 350 passengers and crew lost and 85 survivors of whom only 24 were passengers.

[48] [Anon.] (1864: 496–8); Ridgeley-Nevitt (1981: 157–8) (arbitrariness of departure and arrival points).

[49] Jeremiah 49:23; Brown (1961: 149–50, 167) (listing six clergymen who preached sermons on the disaster, including Beecher); Albion (1938: 171) (Luce).

ship *Pacific*. It is now to be feared that all onboard are lost!' The *Pacific* had gone missing, presumed to have struck ice in the Atlantic.[50]

'We have now been nearly twenty years connected in business, a large portion of our ordinary life, and *we have been most providentially protected*', Samuel Cunard wrote to Charles MacIver in 1858. '– see how many great houses have fallen around us during the last year, who were at the commencement of the year in affluence –'.[51] Not only had the Line weathered financial storms better than any of its competitors, but between 1843 and 1858 at least thirteen transatlantic steamers belonging to rival firms had been lost, several with heavy casualties, compared to one Cunarder with no losses of passengers or mail (Chapter 6).

During the first half of the 1850s, different ship-owners and shipbuilders chose to shape a striking range of steamship systems according to their local contexts and larger geographical ambitions. With capital generated from the supply of steam coal to British mail lines around the globe, Lindsay and his Liverpool friends cautiously invested in screw steamers, often sail-dependent, and employed primarily on cargo services. At the other extreme, Russell and Brunel projected the 'Great Ship' designed to serve as its own coal transporter. Collins, following Cunard, instead opted for high-class wooden paddle-steamers for its rival transatlantic mail service, adding a culture of Collins luxury to Cunard comfort. Everywhere, it seemed, did the press hail the sea-kindliness and passenger satisfaction of the American newcomers. But the sudden, tragic collapse of the former's grand system generated a public narrative of Cunard safety over Collins recklessness that long defined the reputation of the two lines.

[50] GHD (12 March 1856); [Anon.] (1864: 500–1).
[51] Samuel Cunard to Charles MacIver, 1 January 1858, CP, LUL. My italics.

16 'The Engines Were Imperfect'

Pacific Steam's Coal Economy

> It was during the Russian [Crimean] war, when tonnage for the conveyance of coal hence [from Britain] to the Pacific became so scarce, and the cost of the article abroad was thereby more than doubled for a time, that we were led to inquire into the question of a saving of coal. Mr Elder was called in and consulted, and the double-cylinder engine adopted ... with a success far beyond our most sanguine expectations, or the advantages held out by Mr Elder himself. Indeed I am in fairness bound to admit, that his double-cylinder engines never exceeded the promised consumption, nor fell short of the guaranteed speed. On the contrary, the promised results were always more than realised ...
>
> *PSNC's managing director William Just looks back on John Elder's role in the introduction of the marine compound engine into ocean steam navigation (1870)*[1]

Summary

William Just's recollection masked the protracted gestation of the marine compound engine and the resistance against which David Elder's son John and his colleagues struggled in the mid-1850s to win PSNC's trust. At the close of the 1840s, the Liverpool firm had ordered four large steamers from Robert Napier. Witnessed by PSNC directors, the first of these steamers spectacularly failed to meet performance promises as to coal consumption. Indebted to, working within and backed by Glasgow networks of scientific and practical engineering, Elder persuaded the sceptical ship-owner to adopt the new system as one reliably grounded on an experimental knowledge of the principles of heat engines. Yet his task of implementing the marine compound engine in practice, and resisting PSNC's rising stock of doubt, provided a challenge of no small magnitude.

16.1 'It is Easier to Instil Poison Than to Extract it': PSNC's Inauspicious New Beginning

The financial losses of PSNC during its first five years had amounted to two-thirds of its paid-up capital. Largely on the strength of the 1845 mail contract, together with fresh confidence injected by the new Board (Chapter 11), PSNC

[1] William Just to [Rankine], 21 October 1870, in Rankine (1871: 66).

ordered its first iron paddle-steamer. In January 1846 Tod & MacGregor delivered the 323-ton *Ecuador* for the Callao to Panama service that linked to RMSP's mail steamers overland across the isthmus. A second iron steamer, the 694-ton *New Granada* from Smith & Rodgers (sited next to Robert Napier's yard in Govan), entered the same service in August.[2]

In late 1845 PSNC recruited Just, a director of the Aberdeen Steam Navigation Company, and appointed him joint managing director with Wheelwright. The Aberdeen Steam Navigation Company (ASNC), operating passenger steamers between Aberdeen and the British capital, generated healthy profits of around £40,000 per annum. Its 1,067-ton *City of London*, constructed at Napier's in 1844 as one of his earliest iron paddle-steamers, cost £32,700 – almost as much as Cunard's first Atlantic steamers. Evidently well-pleased with the steamer, the Aberdeen line ordered two further vessels from Napier for delivery in 1847. Given the likelihood of Just's close involvement with these contracts, PSNC's next order, for the iron-hulled *Bolivia*, went to Napier in 1849. Meanwhile, the Board had sent Wheelwright to the Pacific coast, leaving Just in command of the Liverpool office.[3]

Gaining in confidence, PSNC ordered four iron steamers (*Santiago*, *Quito*, *Lima* and *Bogota*) from Napier of similar specification to the first transatlantic Cunarder's. At this time, James Robert Napier (one of Robert's two sons and with ambitions to be a scientific naval architect) managed the building yard at Govan, David Elder remained in charge of the engine works at Lancefield and Vulcan and John Elder (one of David's sons) occupied the position of chief draughtsman. The contract specified a speed of 12 knots with 500 tons of fuel and cargo, apparently some 150–200 tons more than the maximum that the vessels were designed to carry while in regular service. The PSNC, however, requested an 'increased tonnage ... over those originally contracted for'. On behalf of his father, James undertook to cost the increases and 'offered to build the sharper one'. The first of the quartet, *Santiago*, emerged with a lower tonnage (961-tons) than the other three ships (1,461-tons) for almost the same length but one foot less beam. The likelihood is that the hull was also 'sharper' at the bow and stern, with the expectation of greater speed and economy.[4]

In August 1851, the *Santiago* made a trial trip from the Mersey with PSNC directors on board to witness the expected performance. Overseeing the trials, James Napier had arranged for 420 tons of coal and 80 tons of other stores to make up the 500 tons weight in the absence of actual cargo. 'The Santiago has done miserably today', he reported to his father as he struggled with the

[2] Haws (1986: 30); Wardle (1940: 70–4). This chapter draws extensively on Smith (2013b and 2014).

[3] Somner (2000: 12–15, 58) (*City of London*); Wardle (1940): (Just's appointment); Haws (1986: 30) (*Bolivia*); Smith (2014: 86–7).

[4] James to Robert Napier, [c. 1851], NP; Napier (1904: 260); Haws (1986: 30–1).

engines (overheating and with no vacuum) and with the paddlewheels (too deep in the water). 'I trust things will come right ... [T]o Dublin & back will be a famous trial. I shall take care they burn or waste plenty of fuel that we may be as light as possible'. The owners, however, refused to accept the *Santiago* until Napier's had completed expensive alterations. Thereafter, they claimed penalties for delays in delivery.[5]

From Valparaiso in January 1852, Wheelwright relayed to PSNC a tabulated report on the *Santiago*'s long-distance runs between that port and Callao. The directors pronounced these results 'very unsatisfactory', especially with regard to 'the extraordinary consumption of coal onboard the vessel', and instructed their Pacific coast manager, Mr Mathison, 'to dispose of her if possible' at a price of £25,000 or even lower 'as it is very desirable to get quit of said vessel'. The Line now seriously considered its choices for future vessels, including auxiliary screw steamers fitted with a low-powered engine and detachable propeller.[6]

Napier admitted to his son that PSNC's reports on the *Santiago* 'were any thing but satisfactory ... either as to vessel or machinery [–] consuming an enormous amount of coals & no speed'. Concerned to protect his reputation, he seized upon anomalies in the log. In particular, the vessel seemed to move at the same speed whether the paddlewheels turned at fourteen or nineteen and a half revolutions per minute. He therefore pointed out to the directors that this and other 'errors' showed 'how little dependence can be placed on their logs'. His annoyance, he told James, derived from 'the evidence contained in the different letters [received by the owners] that the early captains had done their work thoroughly of poisoning the minds of the agents & Mr Wheelwright as to your being the cause of the failure ...' Furthermore, 'I have done what I could to counteract these calumnies but it is easier to instill [sic] poison than to extract it'. By thus 'running down Santiago', directors had 'carried their point against us' and 'were trying to lay the blame on us'.[7]

The large geographical distance between witnesses and PSNC contributed to the weakening of trust between the owners and contractors. In this period the steamers rarely if ever returned to British waters. Engine logs, moreover, were only as reliable as the engineers who compiled them. On the other hand, shipmasters were often intensely loyal to the sea-keeping qualities of the vessels under their command. Robert Napier derived consolation from captains of two other members of the quartet, *Lima* and *Quito*. Most notably, Captain Wills of the *Quito* had reported on the vessel's performance in 'a hurricane which

[5] James to Robert Napier, 7, 8 August 1851, NP; Napier (1904: 184–5); Smith (2014: 87–8).
[6] PSNC Minutes (17 April and 8 May 1852).
[7] Robert to James Napier, 25 May 1852, NP; Smith (2014: 88); Smith and Scott (2007: 481–94) (Napier's reputation).

Figure 16.1. PSNC's coal-hungry *Quito* (1852)

frightened him at first lest the storm swamp her. She however rose to it like a duck & never shipped a drop of solid water & what is to me most singular the consumption of fuel is reported to be moderate' (Figure 16.1).[8]

One solution to the problems of credible witnessing was to despatch an experienced engineer to the Pacific. John Elder's brother David took a passage on one of the new steamers with a view to seeing the problems for himself. His involvement also extended to the whole PSNC system of operations. In March 1852, for example, he enclosed in a letter to PSNC '[a] specification and tracings of a steam engine boiler required to be sent out'. In April he submitted reports from himself and 'from the engineer of H.M. Ship *Virago*, on the quality of the Talcahuano and Coronel Coals'. Another letter was 'accompanied by a statement of repairs done to the steamers *Santiago* and *Lima*, chargeable to Mr Napier, and amounting to \$673.7'.[9]

By August 1852 PSNC had received news 'of the reported improvement in the working of the *Santiago*' but expressed much surprise that they had not also received 'a full explanation of the causes of said improvement'. David Elder subsequently supplied an explanatory letter which is unfortunately no longer extant. He also reported 'favorably of the working of the "Quito" on the voyage from Callao to Panama' and attended to 'the partial derangement' of one of the engines of the sistership *Bogota*. At the beginning of the following year, PSNC raised the sale price of the *Santiago* by some £10,000 'in

[8] Robert to James Robert Napier, 25 May 1852, NP.
[9] PSNC Minutes (1, 29 May and 12 June 1852).

consequence of the great rise in the price of iron and the increased cost of iron ships in this country', but probably also because of the turnaround in the vessel's performance.[10]

The problems with coal consumption for PSNC, however, were far from over. The Napier steamers all continued to consume prodigious quantities of coal compared to their smaller predecessors. With local coal supplies limited and of low quality along much of the west coast, the continued use of sailing vessels to ship Welsh coal halfway across the world was costly in terms of money, men and materials, return guano shipments notwithstanding (Chapter 11). And then, in 1854, the outbreak of war in the Crimea meant high freight rates and a corresponding dearth of tonnage.[11]

16.2 'A Much Greater Amount of Mechanical Effect': Glasgow's Scientific Engineers

In the early 1840s John Elder served a five-year apprenticeship with Napier's firm under the direction of his father in the pattern shop, moulding shop and drawing office. John's only higher education was a brief period studying with Professor Lewis Gordon, occupant since 1840 of the regius (that is, crown-appointed) chair of civil engineering and mechanics at the University of Glasgow.[12] Elder then spent about a year as a pattern-maker with the steam engine-builder Hick of Bolton in Lancashire and as a draughtsman at the works of the Grimsby Docks in Lincolnshire. On his return to Glasgow, probably in late 1848, he rejoined Napier's firm and soon took charge of the drawing office in the iron shipbuilding department (Chapter 2).[13]

On 13 December 1848 Elder became a member of the Glasgow Philosophical Society (GPS) (founded in 1802). Members included J. P. Nichol (professor of astronomy at the university), Thomas Thomson (professor of chemistry), James Thomson (professor of mathematics), Lewis Gordon, Walter Crum (cotton manufacturer), Charles Randolph (millwright) and most recently, James Thomson (engineer and son of the mathematics professor) and his younger brother William (professor of natural philosophy from 1846). Elder had joined a network of like-minded scientific and industrial reformers.[14]

Soon after becoming a member in 1840, Gordon had addressed the Society on a theme that later captured the attention of younger GPS members engaged

[10] PSNC Minutes (7 August, 11 September, 6, 20 November 1852 and 8 January 1853); Smith (2014: 88–9).
[11] Rankine (1871: 66).
[12] Marsden (1998a: 87–117).
[13] Rankine (1871: 4–6).
[14] Morrell (1974: 81–94); Smith (1998a: 37–9); *Proceedings of the Glasgow Philosophical Society* vol. 3 (1848–55) (membership).

in marine engineering science. The subject of his presentation was a little known continental interpretation of the motive power of heat published in France by Sadi Carnot in 1824. A more analytical, better known version by Emile Clapeyron followed a decade later, with a translation for Richard Taylor's *Scientific Memoirs* in 1837.[15]

A close friend and student of Gordon, James Thomson recognized the full significance of this French reading of heat engines while working at the Fairbairn shipbuilding yard on the Thames in August 1844 (Chapter 14). Using Gordon's term 'mechanical effect' (which Gordon had adopted from the German *mechanische Wirkung*) for the work done by an engine, he wrote to his brother William that 'I shall have to enter on the subject of the paper you mentioned to me'. James urgently wanted to know 'who it [is] that has proved that there is a definite quantity of mechanical effect given out during the passage of heat from one body to another'. Having recently studied water-wheels with Gordon, he also noted the resemblances with heat engines. Just as a measurable quantity of water 'fell' between two levels, so a measurable quantity of heat 'fell' between a higher (boiler) temperature and a lower (condenser) temperature. And the greater the difference between high and low levels (or temperatures), the greater the useful work that could be obtained from a given quantity of water (or heat). This letter sparked an intensive research programme, focused on the Carnot–Clapeyron reading of heat engines, developed by the Thomson brothers.[16]

In the autumn of 1846 William Thomson, aged twenty-two, had been elected to the natural philosophy chair in Glasgow College. The University was currently proving attractive to the sons of several Clyde shipbuilding and marine engineering families, including John Scott, William Simons, John Elder, James Napier and John Caird.[17] Without any such family credentials, an unknown twenty-year-old, Robert Mansel, attended Thomson's first undergraduate class from November that year. A native of Glassford, Lanarkshire, Mansel had previously studied at the Andersonian Institution, well known for its industrial and practical ethos, and at the Glasgow Mechanics Institution.[18]

During his first six-month session, the new professor initiated a radical overhaul of the lecture demonstration apparatus and introduced precision experimental research into the venerable rooms that constituted the College territory of natural philosophy. At the end of the session he selected Mansel as his

[15] Thomson (1847: 169–70); Fox (1986); Smith and Wise (1989: 296); Smith (1998a: 47).

[16] James to William Thomson, 4 August 1844, Kelvin Collection (KC), ULC; Smith and Wise (1989: 289–90); Smith (2013b: 517–18).

[17] Smith (1998a: 29 (Caird), 151 (Elder)); Marsden and Smith (2005: 117–19) (Scott); Smith and Wise (1989: 24) (Napier); William Thomson's Natural Philosophy Class register 1847–8, Glasgow University Archives (GUA) (Simons).

[18] [Anon.] (1905–06: 452) (Mansel obituary).

experimental assistant. Mansel continued in the role for the following three sessions. During those years, he and the Thomson brothers laid much of the experimental groundwork for a new science of the motive power of heat that they termed 'thermo-dynamics'.[19]

In clearing out his predecessor's stockpile of apparatus early in 1847, Professor Thomson had come upon a model Stirling air engine which used air rather than steam as its working substance. Originally patented in 1816 but granted a fresh patent in 1840, the Revd Robert Stirling's engine quickly assumed both a scientific and a practical significance for the Thomsons. Moreover, Stirling's commercially minded brother James had recently addressed the ICE on the engine's economic potential for marine use. He explained the value of increased space for freight: 'every ton of fuel that is saved is not only a direct gain to the extent of its market price, but in long voyages an indirect gain is effected of at least twice as much more, from the increased accommodation which it allows for the carriage of merchandise'.[20]

A month after the discovery of the model engine, Professor Thomson addressed a short paper entitled 'Notice of Stirling's air-engine' to GPS members. In doing so, he carefully confined himself to physical matters and avoided any trespass upon engineering or business territories.[21] Explaining the engine's action in terms of the Carnot–Clapeyron reading of heat engines, he inferred that the air engine operated with a much greater temperature difference than a steam engine, thereby delivering in principle 'a much greater amount of mechanical effect ... by the consumption of a given quantity of fuel'.[22] Over the next year, his brother investigated the practical possibilities. With his father, brother and Mansel as witnesses, James discussed the working of the air engine with Robert Stirling himself. Keenly aware of patent conventions, he also told the clergyman–inventor 'not to tell me anything that he did not regard as entirely public, because I had some ideas on the subject myself'.[23]

In his 1847 address, however, William had arrived at an unexpected inference from the engine's working. The Carnot–Clapeyron interpretation *seemed* to point to heat transfer, under certain conditions, from a cold region to a warm one within the engine, such that ice could be produced without expending work. But, as he told James later, since water expands on freezing, the process could be made to do work given the right mechanism. If, therefore,

[19] Smith (1998b: 118–46) (Thomson's reform of Glasgow natural philosophy); Smith (2013b: 518–23, 528) (Mansel); Smith (1998a: 77–169) (science of energy).

[20] Thomson to Forbes, 1 March 1847, Forbes Papers (FP); Stirling (1846: 559–66); Smith and Wise (1989: 294); Marsden (1998b); Smith (1998a: 48).

[21] Thomson (1847); Marsden (1998a: 87–117) (disciplinary sensitivities in Glasgow University).

[22] Thomson (1847: 169–70); Smith and Wise (1989: 296–8) (with explanatory diagram). I thank Norton Wise for the concise analysis of Thomson's paper.

[23] James Thomson, 'Motive power of heat: Air engine', Notebook A14(A), James Thomson Papers (JTP), Queen's University Belfast (QUB); Smith (2013b: 520–1).

ice could be made without the consumption of work while in the process the expanding water generated work, a perpetual source of power would be possible. Committed to the impossibility of obtaining work for nothing, James responded by arguing that the production of external work from the expansion of the freezing of water required a temperature difference. He therefore predicted that ice under pressure would involve a lowering of the freezing point. Were this prediction to be experimentally confirmed, the Carnot–Clapeyron account would win credibility as a central foundation of the new science of heat engines (thermodynamics).[24]

In January 1850 Mansel constructed an exceptionally sensitive (ether) thermometer. Professor Thomson excitedly informed his natural philosophy class in January 1850 that it was 'assuredly the most delicate that ever was made, there being 71 divisions in a single degree of Fahr[enhei]t'. A few days earlier he had communicated to his fellow-professor in Edinburgh, J. D. Forbes, the sequence of experimental work almost as it happened. At first, he expressed pessimism over the great difficulties of making comparisons using existing thermometers, rendering it impossible 'to attain any satisfactory accuracy'. Later in the same letter Thomson admitted that they were not obtaining the hoped-for agreement with James' theoretical prediction. Everything, however, changed while he continued to compose the letter. 'As soon as we got the thermometer (hermetically sealed in a glass tube) into Oersted's apparatus [for demonstrating the compressibility of water] everything was satisfactory', he wrote. 'The column of ether remained absolutely stationary until pressure was applied'. Measurements of the decreases in temperature on Mansel's instrument matched James' theoretical predictions such that 'the agreement is wonderfully satisfactory'. Indeed, he confessed that, having only just become aware of the agreement, it 'really surprises me by being so close'.[25]

Throughout the experiments, Mansel was a full participant – if a largely invisible experimental assistant – in this critical and formative period for the analysis of the motive power of heat. '[D]own to the spring of 1850, we have an interval in which the mechanical theory of heat received its first thorough exposition', Mansel later wrote. 'It was the salient feature of Sir William Thomson's physical teaching during this period, and I have never forgotten the enthusiasm with which the then young professor entered upon and conducted this subject, nor the generous recognition he bestowed upon the labours of his predecessors and contemporaries whether experimental or mathematical ...'[26]

[24] Thomson (1849: 575–80); Smith and Wise (1989: 298–9) (full exposition of James' reasoning by Wise).

[25] Thomson to J. D. Forbes, 10 January 1850, FP; Smith (1998a: 81, 96–7, 1998b: 6–29, 129–30).

[26] Mansel (1882: 91–2).

With an intensity of iron shipbuilding and marine engine-building on the Clyde taking place as never before, the Napier's appointed Mansel to their shipbuilding yard where he 'acquired a practical knowledge of iron ship building' to add to his exceptional skills with precision measurement. It is likely that for a time Mansel served both in Thomson's classroom and in the yard before being appointed in 1850 as naval architect, a position that he retained for some thirteen years. During that period he played a role in designing the hulls of the record-breaking Cunard mail steamers *Persia* (1856) and *Scotia* (1862), as well as the iron-clad HMS *Black Prince* (sister to the Thames-built HMS *Warrior*). On his departure from Napier's, Mansel founded (with James Aitken) his own yard which, over a thirty-year period, built many vessels, including ocean steamers for the Union Steamship Company running to South Africa and cross-channel steamers for the London and South-Western Railway.[27]

From at least 1850 until mid-1852 Robert Mansel (as naval architect), John Elder (as head of the drawing office) and James Napier worked face-to-face with one another at Robert Napier's iron shipbuilding site at Govan. From Thomson's laboratory, Mansel brought to the yard a hands-on familiarity with the experimental practices and mathematical principles of the science of the motive power of heat. Glasgow's scientific engineers now actively sought to embody the new science of thermodynamics into two innovative systems of marine motive power. First, James Napier began a joint project with his friend Rankine (soon to be Gordon's successor in the Glasgow engineering chair) to design a marine air engine in accordance with thermodynamic principles. Those principles were set forth in Rankine's address to the BAAS's Liverpool meeting in 1854. Difficulties with the experimental model, however, precluded the construction of full-scale air engines to their design.[28]

Second, Elder left Napier's yard in the summer of 1852 to become a partner in the firm of Randolph, Elliott & Company based at Centre Street on the south side of the Clyde close to central Glasgow. Formerly known as Randolph, Cunliff, the firm had employed James Tennant Caird in his youth (Chapter 10). Son of a Stirling printer, Charles Randolph had studied arts courses at the University of Glasgow, but found the lectures of Andrew Ure at the Andersonian more to his liking. Apprenticed to Robert Napier and trained by David Elder at the Camlachie and Vulcan works (Chapter 2), Randolph then joined in turn the Manchester engineering firms of Omerod and Fairbairn (Chapter 14) before setting up his millwright business with his cousin Richard Cunliff, yarn merchant, in Glasgow in 1834. John Elliott, a former manager at Fairbairn's, became a partner in 1839. The firm, with a reputation for the

[27] [Anon.] (1905–6: 452); Fox (2003: 163, 190).

[28] Rankine (1855: 1–32); Marsden (1998b: 395–400); Smith (1998a: 150–7); Marsden and Smith (2005: 73–5, 113–16).

accuracy of its gear-cutting and machining practices, supplied a very wide geographical range of textile manufacturers across Scotland, Ireland and continental Europe. It was from the Centre Street works that Randolph, Elder took out their first patent for a double-cylinder marine engine in January 1853. They then fitted this system to the screw steamer *Brandon* whose seagoing performance George Peacock witnessed for himself. The *NM* published a highly favourable report on the fuel consumption. Soon after the PSNC contracted for two sets of paddle-steamer engines to Randolph, Elder's modified design.[29]

16.3 'To Meet All Doubters on Their Own Grounds': John Elder's Powers of Persuasion

Master engineer John Elder, in business with Randolph, quickly forged a good relationship with PSNC's William Just. Alongside shipbuilder John Reid, Elder acted as a consultant to the Liverpool firm, especially in connection with the purchase of secondhand tonnage. The PSNC also finally settled with Napier the long-running dispute over the troublesome quartet. Freed from this millstone, the Board instructed Just to correspond with John Wood, Port Glasgow shipbuilder (Chapter 2), 'with a view to obtaining a suitable model and specification' for a mail steamer to replace the recently lost *Quito*. As the celebrated shipbuilder contemplated retirement, Reid, who had married Wood's niece, gradually integrated Wood's firm into his own.[30]

Once PSNC had considered tenders from five firms, it opted for Reid's offer to build the hull, with engines from Randolph, Elder, at an agreed total price of £31,600. Were double-cylinder engines adopted, the extra price would be 'in proportion to the ascertained saving of fuel, by the use of the said engines, the aggregate amount of such extra price not to exceed the sum of fifteen hundred pounds'. Requiring a trustworthy and independent authority with the skills to evaluate these promised savings in fuel, the PSNC Board invited William McNaught to undertake the role of superintendent engineer during construction of the new engines for a fee of £150. He would also act as independent arbiter of the fuel savings over conventional engines during trials.[31]

McNaught had begun his career as an apprentice under David Elder in Napier's Vulcan Works around 1827. In 1845 he took out a patent (known colloquially as 'McNaughting') for adding a second, high pressure cylinder

[29] On Randolph, see [Anon.] (1886: vol. 2, 267–8); 'Randolph, Charles (1809–1878)', *ODNB*; [Anon.] (1854: 507–9) (*Brandon*); Rankine (1871: 6, 31); Napier (1904: 183–6).

[30] PSNC Minutes (17 September, 15 October and 19 November 1853; 18 February, 11, 18 March and 3 June 1854; 15 March 1855); Shields (1949: 76) (on Wood and Reid); 'Wood, John (1788–1860)', *ODNB*.

[31] PSNC Minutes (29 March, 12, 26 April and 3 May 1855). The 'saving of fuel' arrangement echoed that of Watt with Cornish mine owners. See Marsden and Smith (2005: 57).

to traditional beam engines for the textile industry. Moving to Manchester, he exploited a large market for the McNaughting of stationary steam engines among the cotton mills of Lancashire. His father had earlier designed a steam engine indicator with a small revolving cylinder to carry the paper upon which the pen traced the cyclical pressure within a steam cylinder. From 1838 William worked alongside his father in the manufacture and marketing of these self-registering instruments. The enclosed area of the resulting indicator diagrams would represent work done from which they could readily calculate the indicated horsepower (ihp). The number of pounds of coal per ihp per hour provided a measure of performance among different designs of steam engine. These indicators would be central to the business of evaluating performance claims for PSNC's marine engines.[32]

As the contract for the mail steamer *Valparaiso* took shape, Elder successfully negotiated a second order, this time for a small branch line coastal steamer (*Inca*) to be built by Steele (another builder with strong links to Wood through construction of the first series of Cunard steamers).[33] Laid down in November 1855 and launched in March 1856, the *Inca* completed builders' trials on the Clyde prior to arrival in the Mersey two months later. Exceptionally fast construction coincided with approval of Randolph, Elder's second major patent for a double-cylinder arrangement in the spring of 1856 (Figure 16.2). As the prototype for the new patent, the *Inca* had much to prove. McNaught's verdict, however, was little short of damning. Having 'tried' the engines, he 'found their indicated power considerably less than it ought to be'. After face-to-face meetings with McNaught and Elder, PSNC's chairman reported to his Board that Elder had admitted in the presence of himself, the secretary, the marine superintendent and PSNC's solicitor 'that the engines were imperfect, and that the steam had fallen off to six pounds [psi], and could not be maintained; [and] that Mr Elder proposed to take the vessel back to Glasgow, as there he had the proper means of remedying the defects of the engines'.[34]

The same Board meeting received from Randolph, Elder indicator diagrams that contradicted McNaught's conclusions. Soon after, the Board secretary reported an interview with McNaught who 'considered the diagrams taken by Mr Elder [as] altogether fallacious'.[35] The seemingly inconsistent verdicts hint at a fundamental divergence of understanding between McNaught and Elder. The former increasingly regarded the new engines as flawed in practice.

[32] Hills (1989: 57–9); 'McNaught, William (1813–1881), *ODNB*.

[33] PSNC Minutes (10 May, 12 July, 30 August, 13 September, 4, 18, 25 October, and 22 November 1855).

[34] PSNC Minutes (15 November 1855; 7 March, 24 April, 15, 22 and 30 May 1856); Rankine (1871: 31–2) (patent dated 28 February 1854). The steam pressure probably should have been in the range 20–30 psi.

[35] PSNC Minutes (30 and 31 May 1856).

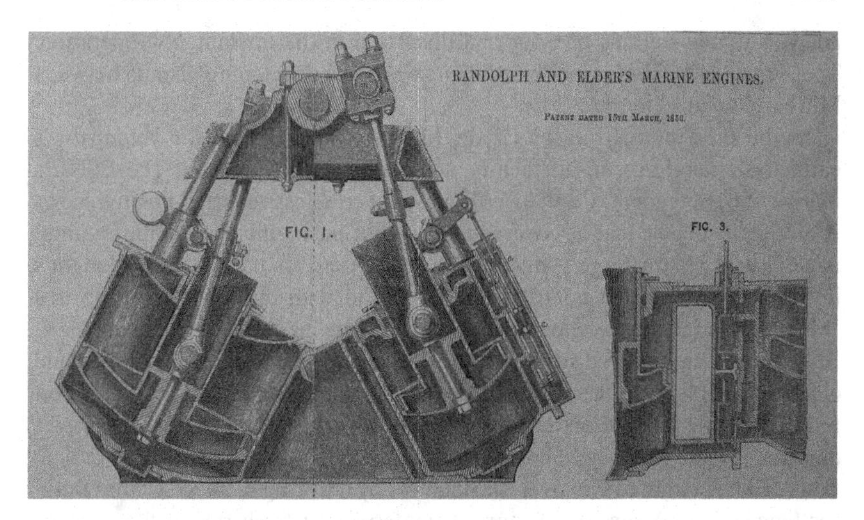

Figure 16.2. Randolph, Elder's double-cylinder marine engine (1856)

Although the latter openly acknowledged that there was a very conspicuous falling off of steam pressure due to causes as yet unknown, he was unwilling to grant unqualified legitimacy to McNaught's measurements. As the protracted trials continued, the differences between the two engineers became centred on the trustworthiness of McNaught's instruments.

Especially in the years immediately following these trials, there is strong evidence that certain groups of engineers questioned the reliability of such indicators. 'For the more rapid running higher-pressure engines of the present day', John Hannan told the Institution of Engineers in Scotland (IES) in 1866, '... [the indicator] is found to be totally useless, and in others to give very unsatisfactory results, arising from the great momentum of the moving parts, causing the marker [pen] to make a series of vibrations which, in some cases, were not completed before another diagram was commenced'. In discussion, James Napier welcomed Hannan's legitimate modifications to McNaught's instrument in order both to reduce the vibrations and increase the accuracy of the recordings. As Napier noted, however, previous misconceived attempts to achieve smooth diagrams by frictional dampening tended to produce results that underestimated the power of the marine engine and overestimated the efficiency of the ship's hull.[36] David Miller has investigated the contested nature of indicator measurements in the later nineteenth-century. His findings highlight the role of the practical engineer's often tacit knowledge in establishing results for the customer's benefit in contrast to the scientific engineer's more critical

[36] Hannan (1866: 75, 78).

analysis of the sources of error and distortion in the instruments' operation. These sorts of tensions manifested themselves in a developing dispute between McNaught and Elder.[37]

As the *Inca* returned to the Clyde, McNaught now found the *Valparaiso*'s early trials similarly unsatisfactory as to engine performance. The PSNC's marine superintendent, Captain Williams, nevertheless reported positive news from Elder: 'that he had discovered the cause of the inefficiency of the engines, namely water in the large cylinders'. But the Board also received McNaught's telegraph message that contradicted this claim with the blunt assertion that 'Mr Elder had not discovered the true cause of the inefficiency of the engines'. Elder, however, believed firmly in the truth of his diagnosis and in the sound character of his designs. He therefore spoke frankly to PSNC 'of the incorrectness of the indications taken by Mr McNaught on the trial of the *Inca*'s engines'. While McNaught continued to affirm 'the correctness of his indications', Elder sought to win over the doubters by allowing them to see the 'truth' for themselves. When the *Valparaiso* underwent a fresh run in the absence of McNaught, Captain Williams pronounced the trial 'satisfactory'. That triggered the despatch of McNaught and a PSNC director to the Clyde, where McNaught again judged the performance of the engines to be 'unsatisfactory'. He therefore proposed that Randolph, Elder consent to his being called to their assistance.[38]

McNaught appeared before the Board to explain 'the results of his endeavours to remedy the defects of the [steamers'] Engines – that steam could not be maintained in the *Valparaiso*'s boilers, which he attributed, chiefly, to the imperfect boring of the large cylinders'. As a practical engineer, McNaught had identified a practical solution. But the implication of imperfect or faulty workmanship clearly stung Randolph, Elder. They assured the Board that 'they were ... making some alterations from which they anticipated some improvement and that, when these were completed, it was their intention to have the performance of the ships certified by several leading practical marine engineers, of which they would give due notice to the Board'. Unconvinced, PSNC asserted that 'the performance of the Engines must be tested and determined in the manner provided for by the contracts, and not otherwise'. In other words, McNaught still stood as the sole independent arbiter in the matter of the steamers' performance.[39]

By late August, the *Valparaiso*'s master and chief engineer sent the owners a full account of the alterations to the machinery, most probably in the form

[37] Miller (2012: 212–50, esp. 222–3). Another critic from the early 1860s cited here was marine engineer John Bourne.

[38] PSNC Minutes (26 June and 3 July 1856).

[39] PSNC Minutes (7 August 1856).

of steam jacketing at the top and bottom of the cylinders. Rankine later noted that in Randolph, Elder's patent (first lodged in March 1856 but revised before final approval) the 'specification fully states the importance of providing each cylinder with a steam-casing or jacket to prevent liquefaction: but this is not claimed; for it was not a new invention, but ... the revival of a practice which had fallen into neglect, though essential to the economical use of high rates of expansion'. He also noted that, unlike the next generation of Elder double-cylinder engines, the *Valparaiso* and *Inca* were 'jacketed at the top and bottom only, and not round the sides'. This difference provides strong evidence that the jacketing was introduced following the initial patent in order to solve the 'defects' involving liquefaction of steam in the large cylinder.[40]

Better insulation of the cylinders, including steam jacketing, seemed like a practical means of preventing the liquefaction of the steam in the larger cylinder. But it also gained legitimacy from the new science. To prevent liquefaction, Rankine explained in 1871, additional heat had to be provided to the steam through the cylinder walls to replace the heat that had been converted into work. Elder's 1859 paper to the Aberdeen meeting of the BAAS, however, offers a glimpse into his own perspective much closer to the time of these early trials. There the master engineer spoke with phrases such as how steam jackets 'saved a vast quantity of heat' and how the 'economy of the machinery [was] realised'. The latter phrase directly echoed the title of Rankine's 1854 address to the Liverpool meeting of the BAAS, 'On the Means of Realizing the Advantages of the Air-Engine'. That address underscored the importance of identifying, and then reducing, the causes of waste in actual heat engines by which they fell short of a perfect thermodynamic engine.[41]

Both Rankine and Elder subscribed to a parallel standard of moral and spiritual perfection in their version of Christianity – with the imitation of Christ as their religious imperative. Elder was a member of the Barony congregation (Chapter 1) under the ministry of Norman Macleod. In 1857 Macleod officiated at Elder's wedding, held in Glasgow's Blythswood parish church, to Isabella Ure.[42] Only twelve years later, after Elder's death, Macleod bore personal witness to Elder's character as that of modern Christian exemplar (Figure 16.3). It was a character, Macleod affirmed, that 'told upon every department of his workshop and building-yard'. He himself had, on a recent visit to 'his great building-yard', heard from 'one of his oldest and most trustworthy men' that '[e]very man trusted him, and knew that he would do all that was possible to

[40] PSNC Minutes (14 August 1856); Rankine (1871: 35, 38).

[41] Rankine (1871: 28); Elder (1859: 291); Smith (1998a: 150–7; 2014). For an account of the causes of liquefaction based on later scientific understanding, see Cardwell and Hills (1976: 11).

[42] The marriage details are recorded in the facsimile copy of the marriages in the district of Blythswood in the Burgh of Glasgow (1857) in www.scotlandspeople.gov.uk. Both David (senior) and John Elder are designated 'master engineers' by profession.

Figure 16.3. John Elder

benefit them in every respect'.[43] The simultaneous pursuit of engineering and spiritual perfection here represented the human engineer. Honest and trustworthy in himself and as well as in his works, he aspired to imitate the divine engineer whose works revealed His trustworthiness through the books of nature and of revelation.

Another clergyman friend, Revd W. G. Fraser, recalled a conversation on the 'infidelity of the age' in which the master engineer, speaking of some of the workmen's unbelief, 'counselled us and all teachers' to adopt a mode of grappling with such doubters. Elder took as an inspirational biblical exemplar 'the doubting disciple', Thomas, 'who did not at first believe that most vital and fundamental truth', Christ's resurrection after the crucifixion. There 'the Saviour did not frown upon him [Thomas] as an infidel, nor sneer sarcastically at him, but came down and met him on his own ground, as if he entered into his doubts, and asked him to examine for himself the unmistakable proofs of the *facts* of his resurrection; whereas, had Thomas been treated coolly, and called hard names for doubting what all the others believed, humanly speaking, he might have turned away in confirmed unbelief'. In this manner 'the great Teacher' led Thomas from unbelief to faith and to declare 'my Lord, and my God' as affirmation that he had witnessed for himself the truth of the claim that Christ had risen after His sufferings and death. As Christian believers, Elder inferred, 'we should endeavour to meet all doubters on their own grounds,

[43] Macleod to [Rankine], in Rankine (1871: 57–9); Smith (2014: 90–4). On Rankine's debt to Thomas à Kempis' *Imitation of Christ* see Smith (1998a: 151). I thank Ben Marsden for this insight.

giving them credit for what they do believe, and striving to furnish evidence for what they have difficulty about. In this way many might be saved from the ranks of unbelief'.[44] But Elder's mode of reasoning had also served him well in persuading a doubting PSNC to witness for themselves the power and economy of his engines.

The PSNC's doubts continued to be fuelled by McNaught's evidence of serious imperfections. In September 1856 McNaught submitted to the PSNC Board formal letters detailing his evaluation of the performance of the steamers according to the respective contracts. In each case he began by grimly asserting that he found 'that the Engines do not consume any less fuel than Engines of the ordinary construction for the indicated or actual power exerted'. In the case of the *Valparaiso* he calculated that the engines 'consumed 4 lb of Scotch coal per hour per horse power', a figure which compared unfavourably with a four-year old conventional vessel, the Clyde-built Isle of Man steamer *Monas Queen*, 'tested by me' and found to consume only 3.75 lb.[45]

McNaught then found that the *Valparaiso*'s 'actual' power (most likely derived from dynamometer readings of the power delivered by the crankshaft) was equivalent to only 629½ ihp compared to his calculation of an equivalent contract specification of 1,024 ihp. He thus concluded that he had 'no award to make in respect of saving of fuel, and consider[ed] that the Contractor, Messrs John Reid & Co., have failed to fulfil the conditions of their contract, in respect of power'. The *Inca*'s engines, in comparison, worked 'to a fraction more indicated horse power than the vessels tested by me' and therefore 'I consider that the engineers have fulfilled their contract in respect of power'. In consequence PSNC reluctantly accepted the *Inca* but declined the *Valparaiso*. McNaught's second report on the latter's performance showed a reduction to 3.13 lb, but only a slight increase in actual power.[46]

Randolph, Elder, however, wrote to McNaught in mid-October 'impugning the correctness of the Award given by him for the steamer *Inca*'.[47] Again they challenged the reliability of his measurements based upon his indicators. The move against McNaught was not coincidental. On the same day, the PSNC secretary read to the Board two communications concerning a much longer trial trip with the *Valparaiso* that immensely strengthened the character of the engines in the face of such doubt. Elder had persuaded William Just (back from a long absence in South America) and John Edward Naylor (one of PSNC's directors) to see for themselves, independent of McNaught's numerical data, the performance of the new steamer.

[44] Fraser to Rankine, in Rankine (1871: 62–3). On Victorian doubt in parallel contexts, see Leggett (2013).
[45] PSNC Minutes (11 September 1856).
[46] PSNC Minutes (11, 27 September 1856).
[47] PSNC Minutes (16 October 1856).

First, the master, Captain Wells, telegraphed from Queenstown (Cobh) in southern Ireland to announce the arrival of the steamer there after a passage of more than 300 nautical miles from the Clyde. Clearly this trial was no brief experimental run within the confines of the sheltered Firth. Second, Just communicated by letter the news that 'the performance of the *Valparaiso* was, so far, satisfactory, in regard to speed and consumption of fuel'.[48] Trusted by the PSNC Board since his appointment a decade earlier, and with an unrivalled knowledge of Scottish marine engineering, Just was a credible witness who had seen for himself (to adapt Macleod's phrase) 'how truly they [the engines] spoke'.[49] The double-cylinder engines did after all deliver on promises, McNaught's authoritative measurements of power notwithstanding.

On the vessel's return, Just reported fully to the Board. The trial trip, while delayed for some thirty-nine hours at Queenstown due to thick weather and a heavy sea, had continued to Cape Clear and round the Fastnet lighthouse off the south-west coast of Ireland before returning to the Mersey. Just and Naylor, as witnesses, were evidently persuaded by the trial. Elder had won over the doubters by entering into their doubts, meeting them on their own ground and finally letting them see for themselves the reasonableness of the case for belief in the double-cylinder engines. The Board therefore resolved, subject to payment for delays in delivery as well as an allowance for the alleged shortfall in power, to accept the steamer.[50]

Reid and Elder, confident in their steamers but uncomfortable with McNaught's reports, now turned the tables by offering to keep the vessel themselves and to repay all PSNC's instalments along with interest. They rejected PSNC's newest proposal to lay the matter of financial settlement in the hands of an English barrister and instead offered 'to leave it to an eminent Engineer, to be mutually chosen, to decide as to deficiency of power and consumption of fuel'. This counterproposal implied that the mutually chosen engineer would be someone other than McNaught. Initially resisting, PSNC received two letters from the master of the *Inca* concerning safe arrival at Madeira and his intention of proceeding for Rio. In accordance with PSNC's usual practice, the master would have accompanied his letters with copies of the bridge and engine room logs, including fuel consumption. This promising news convinced the Board to accept Reid and Elder's terms: it now ordered that 'a cheque be signed for £1,600 [in relation to the promised saving of fuel] to be lodged in the Bank of Scotland' in the joint names of PSNC's chairman and John Reid.[51]

[48] PSNC Minutes (16 October 1856).
[49] Macleod (1876: vol. 1, 279–80); Smith (1998a: esp. 28).
[50] PSNC Minutes (18 October 1856).
[51] PSNC Minutes (21, 23, 27 October and 1, 6 November 1856).

After identifying a suitable steamer of similar power and consumption as a 'standard' against which to test the *Valparaiso*'s performance, the parties received McNaught's verdict in April 1857: 'the fuel consumed in working at a given horse power the said double cylinder expansion engines is less by the rate of twenty five per cent than the fuel consumed in working at the same power engines of the kind mentioned in the said specification'. McNaught thus 'certified that during the run from Glasgow to Liverpool the consumption was ... 2.98lb per indicated horse-power per hour for the *Valparaiso*, while the ['standard'] *Pride of Erin*, during 5½ hours run, consumed 4.27lb per horse-power per hour'. This time a railway engineer, Daniel Kinnear Clark, independently confirmed the results. In the final settlement in September, Randolph, Elder received the £1,600, of which some £983 went back to compensate PSNC for contractors' delays.[52]

Early the following year, PSNC received two letters from their superintendent engineer at Panama 'reporting favourably of the working of the *Valparaiso*'. Their manager on the West Coast also notified the Line of a fresh mail contract with the Chilean Government worth $50,000 per annum, as well as of lucrative shipments of silver bars. Thus the necessity 'of contracting for a new ship, as early as possible, ... [was] very urgent'. The PSNC quickly accepted Reid and Elder's joint bid at £34,000 as the lowest. Within two months, Reid's yard had the vessel in frame. Well within the contract time, the new steamer, *Callao*, undertook a full set of trials on the Clyde. The chairman reported of one trial 'that it was highly satisfactory ... [and] that he saw no reason to hesitate in taking delivery of the vessel'. He further confirmed that 'the performance of the *Callao* from Port Glasgow to Liverpool was equally satisfactory', a claim supported by a letter from the vessel's chief engineer Duncan Morrison. Two older vessels, the *Lima* and *Bogota*, were soon re-engined with the new machinery.[53]

In 1858 *The Engineer* (reprinting a piece from the *Liverpool Albion*) reported the departure from Liverpool of the *Callao* on its outward voyage to the Pacific. Noting the *Callao*'s running of the Clyde's measured distance at a speed of some 13 knots, the journal stated that the 'performances from the Clyde to this port were equally satisfactory, both in regard to speed and economy of fuel, which latter is the great leading feature in these engines'. With them, indeed, 'a ship can steam the greatest distance possible with a given quantity of coals; that a given distance can be performed in the shortest time, on account of the small weight of coals necessary to be carried; that a larger amount of cargo and passenger accommodation is thus obtained, a less expensive ship thus necessary; and that the number of firemen and stokers can also

[52] PSNC Minutes (13, 15 November and 11 December 1856; 2, 8 January, 10, 24 February, 5 May, and 22 September 1857); *The Engineer* vol. 6 (1858: 348). On Clark see Hills (1989: 139).

[53] PSNC Minutes (4, 18, 25 May, 22 June and 19 October 1858).

be reduced'. The following spring *The Times* included in its short report on Liverpool shipping movements the news that the Pacific company's vessels supplied with these engines 'consume about one-half the quantity required by those they formerly had'. And around the same time Randolph, Elder, their credibility much enhanced by its PSNC contracts, took over the 'Old Yard' at Govan from James Napier and extended their business into shipbuilding (Chapter 18).[54]

'What do you believe to be the real economy of the best steam engines?' Professor William Thomson asked James Napier in 1879. 'What does 1½ lbs [of coal] per hour per horse power mean? ... Is the indicator ever trustworthy? Never? Hardly ever? If trustworthy, I suppose it gives really the work done [by the steam] on the piston. This is very interesting scientifically'.[55] Such scepticism on the part of an international authority on physics illustrates dramatically that the trustworthiness of indicator practices long remained open to question.[56] But the story of the *Inca* and *Valparaiso* showed not only the often-fraught relationships among ship-owners, builders and consulting engineers, but also how a master engineer and his associates negotiated their way through a treacherous period of scepticism and doubt over a radical new marine system of motive power. Eventually, however, Randolph, Elder persuaded not only PSNC, but the wider engineering and national press of the superior character of their engine in terms of distance run and internal space saved.

[54] *The Engineer* vol. 6 (1858: 348); *The Times*, 16 May 1859; Wardle (1940: 93); Shields (1949: 60).
[55] Thomson to James Napier, 31 August 1879, NP.
[56] Miller (2012).

17 'A Constant Succession of Unfathomable and Costly Experiments'

Making Credible the Marine Compound Engine

> theory further points out, what practice has partially confirmed, that the work done by an engine for a certain expenditure of fuel is proportional to the difference of the temperatures at which steam enters and leaves the engine. From this principle arises the economy of using high-pressure and super-heated steam ... The economy already effected in this manner is wonderful. ... But it is in steam navigation that the improvement of the engine will have most marked effects. Any extensive saving of fuel, saving its stowage-room as well as its cost, will still more completely turn the balance in favour of steam, and sailing-vessels will soon sink into a subordinate rank.
>
> *William Stanley Jevons' reading of the new science of thermodynamics in The Coal Question (1865)*[1]

Summary

In the wake of Randolph, Elder's double-cylinder engine patent and its eventual adoption as standard by PSNC, other engine-builders devised different compound engine systems that also aimed to address problems of fuel economy. As ocean steamship lines introduced larger vessels, so coal questions assumed ever-greater prominence at AGMs, in the engineering press and among informal networks of shipbuilders and engineers. On the Clyde, Greenock shipbuilder John Scott privately constructed the iron-screw steamer *Thetis* (1859) in order to conduct quantitative trials with a double-cylinder engine deploying steam at very much greater pressure than that of conventional ocean steamers. On the Thames, the Deptford engine-builder Edward Humphrys persuaded P&O to invest in the compound-engined *Mooltan* (1861) as the first of a series of expensive mail steamers for its mainline services. In both cases, trouble-free performances were far from guaranteed. Reflecting on the difficulties, Alfred Holt embarked on experimental trials with his own steamship. Upon the results of these extensive trials, he ordered three large iron-screw vessels from Scott in 1865 which were designed for a new line of cargo steamers from Liverpool to Singapore and Chinese ports.

[1] Jevons (1865: 109–13).

17.1 'Creditable to the Ingenuity of the Clyde Engineers': Scott's High-Pressure Engine

The PSNC's first compound-engined vessels operated at modest steam pressures of around 20 psi. In 1860 *The Engineer* hailed Randolph, Elder's engines as now working with superheated steam of 45 psi with coal consumption about one-half of the usual expenditure in ocean steamships. Three years previously, however, the gentleman-shipbuilder John Scott, the youngest of several John Scotts in the family, publicly demonstrated that much higher pressure steam (at over 100 psi) offered still greater saving of fuel.[2]

Scott followed Russell in pursuing experimental trials using models on the neighbouring Loch Thom and in the firm's dry dock. It was a practice that paid dividends. His iron clipper *Lord of the Isles* beat two celebrated American clippers to arrive in Britain from China in a record-breaking eighty-nine days. On another voyage the ship apparently averaged more than 13 knots over five consecutive days and nights (Figure 17.1).[3] These victorious trials-by-space for pure sail would have been enough to secure Scott's reputation as a naval architect and master shipbuilder, but he was even more resolved to win a reputation in the high-risk field of deep-sea steam navigation.

Scott constructed to his own account the schooner-rigged, iron-screw steamer *Thetis* for experimental purposes rather than as the plaything of a wealthy patron. As with Napier's use of Assheton Smith's steam-yachts, the strategy exonerated Scott from accusations that he was 'experimenting' with the valuable ships of an established owner (Chapter 2). Completed on 13 May 1857, the *Thetis* spent the next two years undergoing experimental trials, most of them witnessed by a variety of authorities including Professor Rankine. Operating at some 115 psi boiler pressure, it embodied Scott's conviction, consistent with his Glasgow University mentors, that high pressure steam was the key to economy in marine engines.[4]

In late 1858 Rankine reported informally to James Napier that 'Rowan's engine burns 1.018 lb of coal per ihp per hour'. Messrs Rowan & Co., whose Atlas Works in Glasgow constructed the boilers, surface condensers and compound engine for the *Thetis*, had commissioned Rankine to draw up a formal evaluation of the working vessel using indicator diagrams. These measurements coincided with a surge of BAAS interest in a range of topics concerned with steamship performance, including tonnage measurement, strength of wrought

[2] *The Engineer* vol. 9 (1860: 9) (editorial). This section draws on Smith (2011b: esp. 235–42). John Scott entered Glasgow University as an undergraduate in the 1846–7 session, at just the time William Thomson and Robert Mansel began building the new science of thermodynamics (Chapter 16).

[3] [Scotts] (1906: 13–14); Lubbock (1946: 82); [Robb] (2009: 69); Napier (1904: 21) (Russell's experiments).

[4] Robb (1993: vol. 2) ship no. 40 (data on ships built by Scotts).

Figure 17.1. Celebrated iron clipper *Lord of the Isles* built by Scotts of Greenock (1853)

iron and steel and fuel consumption. Moreover, as Ben Marsden has shown, in the autumn of 1858 Napier and Rankine promoted in *The Engineer* the Napier-built steamer *Admiral* as an exemplar of science-based naval architecture.[5]

[5] Rankine to James Napier, 22 November 1858, NP; Rowan (1880: 52–9) (published versions of Rankine's 1858 reports); Marsden and Smith (2005: 119) (*Thetis* summary); Marsden (2008: 79–94); Bellamy (2012: 161–77).

The Atlas Works had been the site for rigorous boiler tests to 240 psi in the presence of the Board of Trade surveyor in April 1857. Subsequent trials of both engines and boilers showed 'that there were grave defects inherent in their design', including the development of small holes in the boiler tubes carrying the hot gases from the furnace through square vertical water tubes that made up the boiler. John Martin Rowan and Thomas Horton (works manager) duly incorporated changes in an engine patent taken out in their names one year later.[6]

When *The Engineer* ran an editorial entitled 'Steamship Economy' later that year highlighting the poor fuel economy of ocean steamers, it made no explicit mention of the concurrent *Thetis* trials. 'No other application of steam is yet so unsatisfactory as that to ocean steam navigation', the editor complained. 'It is not so much the power required to drive the vessel as it is the fuel required to produce the power'. He cited the exemplary performance of a stationary, almost certainly compound, Cornish pumping engine as 1.57 lb of coal per horsepower per hour and that of the best South Western Railway locomotives consuming scarcely twice that quantity. In contrast, the editor stated that it was 'very generally known that the consumption of coal in ocean steamers making long voyages is about six pounds'. And he even suggested that 'Dr Lardner was not so far wrong when, in 1837, he estimated the extravagant expenses of ocean steam navigation' (Chapter 3).[7]

The editorial then listed six desiderata for marine engine systems 'most likely to afford permanently successful results'. The first two urged improved combustion of the fuel in the furnace. The third advocated the use of high pressure steam in the range 60–90 psi. The fourth recommended using superheated steam, the fifth expansive working and the sixth advised surface condensation to avoid the problems of encrustation arising from the use of seawater. With 'these plans of improvement there is no doubt that the fuel consumption of ocean steamers might be reduced to 2¼lb. of coal per horse power per hour – thus saving from 1,500 to 2,500 tons of coal now required by a large paddle steamer on an Australian voyage'.[8]

In early summer 1859 *The Engineer* printed an eyewitness account of the *Thetis'* machinery by three Liverpool Polytechnic Society members. Mr Arnott declared the system 'creditable to the ingenuity of the Clyde engineers'. It 'promised to revive the progress and prospects of our mercantile steam marine, which was held in check by the cost of fuel, and consequent displacement of cargo at much sacrifice, especially on long voyages'. Messrs Lamont and Gray followed up with explanations of the system. Most notable features were the

[6] Rowan (1880: 52–3); [Scotts] (1906: 34–5); *The Engineer* vol. 10 (1860: 210) (Rowan and Horton's patent).

[7] *The Engineer* vol. 6 (1858: 357–8).

[8] *The Engineer* vol. 6 (1858: 357–8).

boiler design ('certified to be safer with 120 lb. pressure than nine-tenths of the common marine boilers with 15 lb.'), superheating of the steam, the use of a surface condenser, the arrangement of the twin engines (each with one high pressure and two low pressure cylinders with the three pistons connected by a cross-head) and the use of expansive working of the steam. In effect, the system met most of *The Engineer*'s desiderata for the economy of marine steam engines.[9]

In the first of two printed reports on the *Thetis* labelled 'For private circulation only', Scott focused on the steamer's trials within the upper Firth of Clyde between the Cloch and Toward Lighthouses (a distance of 5.8 nautical miles). It recorded every quantitative detail of the vessel, the engines and boilers, particulars of the trial, power developed (using four sets of indicator diagrams taken at half-hourly intervals during the trial), fuel consumed and performance. At a boiler pressure of 90 psi, with engines delivering 256-hp and the vessel making 9.66 knots, the calculated coal consumption came to 1.202 lb per ihp per hour. Witnesses included M. Forquement (a French Government naval engineer) and Thomas Stirling Begbie who acted as London-based agent for a number of Clyde shipbuilders (among them Scott, Denny and Caird) (Figure 17.2).[10]

The second report concerned the results of some fourteen sea voyages under commercial conditions, transporting in total 7,622 tons of cargo and consuming just over 121 tons of coal. The distinctive language in the document of 'dynamical power' and 'dynamic results' echoed William Thomson's and Rankine's terminology, especially in Thomson's 'Dynamical theory of heat' and 'Thermo-dynamics'. Explaining that because the vessel had on occasions been steaming light with the screw not fully covered, the document claimed that 'the total quantity of cargo transported does not convey an accurate idea of her dynamical power'. But 'experience in her working amply proves that when the Screw propellor [sic] is properly covered the dynamic result is very superior'. The quantitative results nevertheless showed, Scott insisted, that the system 'still surpasses to a very marked extent the best performances hitherto obtained from ordinary marine engines of the most approved construction'.[11] Both documents were clearly intended to persuade prospective customers who could be trusted to treat commercially sensitive knowledge with due care and respect.

In contrast its earlier pessimism, *The Engineer* headed its January 1860 editorial 'Progress of steam navigation'. 'In regard to steam economy', it enthused, 'some of the best results have been obtained within the year, and the direction

[9] *The Engineer* vol. 6 (1858: 381).

[10] John Scott, 'Notes of an experimental trial made with Rowan's patent expansion steam engines', NP. See also Robb (1993: vol. 1, 508–10) (Begbie); Graham (2006: 71–102) (Begbie, Scott and the 1860s 'blockade runners').

[11] John Scott, 'Abstract of the performances of screw steamer *Thetis*', NP; Smith (1998a: 107–25) (dynamical terminology).

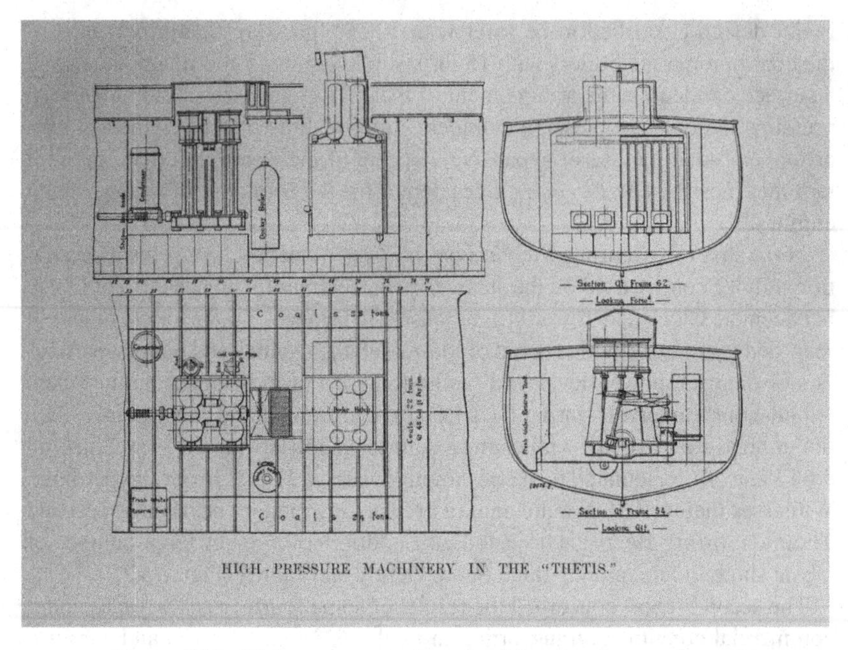
HIGH-PRESSURE MACHINERY IN THE "THETIS."

Figure 17.2. High-pressure engines for John Scott's experimental steamer *Thetis*

of improvement in this respect has been precisely that which we have so often pointed out'. Remarking on the gains in economy from Randolph, Elder's use of 45 psi, the journal cited the *Thetis*' performance in which the engines 'obtained the best result yet recorded, and have probably maintained a rate of about 2lb. of coal per horse power per hour, in regular working, for a period of several months'. The editorial therefore concluded that '[w]e believe there was never before a time when such really earnest endeavours were being made as now, to reduce the fuel consumption of steam engines to a minimum'.[12]

Not every engineer, however, welcomed the new designs. Writing to *The Engineer*, Frederick Spencer of London Bridge denounced 'combined cylinders or other complications' as tending 'to retard and not advance steam economy'. As with the Victorian fashion for impractical styles of female clothing, he asserted, such a system was 'the "crinoline" of our steam engineers, and will have its day'.[13] Provoked, John Scott insisted that he was 'not aware that Mr Spencer has had any experience in the construction or working of combined

[12] *The Engineer* vol. 9 (1860: 9).
[13] *The Engineer* vol. 9 (1860: 22).

cylinder engines'. Scott declared that he himself had had 'the opportunity of testing on a large scale the merits of both [single and combined cylinder] systems, and am therefore in a position to speak on the subject'. And if Spencer could not realize his promises by producing results that 'surpass or equal those already for many months past obtained by us', Scott could not consider Spencer's claims of much value. Moreover, he affirmed, the fact that some fourteen pairs of engines were currently under construction on Rowan's system was 'no mean indication of the attention which it is attracting'.[14]

Such attention, however, was often critical. Built by Robert Stephenson's Newcastle works under license, four sets were fitted to River Ganges steamers. In the salt-free waters the engines gave satisfaction. In contrast, two other sets for coastal steamers running between Sydney and Brisbane were beset by leakage of seawater into the system. Two deep-sea vessels for the Mediterranean trades experienced similar problems. In an historical account presented to the Institution of Engineers and Shipbuilders in Scotland (IESS) in 1880, Rowan's son Frederick exonerated his father from any charge of faulty design. Instead, he attributed the failures to '[t]he evils of faulty workmanship, insufficient arrangements, boiler corrosion, and the numerous minor difficulties of pistons and piston rod packing, and of lubrication in presence of steam of a high temperature'. He even blamed one of the vessel's masters and engineers for their timidity in cutting short the maiden voyage.[15]

Not every member of the Institution accepted Rowan's account. Master boilermaker and engineer James Howden's explanation of the failures of Rowan's system focused on the three principal components. First, the boilers, although designed for strength, safety and apparent simplicity of construction, 'were necessarily complicated ... and in the event of salt water getting inside, and causing incrustation, it was quite impossible to remove it'. Second, the surface condensers were of similar complexity and, by leaking, acted as the source of saltwater contamination, except, of course, in the case of river steamers. Finally, the engine arrangement, with six cylinders for each pair of engines, 'did not tend to economy in small engines such as those fitted in the steamers referred to', due to a very great loss of pressure as the steam passed through the various passages and valves between high and low pressure cylinders. His own firm, he concluded, had lately replaced the problematic boilers with tubular ones in two of the seagoing steamers and had shown to the owners' satisfaction that a high pressure steam system could be made durable.[16]

[14] *The Engineer* vol. 9 (1860: 74).
[15] Rowan (1880: 68–76); Smith (2011b: 239–41).
[16] Rowan (1880: 88–93).

17.2 'Economiser General': P&O's Costly Experiments with Power and Economy

Responding to the toast after the launch of the *Bengal* in 1852 (Chapter 14), Arthur Anderson told guests that P&O now had 'in active service and in progress of construction a fleet of 41 steamships, of the aggregate tonnage of 52,000 tons, and of about 16,000 horses' power of machinery, and being in value upwards of 2,000,000l. sterling. (Hear, hear)'. The ships annually navigated a distance of almost 1 million miles, carried upwards of 3,000 seamen, and required some 60,000 tons of sailing vessels with another 3,000 men to transport coal to its various stations at home and abroad. The Company's current annual expenditure, including capital invested in new steamers, exceeded 1 million pounds, while he estimated that the Line 'afford[ed] subsistence to 100,000 persons employed and their families (Applause)'. Small wonder that he claimed for P&O 'the rank of the first of private maritime enterprises which the world had yet seen'.[17]

Throughout the 1850s, P&O's voracious appetite for new tonnage far exceeded the capacity of Tod & MacGregor's yard, which in any case was also heavily engaged in constructing iron-screw steamers for the expanding Inman Line. The Company therefore ordered its principal vessels from a variety of yards (Table 17.1).

Of all the P&O screw steamers from the Meadowside stable, the 2,441-ton *Simla* (1854) was, by common consent, the finest. In March 1867 the IESS President told his members that Tod's son William had worked alongside Archibald Gilchrist (draughtsman at the yard since 1842) in designing and

Table 17.1. *P&O's shipbuilders (1853–60)*

Tod & MaGregor (Glasgow)	8
Robert Napier (Glasgow)	1
Mare (Blackwall)	3
Thames Iron Works (Blackwall)	1
Samuda (Poplar)	2
Summers, Day (Southampton)	2
Lairds (Birkenhead)	3

In addition, P&O purchased several vessels while under construction (Lairds (four), Denny (one) and Bourne of Greenock (one)) (Rabson and O'Donoghue (1988: 46–61)).

[17] *The Times*, 3 November 1852.

constructing the machinery of the vessel which 'has proved to be one of the most successful that ever was built, either before or since, and yet holds an unsurpassed reputation in the numerous fleet of the great company to which she belongs'. A remarkable builder's model of the engines is currently displayed in Glasgow's Riverside Museum and shows very clearly how the earlier 'steeple' engine had been transformed into a compact design in which the crank shaft drives a large gear wheel situated directly above (and not to the side of) the smaller gear wheel on the propeller shaft (Figure 17.3). 'The *Simla* is a most magnificent ship and, independent of her steam power, is fully rigged as a clipper bark [sic]', *The Times* observed of its maiden voyage, 'so as to take the greatest possible advantage, on all occasions, of a leading wind'. After serving P&O for twenty-one years, the *Simla* was eventually converted into a pure sailing vessel in 1878 and continued to trade until lost by collision in 1884 while owned by sailing ship-owners Devitt and Moore.[18]

In stark contrast, the 3,438-ton *Himalaya* (1854), longest and largest steamship afloat, barely lasted six months with P&O. Ordered from the Blackwall yard of Charles Mare with engines by John Penn at Greenwich, the steamer had begun as part of a much larger project. Around 1850–1, *The Times* explained, the P&O Company faced the expiry of its mail contract on the route to India and 'powerful competitors were understood to be preparing to contest ... [its] commercial supremacy'. The original paddlewheels and 1,200-hp engines promised 'a speed of 20 miles an hour' to enable 'the run from Southampton to Alexandria to be effected in eight or nine days'. Five further ships of the same or greater tonnage were envisioned for employment on the other side of Suez. 'If this project had been brought to perfection', the paper lamented, 'the communication between England and our Indian empire would have been facilitated to an extent that even now can be hardly realized, and Calcutta brought within 25 days of Southampton'. Intended to out-manoeuvre Eastern Steam's planned system (Chapter 15), however, P&O abandoned its extraordinary project after it successfully retained the mail contract. In place of the paddle-engines, the *Himalaya* now received a screw arrangement, 'tried with favourable results on a small scale by the company'.[19]

The 'small scale' trial referred to an experimental steamer constructed to Andrew Lamb's own account in Glasgow 'for the purpose of experimenting upon boiler, paddle wheels &c.' Penn constructed the new 700-hp engines for £36,750, over £20,000 less than the original 1,200-hp engines. Assigned to the Alexandria mainline, the ship's master assured *The Times* that under steam and canvas 'there would be no difficulty in getting 18 knots or 20 miles an

[18] General Minute Book (27 March 1867), IESS, GUA; Fox (2003: 177) (Gilchrist); *The Times*, 15 May 1854; Rabson and O'Donoghue (1988: 49).
[19] *The Times*, 19 January 1854.

Figure 17.3. Exhibition model of Tod & MacGregor's engine for P&O's *Simla* (1854)

hour out of the ship'. Within a few months, however, the steamer had been withdrawn from commercial service and chartered to the Government to carry troops to the Black Sea. The vessel was soon sold to the Government for £130,000, a figure considerably more than the construction price of £115,000.

At a consumption of over seventy tons of coal per day and at a time of rising prices, the Company had disposed of their most costly experiment to date.[20]

The deaths of David Tod and John MacGregor in 1858 resulted in the transfer of control of the business to David's son William. The close relationship with P&O, however, did not continue to its former extent. Instead, P&O turned to a relatively new London engine-builder, Humphrys, Tennant & Dykes, established in 1852 next to Deptford Pier on the Thames. Edward Humphrys had a powerful engineering pedigree. He and his elder brother Francis were closely involved with the machinery of both the *Great Western* and *Great Britain* (Chapter 6). After managing Rennie's engine works in London, Edward was appointed around 1848 as chief engineer of the Woolwich Dockyard. By 1851 he was briefly with Nasmyth in Manchester before establishing his engineering firm at Deptford the following year. Much favoured by the Admiralty, 'he had all the work he could do, and good work he made of it'. Described in his obituary as 'a lion in energy and will', he apparently 'drew his own plans, was always the first man in the factory in the morning and the last out at night, and ... was wonderful for his economical "dodges", as engineers understand the term'. The engineering work took place in a shed constructed of boiler iron, with the engines and heavy components moved by a steam traveller and lifted aboard ships by a massive sheerlegs on the wharf.[21]

Humphrys' initial work for P&O involved the fitting of a 2,054-hp direct-acting two-cylinder (but not compound) engine to the Samuda-built *Ceylon* in 1858. He also supplied a smaller engine to the 796-ton steamer *Nepaul* [sic], built by the Thames Iron Works (successor to Mare). Favourable reports on the *Nepaul*'s trials augured well for the increase in mutual trust between shipowner and engine builder.[22]

As he won favour with P&O, Humphrys received from the engineering press the epithet of 'Economizer General' to the mail line. High pressure and high speed he avoided, while he 'superheated, and he surface-condensed, and he expanded' in pursuit of a reduction in coal consumption. His range of screw engine designs, characterized by short connecting rods around 1.75 times the length of the stroke, were compact. Humphrys' 'double-cylinder' engines exemplified this arrangement by having two small high pressure cylinders above and below the piston rod of the large cylinder. The piston rods of each small cylinder were fixed to the piston of the large cylinder in such a way as to

[20] *The Times*, 19 January 1854; P&O Minutes (28 October 1851; 25 January and 22 March 1853) (Lamb's steamer); (30 April 1852) (Penn); Rabson and O'Donoghue (1988: 48–9); Harcourt (2006: 182–4).

[21] P&O Minutes (3 December 1858); Banbury (1971: 189–90) (Humphrys); [Anon.] (1867: 528); John Bourne to the Editor, 31 May 1867, *Engineering* vol. 3 (1867: 559); [Nasmyth] (1883: 238–43).

[22] P&O Minutes (14 December 1858); Rabson and O'Donoghue (1988: 56, 58).

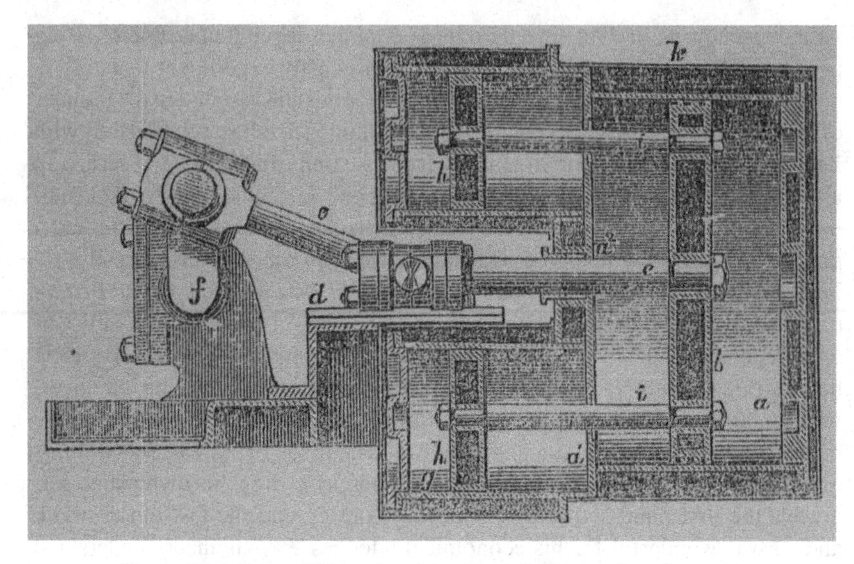

Figure 17.4. Edward Humphrys' double-cylinder engine plans

move back and forth with it. As the high pressure steam entered and expanded in the small cylinders, it drove the low pressure piston back; and as the low pressure steam next entered and expanded in the large cylinder, it drove the low pressure piston forward. The whole system of cylinders was surrounded by a steam jacket (Figure 17.4).[23]

On Christmas eve 1858, P&O received a communication from Humphrys' works describing a new 'screw engine which they could guarantee should not consume more than 3lbs of coal per indicated horse power per hour [and] stating that the average present consumption of the Company's engines was 4½ lbs per hour'. With Humphrys' design for a double-cylinder marine engine before them, the directors considered 'the plan was worth a trial' and authorized the MDs to obtain more precise information as to cost. An undated document by Humphrys included a practical argument on the advantages of a double- over a single-cylinder system. Assuming both types to be equally well insulated, he wrote, the heat loss would be less in the double-cylinder engine because the high pressure steam admitted into the smaller cylinder 'never comes in contact with the great piston [of the large low-pressure cylinder] which is always in communication with the condenser'. He further insisted that '[t]he double cylinder engine is the most simple machine [and] indeed the [ship's] engineer need not know, so far as working the engines are concerned, that he has a

[23] [Anon.] (1867: 528); [Anon.] (1860: 73).

double cylinder engine in his charge'. And Humphrys' prototype engine had made '60,000,000 of revolutions without any apparent wear'.[24]

Early in 1859 the Board directed the MDs to arrange with Humphrys for the construction of a pair of engines 'on the principle advocated in the correspondence ... of such power as they may consider best to test the ability of the engineers to effect the economy of fuel they have promised therein'. Humphrys expressed a strong desire 'that these engines might be put into a new ship of about 310 feet in length and 40 feet beam'. Only after another six months did P&O place an order with Thames Iron Works for a 350-foot hull that became the 2,257-ton *Mooltan*.[25]

In an 1858 editorial *The Engineer* observed that P&O was currently receiving £430,000 per annum in Government subsidies. Two years later, it noted that nearly £650,000 (or very roughly 220,000 tons at £3 per ton) was spent on coal to maintain annual voyages totalling 1.30 million miles. In addition to the move to compound engines, P&O's imperative to reduce that enormous coal consumption was also reflected in its concern to install superheating apparatus on selected vessels. Lamb thus received approval to fit his (and Summers) patent superheating apparatus to the boilers of the long-serving *Sultan*, converted to screw propulsion by Caird in 1855.[26]

The related strands of superheated steam, steam jacketing, high pressure steam and the advantages of compound over simple engines not only permeated the board rooms of mail lines and the workshops of engineers, but also generated intense discussion at the first meeting of the new INA held in London in March 1860. The meeting brought together several of the protagonists involved in naval architecture and marine engineering economy, most notably Scott Russell, Humphrys and James Napier. With Russell in the chair, Robert Murray presented a paper 'On various means and appliances for economising fuel in steam-ships'. Murray, with whom James Thomson had worked on drawings for the *Pottinger* at Fairbairn's shipbuilding yard in the mid-1840s (Chapter 14), was now engineer–surveyor to the Board of Trade. His brother Andrew, former manager at the now-defunct Fairbairn yard, had become chief engineer and inspector of machinery at HM Dockyard, Portsmouth. Both were active in writing on shipbuilding and marine steam engines.[27]

The second half of Murray's paper focused the advantages of superheated steam, steam jacketing and double-cylinder engines. He drew particular attention to P&O's plan, based on the patent by Lamb and Summers, to superheat boiler steam from about 250° to 400° Fahrenheit by passing the steam

[24] P&O Minutes (24 December 1858); 'Memo by Mr Humphrys on combined engines', P&O Collection, Caird Library. Harcourt (2006: 188) quotes an extract.

[25] P&O Minutes (28 January, 28 June, 19, 22 July and 22 November 1859).

[26] P&O Minutes (19 August 1859).

[27] For example, Murray (1861).

through a gridiron box located at the base of the funnel. 'A great many reliable experiments have been made at Southampton in the vessels of the Royal Mail Company, the Peninsular and Oriental Company, the Cape Mail Company, and others', he stated, 'to test the actual economy of the process ... and in every instance that has come under my observation there has been a perceptible improvement'.[28]

Turning to compound engines, Murray accepted the 'very remarkable economy of fuel' achieved by Randolph, Elder's engines in which steam entered the small cylinder at 42 psi and the large cylinder at 14 psi. Their current system, he noted, also included steam-jackets and superheating to nearly 400°F. Murray, however, favoured single-cylinder engines on the grounds that they were more reliable and just as, if not more, economical.[29]

An intervention from James Napier, on the other hand, represented marine steam engines not as self-contained machines but as components in engineering systems whose economy was governed by energy losses. He cited a recent trial between Randolph, Elder double-cylinder engines and Neilson single-cylinder engines in which the consumption of coal per ihp per hour differed by just 0.06 lb. In deciding the question in practice, 'the criterion of efficiency' that mattered was the 'fuel consumed for the *Useful* Work developed ..., and not the fuel per *Indicated* Work' determined by the indicator measurements. 'Useful work' in this case was the work done in moving the steamship through a given distance whereas the 'indicated work' estimated the total quantity of work done by the engine, a proportion of which would be wasted in the friction of the machinery and the turbulence of the paddlewheels or screw. The question, therefore, was one of friction, space and cost in the steamship as a whole, and not one that could simply 'be determined by Watt's indicator'.[30]

Later in the discussion, Humphrys endorsed Napier's reasoning that the choice was practical. In his summing up, Russell agreed that Humphrys had identified the vital point of 'inconvenience' with regard to the single-cylinder engine and accepted that 'they have been rather better got over by Randolph's engine than by any other, thus far'.[31] Whether the *Mooltan* would fulfil a promise of achieving still greater economy with Humphrys' engines, however, was soon to be tested.

The *Mooltan* slipped into the Thames at Blackwall in October 1860 and began to undergo engine trials at the moorings the following spring. Steaming trials, however, did not take place until late July, implying problems with the machinery. Averaging the speed over four runs and allowing for the current,

[28] Murray (1860: 177–80).
[29] Murray (1860: 179–82).
[30] Murray (1860: 182).
[31] Murray (1860: 183).

Table 17.2. *P&O's double-cylinder engined mainline fleet (1861–5)*

Mooltan	Thames Iron Works	1861
Poonah	Thames iron Works	1863
Carnatic	Samuda	1863
Rangoon	Samuda	1863
Golconda	Thames Iron Works	1863
Baroda	Thames Iron Works	1864
Delhi	Money, Wigram	1864 (engines built by Ravenhill)
Tanjore	Thames Iron Works	1865 (engines built by Ravenhill)
Ceylon	Samuda	1858 (refitted by Humphrys (1865))

Rabson and O'Donoghue (1988: 56–67) (data).

Lamb's computation gave 12.022 knots while the Admiralty mode of calculation gave 11.876 knots. As 12 knots was the contract speed, the Postmaster General approved one mail voyage only. That voyage resulted in detention at Gibraltar 'owing to a derangement of her machinery'. In October the *Mooltan* ran the measured mile at 12.018 knots with Government surveyors aboard and so finally received sanction.[32] Evidently satisfied with the performance despite the teething problems, P&O placed orders for another seven new mainline steamers with Humphrys' engine design (Table 17.2).

Reports of trial trips in 1863 were favourable. The Board received news of the *Rangoon*'s run to Queenstown (Cork Harbour) with the directors aboard. The speed of the ship, the consumption of fuel and the working of the engines was 'very satisfactory' as was a down-river trial of the *Golconda*. But the *Mooltan*'s engines seemed not to improve with time and only five years after the vessel's introduction the Board heard that the steamer was about to be ordered to Britain for repairs that would cost an estimated £15,000. Given the low value of the existing set (£4,000), the Board accepted a Summers, Day tender (£21,000) for replacement engines.[33]

At a meeting of the INA in April 1865, Murray returned to the question of double-cylinder engines. The performances of the new P&O steamers, he believed, 'verified the observation that a complicated machine, however cleverly it may be designed, does not afford the same immunity from accident that a simpler and less sophisticated engine does'. He contrasted their performances

[32] P&O Minutes (5, 23, 30 April, 7 May, 19, 30 July, and 15 October 1861); Harcourt (2006: 188–9).

[33] P&O Minutes (28 July and 22 September 1863 (trial trips); P&O Minutes (13 July 1866) (*Mooltan*'s engines)).

(consuming about 3 lb) with two Samuda-built vessels, *Saxon* and *Roman* (consuming a very similar or even lower quantity). These ships operated the mail run between Southampton and South Africa. Using steam at 20 psi in a single-cylinder, Murray claimed, they '*never break down*'. In contrast, he told INA members that on one voyage the *Mooltan*'s engineers bypassed a faulty high pressure cylinder and found that use of the low pressure cylinder alone occasioned 'the most economical voyage that the vessel had made'. The result, the speaker added, 'was looked upon at the time as rather a good joke against one of our most talented marine engineers'.[34]

Mounting a more general defence of combined-cylinder engines, Maudslay-trained Danish engineer Lewis Olrick used his personal acquaintance with early Randolph, Elder engines to assert that when the high pressure cylinder failed for a short time, use of the large cylinder alone pushed up coal consumption by some 50 per cent. He also neatly contrasted these Glasgow engines with the London ones. In the former, the pistons worked in opposite directions, minimizing the distance passed by the steam from small to large cylinder: 'a considerable waste of steam in the large steam-passages is avoided and a great deal of steam saved'.[35]

The 'joke' against Humphrys was costly for the talented engineer, already struggling against the prospect of a damaged reputation and even failure. Edward Humphrys died at Nice in May 1867 after a short illness. He was fifty-seven years of age. In the words of his obituarist in *Engineering*, 'he worked himself to death'. More particularly, 'the *Mooltan*'s engines, once so famous, were taken out months ago, and the highly economical engines appear to be all in limbo – the limbo of failure, that one fatal word to the engineer'. As the engines had worked on, 'his economical engines broke down, and gave so much trouble that some of them had to be replaced by single engines, and all, or nearly all, are, we believe, doomed'. That it was 'this want of success [that] weighed him down, and hastened his death, none who knew his proud, resolute nature can doubt'.[36]

17.3 'To Establish a Reliable Line of Steamers': Alfred Holt's Ocean Steam Ship Company

On the day before Scott's *Thetis* was officially completed in spring 1857, Alfred Holt arrived at the Greenock yard. Since overseeing work on auxiliary

[34] Murray (1865: 161–2).

[35] Murray (1865: 167–8); Institution of Mechanical Engineers Obituaries, 1881 (Olrick).

[36] *Engineering* vol. 3 (1867: 528). Of the eight major P&O mail steamers fitted with Humphrys & Tennant double-cylinder engines during the period 1861–4, one (*Mooltan*) had the original engine replaced in 1865, three by the mid-1870s, two were laid up in the mid-1870s and two were wrecked in 1869 and 1871 (Rabson and O'Donoghue 1988: 62–7).

steamers for Lindsay earlier in the 1850s, Holt had been a frequent visitor and had become a close personal friend of John Scott. His official task this time was to witness the trials and delivery of Lamport & Holt's first steamer, *Zulu*. Since its foundation in the mid-1840s, his elder brother George's firm had remained resolutely loyal to sail. It owed much of its prosperity, indeed, to the steam coal trade that Lindsay had developed in relation to the demands of the mail lines (Chapter 15). Hitherto, Lamport & Holt had provided Alfred with invaluable experience in the practices of ship-owning; now his skills as a mechanical engineer, honed while apprentice with the L&MR, began to assume a central role for Lamport & Holt.[37] And with the *Thetis*, Alfred could witness for himself the power and economy of high pressure steam at sea.

Alfred strengthened his connection with Scott's firm when he placed an order for a steamer of his own. With the Mersey-built *Saladin*, he entered the West Indies trade in the mid-1850s and, thanks to his trusted shipmaster Isaac Middleton, transformed early losses into profit. Recognizing the inadequacies of a single vessel, Holt took advantage of low prices to contract with Scott for a second steamer, the 695-ton *Plantagenet* (1859). Three more steamers, each one of which was slightly larger in tonnage and power than the one before, followed from Greenock at yearly intervals. His line of steamers also secured a monthly mail service, funded by the Jamaican and Haitian Governments, between Jamaica, Haiti and New York, but failed to win the West Indies contract from RMSP in 1862–3. Just how bold the venture was in practice, Holt witnessed for himself early in 1862 when he took a passage on the *Talisman* under Captain Kidd and encountered the full force of an Atlantic storm.[38]

In 1861 Holt travelled to Newcastle to examine on behalf of Lamport a screw steamer under construction at Andrew Leslie's yard downriver at Hebburn. With Leslie, he quickly forged a durable friendship. Leslie's very first order had come around 1854 from Lindsay, then MP for Tynemouth, for a 1,000-ton auxiliary iron steamer. Lamport & Holt named their purchased vessel *Copernicus* and soon ordered a second steamer from Leslie for the Brazil and River Plate trade. Having sent home a draft specification, Alfred on his return from the West Indies watched over the construction of the *Kepler* at the Hebburn yard and took passage on the first voyage.[39] *Newton* and *Galileo* followed from McNab of Greenock and Leslie of Hebburn, respectively, in 1864 and *Ptolemy* and *Halley* from Leslie in 1865. Profits accrued and around

[37] Scott & Co., to Lamport & Holt, 12 May 1857, in Letterbook (1857–8: 24–5), Scotts of Greenock Papers (SP), GUA.

[38] Holt (1911: 33–7); Kidd Journal, Ocean [Steamship Company] Archives (OA), MMM, pp. 16–17; Anne Holt Diary (27 February and 3 June 1863), Alfred Holt Papers (AHP); Smith et al. (2003b: 402–4).

[39] Holt (1911: 42–3); Clarke (1979: 33, 46) (Lindsay).

seven more 'astronomers', all in the range 1,000–1,600 tons, were delivered from Leslie in 1866.[40]

Selling his own line on account of intense competition in 1863, Alfred now talked steamers with his younger brother Philip (who had just left the service of Lamport & Holt) morning, noon and night until they had identified a trade 'which we should, as we thought, have to ourselves'. In the 1850s and 1860s, the seasonal trade in tea from Chinese ports to Britain had been made famous by the tea races among fast sailing clippers, first American and now British, including vessels built by Scott's, Steele and other yards on the Clyde and in Aberdeen. Towards the end of 1864, the Holts settled on the China trade 'mainly because tea was a very nice thing to carry' and partly because Alfred now confidently believed that 'cargo steamers could be engined so as to go, and pay, on much longer voyages than anyone at that time thought possible'. The brothers would also have known that a year earlier Lindsay's auxiliary steamer *Robert Lowe* had navigated far up the Yangtze to load tea at Hankow for London, a voyage that earned £10,315 in freight alone.[41]

Alfred acted swiftly to circulate shareholders from his previous West Indies venture with a promise 'as soon as the present steam ship mania has subsided, and good vessels can be built at moderate prices, to construct more steam-ships and enter upon a new trade'. The trade was 'a distant one and success or failure is simply a question of consumption of fuel'. Explaining further his meaning, he insisted that the 'steam boat engine as at present used is an exceedingly wasteful machine'. This had 'long been evident to engineers, and many attempts have been made to improve it, some with very considerable success, but the margin of saving still left by the best engines is probably 50% of their actual consumption'.[42] He chose to refit his one remaining steamer *Cleator* (1854) with an 'experimental' compound engine which would, in contrast to Rowan's system, be as simple as possible. It consisted of one high and one low pressure cylinder, the former vertically in line below the latter. The two cylinders shared a common cross-head positioned between them. The cross-head drove two connecting rods with a common crank pin and single crank. A heavy flywheel and other balancing arrangements maximized the smooth running of the crank shaft. Opting for tried-and-tested locomotive engineering, Holt employed a tubular boiler operating at 60 psi.[43]

'Having completed the change of machinery in the *Cleator* we (ie P[hilip] ... and I) went off for a cruise into the channel in her ... [for] about two days',

[40] Anne Holt Diary (AHD) (4, 18 July 1863), AHP; Heaton (1986: 24–30).

[41] Holt (1911: 45–6); Lindsay (1874–6:vol. 4, 468n); Marriner (1961: 25–50, 85–98) (Rathbones in China).

[42] Holt, Circular regarding *Cleator*'s new engine [1854], AHP.

[43] Holt, Circular [1854]; Hyde (1956: 173); Smith et al. (2003b: 411) (plan of Holt's original engine arrangement).

Alfred later recalled, 'The experiment left nothing to be desired, the engine worked perfectly, the vessel's speed was improved, and the coals burnt reduced, so that 5 tons did the work for which 8 tons were previously required'. Added credibility stemmed from the fact that Holt knew every aspect of the vessel's old engines which he had 'designed, and actually drawn, myself, [and] whose speed and coal consumption and performances generally, I knew with absolute accuracy'.[44] With this trial rated a success, Holt placed an order with Scott for three large iron-screw steamers (costing altogether £156,000) for the China trade.

Alfred persuaded a reluctant Lamport to adopt larger versions of his compound engine to be constructed by Tyne engineers R. & W. Hawthorn for the steamers *Arago* and *Cassini* built by Leslie in 1865–6. While at the Hebburn yard in February 1866, Alfred and George sent a third compound-engined steamer, *Humboldt*, to sea under the command of Captain Kidd. 'I took her from the Builders, Leslie's, on the Tyne with a cargo of coals to Gibraltar & returned to Liverpool via Lisbon', wrote Kidd in his diary with characteristic frankness. 'This was a test & trial trip with Alfred Holt's new engine. Mr Lamport thought the engine was a failure. Alfred Holt thought otherwise and sent me to prove it was all right in the *Humboldt*'. This trial allowed the Holt's to assure the Ocean Steamship Company (OSS)'s first GM that 'the leading principle that the same speed can be obtained on about half the usual consumption of fuel is established beyond a doubt'.[45]

At the beginning of the year, Holt had circulated a printed letter announcing to prospective shippers that he was 'about to establish a line of Screw Steamers from Liverpool to China'. The communication promised that the *Agamemnon* would be commanded by Captain Middleton and that the service would involve a non-stop passage to Mauritius via the Cape of Good Hope with an expected time of thirty-nine days out from Liverpool. The passage east would continue with calls at Penang (fifty-four days), Singapore (fifty-seven days) and Hong Kong (sixty-six days) before ending at Shanghai (seventy-six days, all inclusive of port detentions).[46] Here was the projection of a precisely laid down geographical steam navigation system to be initiated by the three identical vessels.

In order to make this promised schedule credible, Holt stressed that 'all the steamers have been built on the Clyde; they are of full power; and will steam the whole passage, both out and home'. In addition to their Clyde-built pedigree and their low dependence on sail, these new steamers had been designed to embody '[e]very precaution which experience suggests ... to fit them for the safe conveyance of valuable cargo'. Moreover, each captain had 'been many

[44] Holt (1911: 46–8).

[45] AHD (16 February 1866); Kidd Journal, OA, MMM (24); OSS 'General Book Vol. 1 (1865–82)' (1); Smith et al. (2003b: 410–14).

[46] Alfred Holt, 'Circular letter dated 16 January 1866', AHP; Falkus (1990: 96–7); Smith et al. (2003b: 410, 414).

years in my employ'. Holt undertook 'to establish a *reliable* line of steamers, which will carry Cargo [and about 40 cabin passengers], at moderate rates of freight, both safely and at *tolerable speed*'.[47]

On 31 March the *Agamemnon* left the builders for an overnight trial run that would take the vessel to the Mersey. Holt privately noted that on departure the engine was 'heating slightly and not working very well for a few hours but after passing the Cumbraes this passed off, and shortly after sighting the Mull of Galloway we were able to allow her to have full speed whereupon she at once went away 10½ knots which she easily maintained during the trial'.[48]

Measurements of fuel economy permeated Holt's diary entry. 'We found her consumption by very accurate experiment to be 6½ tons in 7 hours & 40 mins = 20 [tons]. 6 [hundredweights]. 3 [quarters]. 23 [lbs] in 24 hours, and this result as far as I know is not approached by any vessel afloat. I congratulate myself exceedingly thereon and think I have got a very fine vessel indeed'. This consumption approximated to just over 2 lb of coal per ihp per hour (Figure 17.5).

Eight days behind its original schedule, the *Agamemnon*'s first voyage ended in the Thames on 24 October 1866, a few days over six months since departure from the Mersey. Alfred hastened to meet the ship just entering the Victoria Docks, North Woolwich, where he 'rejoiced to receive an account of her performances which I consider eminently satisfactory'. Had the Holt's been given to showmanship, they would have felt overshadowed by a remarkable triumph of sail that occurred several weeks ahead of the *Agamemnon*'s arrival. Under the heading 'The great ship race from China', *The Times* announced in early September 1866 that three sailing ships (*Ariel*, *Taeping* and *Serica*) had docked on the same tide. What captured the public imagination was that all three of these clippers had raced some 16,000 miles from China, each loaded with around 1 million pounds weight of the new season's tea, to arrive almost simultaneously off the Kent coast with a passage time of ninety-nine days. Two years later, however, Holt noted that although the *Ariel* was first to arrive with the season's tea after a passage lasting ninety-seven days, his steamer *Agamemnon* took only seventy-six days from Hankow and his new *Diomed* only eighty days from Shanghai. By 1869, the *Achilles* made the fastest passage on record to that date, sixty-one days from Foochow to the Kent Coast.[49]

'Ship-owning is a fluctuating, precarious business', Holt cautioned the OSS shareholders at their 1872 meeting. But by 1875 OSS had a working fleet of some seventeen deep-sea vessels, seven of them from Scott and ten from Leslie. Lamport & Holt had built up an even more impressive fleet of ocean

[47] Holt, 'Circular'. My italics.
[48] AHD (c. 23–31 March 1866); Smith et al. (2003b: 414–15).
[49] *The Times*, 12 June, 16 July, 7, 12 September 1866; AHD (29 June 1868; 13 October and 18 September 1869); Falkus (1990: 100–02).

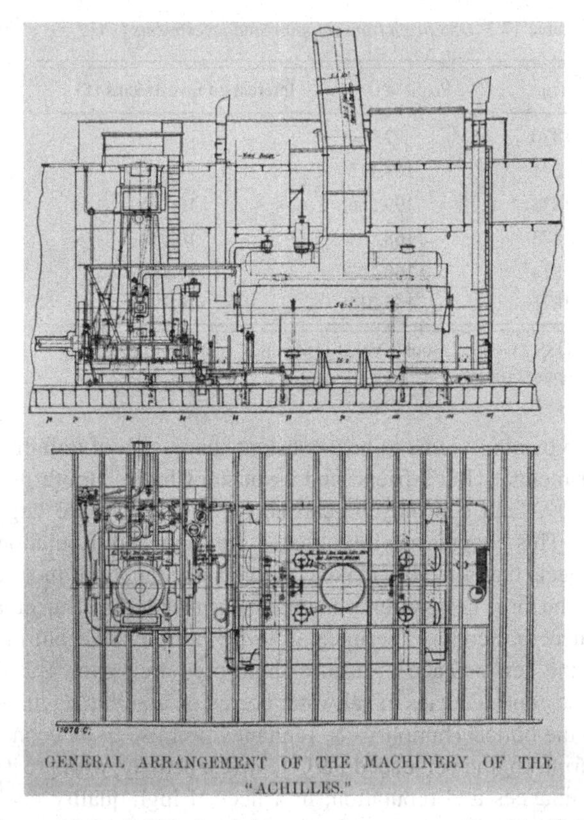

GENERAL ARRANGEMENT OF THE MACHINERY OF THE "ACHILLES."

Figure 17.5. *Achilles'* engine room arrangement with Holt's compact compound engine

steamers since the early 1860s. Over thirty had been new builds, some twenty-five by Leslie.[50] Neither company had Admiralty mail contracts, nor had the shareholders limited liability. Shares were not publicly traded on any stock market, making them private partnerships similar to Cunard. Table 17.3 shows OSS's annual profits over the period 1870–5 inclusive. As will be obvious from the table, there were also income interest payments from the reserve fund, deriving from investments in railway and other shares or bonds at home and overseas. The fund had been set up primarily to allow self-insurance on the ships.

[50] Middlemiss (2002: 41–6); Heaton (1986: 108–13) (fleet data).

Table 17.3. *OSS profit from shipping and investments (1870–5)*

Year	Profit (£)	Interest on investments (£)
1870	92,964	–
1871	164,115	–
1872	199,256	10,705
1873	168,568	16,034
1874	160,197	17,380
1875	145,216	19,391

OSS, 'General Book', Vol. 1 (1865–82)' (data from each annual report).

'Steamship owning seems to be a constant succession of unfathomable and costly experiments', Holt's friend and associate Charles Booth ruefully concluded in 1869, 'and can only be carried on when there are large earnings coming in'.[51] This chapter has investigated the trials and tribulations of three different projects that aimed to make credible an economy of motive power at sea. None of the three new systems maintained a trouble-free trajectory towards inevitable success. Scott's experimental *Thetis* initially held out the promise, founded on the new science of heat, of high pressure steam. But the project faltered on the contingencies of saltwater corrosion combined with difficulties of access to the boiler. Humphrys & Tennant's promise of an economical and compact marine engine persuaded P&O to invest heavily, with great risk to the Company's finances and reputation, in a fleet of high-quality mail steamers which themselves well illustrated the force of Booth's judgement. Cautious in the extreme, Alfred Holt initiated his project in a tried and tested steamer of his own, replacing the simple with a compound engine, and running experimental trials at sea.

[51] Quoted in John (1959: 57).

18 'The Modern Clyde Ships'

Economy and Power for Ocean Steam Navigation

> Now about our being slow in adopting ... those peculiar engines which use the steam twice over. The Cunard Company have not adopted that yet, the P and O Company have not adopted it, and I believe the only other Company who have is the Pacific Navigation Company. [T]herefore it was necessary that we should have a little information before we went into the thing, it might have turned out a great failure; although I believe it to be a most successful and excellent thing.
>
> *RMSP's Chairman, Captain Mangles, defends his Company's caution in adopting a compound system of motive power (1869)*[1]

Summary

When the Holt brothers launched their 'Blue Funnel' system, the principal mail steamship companies had been in business for around a quarter of a century. While each of them faced different challenges to the smooth working and effective control of their lines of steamers, they all shared problems of coal supply and consumption. The RMSP exercised much caution before ordering, in the face of PSNC competition, three large compound-engined mail steamers from Elder in 1870–1. Still burdened by its heavy investment in double-cylinder engines in the early 1860s, P&O revived its old connections with Caird's of Greenock at the end of the decade and, amid internal controversies, began an unbroken owner–builder relationship that lasted into the next century.

18.1 'Economy in Coal': A Troubled RMSP Turns to Clydeside

Under the chairmanship of Captain Charles Mangles between 1856 and 1873, RMSP entered a relatively stable period following the troubles of its first fifteen years (Chapters 7–10). With renewal of substantial mail contracts, its annual surpluses in the 1850s and 1860s typically stood in the region of £250,000, rising to £463,955 in 1855 due to Crimean War earnings. Proprietors, however, frequently voiced complaints that directors were unnecessarily allowing the insurance and replacement funds to rise at the expense of larger dividend and bonus payments to shareholders. The Court's caution reflected its views on the

[1] RMSP GM (27 October 1869: 84–5).

demands of the trade: the Company, for example, attributed the heavy wear and tear on the larger ships to their combined transatlantic and round-Caribbean voyages.[2]

Concerned about a £10,500 increase in coal costs, critical proprietor Dr Alexander Beattie (Chapter 10) reminded the GM in April 1857 that coal remained 'one of the most important and expensive [items] in regard to a steam navigation company'. It was therefore 'most desirable that every advantage should be taken to economise the cost of fuel'. Since at least 1853 RMSP's annual coal costs had been in the range of £195,000 to £225,000 per annum. Deputy chairman Russell Gurney tried assuring Beattie that the Company had 'secured the monopoly' on the very cheapest coal available, that from the Newport, South Wales colliery. After shipment, the cost at Southampton was eighteen shillings per ton. Furthermore, Gurney insisted that they were 'the best coals to send to tropical climates – it is a peculiarly hard coal and not subject to the same amount of waste that others are'. At the following meeting, Mangles claimed that the coal price increases resulted 'from a fluctuation of circumstances over which the Directors have no control'. Dr Beattie, however, remained unimpressed with this familiar Court fatalism, 'seeing that our coals are contracted for at a specific sum'.[3]

For both mainline and branchline services through the 1850s well into the 1860s, RMSP trusted in simple, reliable side-lever engines. In 1853–4 the Line took delivery of their last two wooden paddle-steamers from White of Cowes and Pitcher at Northfleet. At the end of the decade, three large iron paddle-steamers of the size of the *Atrato* (Chapter 10) entered service. The first, *Paramatta*, was wrecked by grounding in the West Indies during the maiden voyage. At the subsequent GM, Dr Beattie voiced a strong suspicion of compass error linked to 'the overweening confidence on the part of those in charge of ships from the large experience that they have had in navigation (hear, hear)'.[4]

Well known for their voracious appetites for coal, these paddle-steamers nevertheless exhibited notable sailing qualities. Built and engined by Napier, the second of the trio, *Shannon*, suffered a serious machinery failure at St Thomas soon after completion. When efforts to tow the steamer to Britain proved impossible, the *Shannon*'s commander ordered the sails to be set for an Atlantic passage under canvas alone. Mangles told his proprietors that, contrary to the views of some competent judges, 'these paddle steamers are able to find their way through the sea as well as other sailing ships'. According to

[2] RMSP Reports and Accounts (1843–4; 1846–59). These printed documents provided the Court's official narrative which often provoked lively proprietorial criticisms over the years.
[3] RMSP GM (16 April 1857: 14–17, 25–6); (28 October 1857: 17–18); RMSP Reports (1850–9).
[4] RMSP GM (26 October 1859: 27–9).

the vessel's master, 'he beat all the other vessels he fell in with under canvas (hear, hear)'. Compared to ordinary sailing ships, Mangles claimed, RMSP's paddle-steamers had a superior fineness of lines. A ship-owning proprietor, Mr Simpson, paid special tribute to the commander, Captain Abbott, for his feat of bringing the *Shannon* unharmed to Southampton amid wild winter seas despite the restrictions on manoeuvrability from the massive paddle-boxes. Such a long vessel, he observed, generally used sails to steady the ship but not 'to give her aid in getting from a lee shore'. Nor indeed were the sail arrangements primarily designed to enable a paddle vessel to cross an ocean without steam power.[5]

Under much less strenuous circumstances, however, sail frequently aided steam in reducing coal consumption and maintaining regularity.

> [T]hrough the cabin port the sunrise shone in, yellow and wild through flying showers, and great north-eastern waves raced past us, their heads torn off in spray, their broad backs laced with ripples, and each, as it passed, gave us a friendly onward lift away into the 'roaring forties', as sailors call the stormy seas between 50 and 40 degrees of latitude.

Revd Charles Kingsley wrote thus of the *Shannon*'s transatlantic crossing to St Thomas one December. Such was the benign influence of the north-east wind that the steamer hurried 'straight as a bee-line' and covered some 260 miles a day in the run under steam and sail towards the Azores.[6]

At the beginning of the 1860s, Admiralty demands for accelerated mail services accentuated RMSP's concerns with coal consumption. 'You cannot drive very old ships fast without consuming a great quantity of coals', the chairman stated at the GM of April 1861. For the first time, however, he drew proprietors' attention to something that 'is now going on and very naturally attracting the attention of all steam companies'. There was, he explained, 'a new Engine called the Double Cylinder Engine by which they work the high and low pressure cylinder, that is to say they can use the steam a second time instead of blowing it off and wasting it'. Mangles evidently had little knowledge of Randolph, Elder's earlier patents. But the timing suggests a loose awareness of the *Mooltan*'s earliest trials (Chapter 17). 'Many people think very highly of the new description of Engine', he continued, 'and they think it will effect a great reduction in the consumption of Coals'. For his own part, he was 'not very strong in my belief for I have met with so many improvements that have ended in smoke that I do not put faith in it'. Indeed, he warned that if the claims

[5] RMSP GM (24 April 1861: 5, 29–30, 43).
[6] Kingsley (1887: 2–4) (first published 1871).

for the new engine were 'found to be true we shall be put to great expense in taking out our present Engines'.[7]

When the European & Australian Royal Mail Steam Packet Company collapsed while still in its early stages RMSP acquired from them in 1858–9 two large and relatively new iron steamers, *Oneida* and *Tasmanian.* Constructed respectively by John Scott at Greenock and by Laurence Hill at adjacent Port Glasgow, they became RMSP's first mainline screw vessels. It is likely that their performance persuaded the Company to order for the Southampton to Rio line two new iron-screw vessels, the *Douro* (1864) from Caird in Greenock and the *Rhone* (1865) from the Millwall Iron Works. After failing at Blackwall in the mid-1850s, but now backed by bankers Overend Gurney, C. J. Mare took over Scott Russell's yard at Millwall (Chapter 17) at the end of the decade and spent £100,000 in redeveloping the site under the new name, including shell plate and armour plate production in rolling mills.[8]

Both vessels incorporated superheating apparatus. Trials on board RMSP steamers had been under way for some years. In 1860, Murray's INA paper had alluded to experiments with superheated steam at Southampton on RMSP (and other owners') vessels (Chapter 17). In 1867 Murray presented to the INA testimony from the *Douro*'s chief engineer. 'The pressure of steam in the *Douro* is 20 pounds', the chief wrote. 'It is super-heated, and admitted into the cylinder at about 304°'. Murray himself attributed the two steamers' 'remarkable economy' of thirty-five tons of coal per day, however, not to the superheating but to ample boiler power and moderate speed (10½ knots). From the chief's letter, he might also have added the exceptional attention to every detail, ensuring, for example, that the usual problem of boiler corrosion linked to surface condensers was largely vitiated by the engineers' precautions to coat the boilers with a sufficient protective scale of brine. As Elder had recently told the IESS, marine engineers faced an enormous challenge of separating out the different strands that entered into questions of coal consumption at sea (see Epilogue).[9]

In his 1865 INA presentation, Murray had observed that screw vessels were 'often placed at a disadvantage as compared with paddles by being underpowered, their great handiness under sail being the temptation to do this'. His view was that the question of screw *versus* paddle for ocean steaming was still far from settled. As evidence, he pointed out that RMSP had recognized the difficulty by recently ordering two large steamships, one screw (*Rhone*) and one paddle (*Danube*), from the Millwall Iron Works.[10]

[7] RMSP GM (24 April 1861: 13–14, 40).

[8] Bushell (1939: 90–7) (RMSP's Australian projects); Haws (1982: 40–3) (data on vessels); Banbury (1971: 211–13) (Millwall Iron Works).

[9] Murray (1867: 157–8) (chief's letter); Napier (1864: 91–2) (Elder's comments).

[10] Murray (1865: 163).

The *Danube* project soon exhibited major shortcomings that threatened the reputations of both owners and builders. Launched in early 1865 and due for delivery a few months later with engines by Humphrys, the hull floated with a draught two feet more than the specification. Expensive lengthening by the builders failed to solve the *Danube*'s problems. In 1870, a proprietor (Mr Carpenter) told the GM that the *Danube* was currently lying inactive in Southampton Docks, 'positively useless' and 'a great waste of money'. 'All tinkering with ships is bad', he proclaimed, 'and worst of all tinkering is the tinkering of a new ship'. He further asserted not only that 'she is a failure' but also that 'she is not considered a safe ship, and is by no means a favourite'. In reply, Captain Mangles denied the safety allegation but confessed that after lengthening 'she has not succeeded in the way that we and other ... more competent judges thought she would'. The builders, he claimed, had admitted to errors and had paid a high proportion of the cost. They were, however, now no longer in business, having been brought down by the Overend Gurney banking crash in 1866 and, presumably, by the damage to their reputation from the *Danube*. But in turn, shareholders indicted management for its fatalistic narrative as an excuse for loosely worded contracts.[11]

When a powerful hurricane struck St Thomas in October 1867, RMSP lost around 200 lives and three steamers, including its prestigious two-year-old *Rhone*. The Company turned for help to their old contractor in Greenock. Since 1856 James Caird had won contracts for an almost continuous series of large iron-screw steamers for the Hamburg America Company (HAPAG) since 1856 and the rival Norddeutscher Lloyd (NDL) since 1858. In all, these contracts amounted to some twenty-two steamships for the former and twenty-nine for the latter company. Indeed, both these Hamburg- and Bremen-based lines hardly looked anywhere else for their new buildings until the mid-1870s. Caird, however, agreed to transfer to RMSP ownership the iron-screw steamship *Rhein*, one of several vessels under construction in 1868 for NDL. Renamed *Neva*, the 3,025-ton vessel had a simple, direct-acting engine that drove the ship at 14¼ knots on trials and at an average speed of 12½ knots on the first round voyage. The chairman reported to proprietors that 'she is a magnificent vessel in every way and I do not think there is a faster or better ship afloat'. So pleased were the owners that (Caird's being busy with their North German contracts) RMSP contracted with Summers, Day at Southampton for the *Nile*, a ship 'equal to and indeed upon the lines as near as can be of the *Neva*'.[12]

[11] RMSP GM (31 October 1866: 18–19); (26 October 1870: 46–52); (26 April 1871: 11–13); Banbury (1971: 212–13) (Mare).

[12] RMSP GM (28 October 1868); Bonsor (1975–80: vol. 1, 387–91) (Hamburg America), (vol. 2, 544–51) (North German Lloyd); Haws (1982: 38–47).

The Revd Kingsley, however, gained a different perspective on the *Neva*'s qualities on his homeward passage from St Thomas around 1870. 'The wind was, of course, right a-beam; the sea soon ran very high', he explained. 'The *Neva*, being a long screw, was lively enough, and too lively; for she soon showed a chronic inclination to roll'. Two female passengers, he observed, had made it clear to those around that had they but known these horrors, they would have gone to Europe in a sailing vessel. As the gale grew worse, a heavy sea 'shook the *Neva* from stem to stern, and made her stagger and writhe like a living thing struck across the loins'. At daylight, 'we found that a sea had walked in over the bridge, breaking it, and washing off it the first officer and the look-out man – luckily they fell into a sail and not overboard'. The water also 'put out the galley fires ...; and eased the ship; for the shock turned the indicator [engine-room telegraph] to "Ease her"'.[13]

These larger and faster steamers of the late 1860s satisfied the terms of the mail contracts as well as providing enhanced space and comfort for passengers, but their simple, one-stage expansion engines did little or nothing for coal consumption either to the West Indies or to Brazil. As for the decade-old paddle-steamers *Shannon* and *Seine*, the language at GMs became harsher, describing them not only as 'large consumers of coal', but also as 'those robbers'. The unspeakable *Danube*, on the other hand, was simply 'a slow ship and a great consumer'.[14]

Since the mid-1860s, independent 'Liverpool' steamship owners (especially Lamport & Holt) intensified the competition for valuable freight on the South Atlantic (Chapter 17). Unforeseen was the challenge to RMSP from PSNC, hitherto serving only the western seaboard of South America, in launching in 1868–9 a monthly line of compound-engined steamers from Liverpool to Valparaiso by way of Rio, the River Plate and the Straits of Magellan, a round voyage of some 19,000 miles that constituted the longest steamline in the world. Having launched the line with four 1,600-ton paddle-steamers, the Liverpool company had on order from Elder at least four 2,800-ton screw steamers, beginning with the *Magellan* (1868) costing £74,500.[15]

At the autumn 1869 GM, Mangles announced RMSP's response. The Company had ordered from Elder the *Elbe*, a 3,108-ton mail steamer to be fitted with compound engines working at around 60 psi. Mangles admitted rather lamely that PSNC 'have adopted those engines for some time, and they report most favourably of them; by which the consumption is very much reduced'. As though addressing sceptical rather than supportive shareholders, he expressed the view of the Board 'that we should join, not as pioneers exactly, but that

[13] Kingsley (1887: 396–8).
[14] RMSP GM (27 April 1870: 33); (26 April, 1871: 11–13, 26).
[15] RMSP GM (27 October 1869: 8); Haws (1984: 36–40).

when a thing has been proved to be really advantageous and leading to a reduction of expense, we should adopt it'.[16]

This news prompted considerable RMSP shareholder disquiet. 'You say we should follow suit'; one unnamed proprietor stated, 'I say we should take the lead (hear, hear)'. A Mr Ward probed further by asking '[w]hy are we always lagging behind other Companies in introducing improvements and new engagements'[?]. His point was that fellow-shareholder Dr Beattie had for long pressed such matters upon the Board, but that it was only when they 'are brought forward by other companies that the Board seems to take advantage of them'. He therefore urged that RMSP instead 'be the first to introduce improvements, by looking in advance, seeing what the public require, what increased commerce requires ... and then introducing improvements, and so to hinder others from coming in and introducing them and turning us out'. And he also told the meeting that many passengers 'do not want splendour and magnificence in the vessels, what they want is comfort; they would rather have plainness with comfort than gilt and velvet without it'.[17]

In response, the chairman told proprietors that hardly a day passed at RMSP's office, or indeed at that of any large company, 'but some inventor does not appear with some new engine which is [promised to require] only about a quarter of the coal that other ships consume'. He vividly expressed 'quite a horror of a man arriving with a green bag under his arm ... [with] some plan or other in view by which he is going to save [us] a great deal of money'. The policy at RMSP was thus 'for a certain time [to] watch and see how a thing works and when we have convinced ourselves that the thing is sound and good and will produce a reduction [and] then go at it as hard as you say we should'. It was therefore wise to let 'other companies test it ... because if we undertook to do so in the first instance I am sure we should lose a great deal of money'.[18]

Further disquiet emanated from a familiar shareholder voice. Mr Simpson targeted a decision to refit the eleven-year-old *Tasmanian* with new, compound machinery and boilers. 'What is the consumption of coal in the Tasmanian'? he asked rhetorically. From a previous meeting he was well aware it was of the order of eighty tons per day. 'We know that it is very large compared with the modern Clyde ships', he answered. A ship built according to modern construction would, he insisted, save £2,000 per voyage in coal alone. And in order to enhance the freight side of the service, 'if we get these more modern ships we should have more room for freight than we have in our present ships'.

[16] RMSP GM (27 October 1869: 16–17); Haws (1982: 45) (*Elbe*).
[17] RMSP GM (27 October 1869: 12–14, 69–78).
[18] RMSP GM (27 October 1869: 79–80).

Indeed, not only would there be less space set in place for coal bunkers, but the number of firemen could also be reduced.[19]

Picking up on his old concerns over coal consumption, Dr Beattie reminded fellow-shareholders that trustworthy masters were as important as trustworthy engines. He commended to the Court P&O's policy of awarding bonuses to commanders 'for saving a large quantity of coal'. Mangles, however, recalled that RMSP had tried the plan in the 1840s with the result that 'there was a constant squabble between the agents abroad and the engineers as to the quantity of coal which was put aboard'. The agents, he claimed, 'had always said it was more, and the engineer always said it was less, and we were never able to arrive at the real truth of the matter; so that we were obliged in self defence to give up the plan'. From his own experience of serving on various railway company boards, he cited cases 'where there have been two engines lying [so] close together that the fireman has stepped over from one engine and stolen coals from the other (a laugh)'. Here the gentlemanly Captain Mangles, MP, was placing the blame on the cunning fireman who sought to trick his superiors by proving 'how little coal he consumed' and thereby claim the bonus.[20] By implication, an RMSP steamer in a distant port could likewise be victim to a variety of fraudulent practices.

None of these reservations dampened RMSP's enthusiasm for the new system which was, as the chairman reported in April 1870, 'certainly producing wonderful results'. He announced that two sister ships had now been ordered from Elder. Even Simpson joined in the mood of optimism by urging the building of still more new ships. The deputy chairman, however, explained that the Government's policy of giving very short-term mail contracts restricted the Company's room for manoeuvre in the five years remaining of the present contract. In contrast, Gurney claimed, the French Government's policy of awarding very long-term contracts allowed those companies to 'adopt every improvement, because they have a length of term sufficient to gain back the expense ... [from] the increased economy of the service'.[21]

The compound-engined trio (*Elbe*, *Tagus* and *Moselle*) entered service for RMSP at yearly intervals between 1870 and 1872. 'I do not believe there are any screw ships that can carry as much cargo [over 1,000 tons] which offer as much accommodation to passengers and steam as quickly as our [new] ships do', the chairman told proprietors in October 1871. '... [The *Moselle*'s] trial trip has been a very satisfactory one, and proves not only great capacity, but great speed'. He did not mention the generous provision of square sails on both masts together with fore-and-aft sails to dampen rolling in a beam sea.

[19] RMSP GM (27 October 1869: 54–7).
[20] RMSP GM (27 October 1869: 83–4).
[21] RMSP GM (27 April 1870: 7, 41–2).

Table 18.1. *RMSP annual coal costs and reductions (1868–70)*

1868	£200,000
1869	£179,000
1870	£159,700
1871	£148,700

Testimony from British and foreign passengers, he claimed, showed 'that our popularity is well established' on the grounds of both comfort and safe navigation.[22]

With the trio operational by spring 1872, Mangles, however, felt that the fuel savings were less than they had been 'at the beginning of our alteration of our system of ships and engines'. From the available figures in Table 18.1, it was apparent that the large savings in earlier years (1868–70) could not be due to the new compound steamers. At an earlier GM, Dr Beattie had again pointed to a probable explanation when he hailed the award of a bonus to officers and engineers for both care in navigation and for 'economy in coal by using their sails at every possible opportunity ... Nothing conduces to the saving of coal than that the officers and engineers should take care when there is any breeze that is at all favourable to use their sails'.[23]

The introduction of the three new iron-screw compound-engined steamers to RMSP's fleet nevertheless heralded a reinvention of the Line's public image that projected comfort, safety and economy underwritten by a distinctly Clyde-built pedigree. A fourth similar steamer, *Boyne*, followed from Denny in 1871 while the Company by 1875 had acquired another five large Clyde-built vessels either while building or with only a couple of years' service. The former European & Australian steamers were re-engined by RMSP with compound machinery and the *Shannon* was converted to screw with a clipper bow. The Line also introduced several compound-engined steamers of over 1,000 tons to inter-island services. But it disposed of the unloved *Danube* to the Union Steamship Company for the Cape run after only six years' service.[24]

18.2 'A Very Foolish Game': PSNC's Mania for Extension

Nothing that RMSP did to modernize its fleet compared with PSNC's subsequent construction programme. 'This Company has for some time past been

[22] RMSP GM (26 October 1870: 4–8); RMSP GM (25 October, 1871: 2–3, 6–8).
[23] RMSP GM (24 April 1872: 3–4) (Mangles); RMSP GM (25 October 1871: 17–18) (Beattie).
[24] Haws (1982: 38–50).

playing what I think is a very foolish game', Alfred Holt wrote in his diary in March 1869, '[for] they had that beau ideal of steam boat enterprise, a profitable obscure trade, and tho' possessing rather an extravagant fleet extravagantly managed were working themselves into a very profitable concern, when the unfortunate mania for extension seized them, and nothing would answer but a line from Valparaiso to Europe'. What was happening, Holt reflected, was that 'money began to burn in their pockets and the vessels they built for the service are as unsuitable for remunerative working as one could well conceive'. The *Magellan* was 'said to have cost £85,000, and tho' a beautiful vessel to look at is probably the most difficult to make money out of that has been built for some years'.[25]

Over the next five years, PSNC took delivery of twenty-one ocean steamers, culminating in a class of six ships of well over 4,000 tons and costing around £150,000 apiece! Sixteen were Elder products, four Lairds and one Napier's. Along with its West Coast vessels, the PSNC fleet totalled almost sixty steamships by 1874 making it the largest such fleet in the world, with the largest active steamship, the 4,671-ton *Iberia*. From 1870, the Valparaiso service, now three sailings a month from Liverpool, also extended north to Callao. From 1872, the line introduced weekly departures, but reduced these to fortnightly in 1874. By the following year, eleven of the ocean steamers were laid up in Birkenhead, with two going to RMSP to replace the wrecked *Shannon* and *Boyne* in summer 1875.[26]

'There is no witchery in the West Coast trade which will prevent misfortune dogging the steps of imprudence and extravagance', *The Porcupine* commented in June 1874, 'and when it does come it comes with sweeping severity'. In a scathing editorial, the satirical paper sought answers as to why PSNC's chairman, Charles Turner, MP, had deserted 'a commercial undertaking in its hour of need' by resigning his position and selling out all his shares. *The Porcupine* also blamed PSNC's downfall on the directors' 'loss of control'. Heads of departments were left to themselves; the accounting system 'prevented an accurate testing of the profits of particular voyages; and stupendous losses were being incurred when the shareholders and even the directors imagined that the concern was a prosperous and remunerative one'. Privately, Alfred Holt admitted that around this very time of crisis, he had been offered the 'managership of that huge concern the Pacific Steamship Company'. He confessed moreover to being 'tempted but principally owing to [his brother] Philip's wise advice I declined'. Instead, PSNC's recovery owed much to a group of London shipowners, including members of the Green family of Blackwall fame (Chapter 10), who from 1877 chartered a number of the laid-up steamers for a new line

[25] AHD (20 March 1869).
[26] Haws (1982: 41,47); Haws (1984: 16–17, 39–50); Newall (2004: 13).

to Australia that in due course became the celebrated Orient Steam Navigation Company ('Orient Line').[27]

18.3 'Almost a Revolution in Steam Navigation': P&O Returns to Greenock

'[T]he Royal Mail Company has nothing like so large a Company as the Peninsular and Oriental who have no competitor treading on their heels pending a negotiation for the renewal of their contracts', Mangles told RMSP proprietors in October 1868. 'The Peninsular and Oriental Company have been wise enough to make their Company so large that the Government Authorities cannot knock them on the head'. In short, P&O were thus deemed too big to fail. From a very different, independent Liverpool ship-owner's perspective, Alfred Holt confidentially deemed the reported renewal of P&O's eastern contracts in late 1867 (£400,000 for twelve years) as 'uselessly enormous'. He calculated that the Government could have saved directly for the country £250,000 per annum for that period together with an unknown sum 'by preventing the Company discouraging enterprise [and] by the commercial weakening of the country'. Wider public concerns over the stability of the Company affected the share price, which fell from over £80 per £50 share in 1863 to under £45 by 1870.[28]

In February 1868, P&O's founding MD and chairman, Arthur Anderson, died. Three years earlier, Anderson had announced a new category of 'management staff' to replace assistant managers and duly proposed the elevation to MD status of the last assistant manager, Henry Bayley, who had begun as a P&O clerk in 1848. Captain George Bain, a P&O master, joined the management staff in the London office. After Anderson's passing, Allan recommended the addition of Thomas Sutherland, until recently superintendent in Hong Kong. A period of intense controversy ensued between Bain and Sutherland, threatening instability within the Company.[29]

Captain Bain, according to the MDs recommendation, 'had been in the Company's service [since about 1853] and commanded with great efficiency various ships on the India, China, Australian and Mauritius services; he has, besides, extensive experience in nautical matters, on education, knowledge and abilities for office business'. His principal command had been that of the

[27] *The Porcupine* vol. 16 (1874: 179); AHD (10 January 1874); Newall (2004: 15). Falkus (1990: 105), misreads Holt's entry as referring to 'the Pacific Mail [sic] Steamship Company', the title of the famous California-centred line. See Lawson (1927: 15–28) (Pacific Mail).

[28] RMSP GM (28 October 1868: 45–6); AHD (29 December 1867); Harcourt (2006: 208) (share price), (198–22) (contract renewal in the context of public criticism and financial turbulence).

[29] P&O Minutes (31 October 1865, 30 October 1866 and 17 March 1868); Harcourt (2006: 194) (Bayley).

970-ton P&O steamer *Norna* built by Tod & MacGregor in 1853, but he had also been master of Anderson's steam yacht in 1857–8. Bain married a niece of Anderson who, in his final years, held him in high esteem and used him as a personal assistant. Bain's roles included an extended tour of inspection of Company establishments and vessels, especially at Suez just as the Canal officially opened for business in late 1869. Continuing East, Bain reported on the Calcutta and Bombay agencies in 1870–1. He returned to resume his duties of reporting on ship construction, trial trips and launches in Scotland.[30]

Born in Aberdeen in 1834 as the son of a house painter, Sutherland had a strict Scottish Presbyterian upbringing that consisted of 'plain living and high thinking, of long graces but not long meals'. He combined unfailing Sabbath attendance at the old West Kirk in the presence of the city's Provost and magistrates with an undeviating adherence to the Church's 'Shorter Catechism' and its unambiguous opening claim that 'Man's chief end is to glorify God and to enjoy Him forever'. As he grew older, he regarded the leading Free Kirk clergyman and deputy to Chalmers, Robert Smith Candlish, as his 'own ideal of an eloquent preacher'. Sutherland's attendance at Aberdeen's Marischal College (one of Aberdeen's two universities at the time) suggests that his initial vocation was the ministry of the kirk. But he left after one year, probably on financial grounds. His aunt Jane Melville, however, wrote to Aberdonian James Allan and in 1852 Sutherland joined P&O's London office as a junior clerk. Far from abandoning his enthusiasm for the kirk, he regularly attended services conducted by famous Scottish preachers based in London's Presbyterian churches, including the zealous Revd John Cumming (whose opposition to the Free Kirk, Catholicism and Tractarianism complemented a firm belief in progress embodied in railways and steamships) and the Revd James Hamilton (who leaned towards the Free Kirk).[31]

Just two years later, the Company appointed Sutherland to the prestigious Bombay establishment. But Bayley apparently sacrificed Sutherland for a favourite of his own, despatching the *Scotsman* to the lesser Hong Kong posting. Sutherland, however, seemed to relish the challenges offered by a vibrant Asian maritime world. During his twelve years of service in Hong Kong, he became a member of the legislature and played a leading role, unbeknown at first to P&O, in the founding of HSBC. Within ten years he had risen to the post of P&O's Hong Kong superintendent with a salary by 1868 of £1,500 per annum, one of the highest in the Company. He took a leading role in negotiating with Japanese authorities, suspicious of westerners' motives, for an extension from Hong Kong to Japan. As a measure of his success in building trust, the Board heard that 'the Japanese authorities had applied to the Company's

[30] P&O Minutes (6 November 1863, 30 October 1866, 30 July 1869, 21 January and 25 November 1870, 13 January and 28 April to 18 July 1871); P&O Report (5 June 1873).
[31] Jones ([1838]: vol. 2, 151–5).

agents at Yokohama for permission to send a few young men backwards and forwards in the steamers to Shanghai in order that they might learn navigation, engineering &c ... and it was agreed that Mr Sutherland be authorized to use his discretion in acceding to this application'.[32]

When Company servant Franklin Kendall took up a posting to P&O's Hong Kong agency late in 1863, he provided a very favourable portrait of Sutherland's character. 'He is a thorough gentleman, as well as being a very clever man and good man of business', he wrote privately to his mother in England. 'His dinners are a by-word in Hong Kong ... [and] in such a style seeking place as Hong Kong is, you must do things well, if you want to hold your own'. To a great extent, he believed, Sutherland had made it possible for P&O to take 'a far higher standing here than they do anywhere else in the East' thanks to his having 'some good men with him since he has been in charge of the Agency'. He was 'the right man in the right place' who took 'a liberal and practical view of everything' and who thereby managed to maintain good relations 'with all the mercantile and official world'. In short, Sutherland was 'the best man whom the Company have out here'.[33]

Sutherland constructed for himself a self-image that in key respects mirrored that of the young Anderson. Presbyterian values of intensely hard, productive work, self-discipline and belief in Providence went hand-in-hand with a Scottish romanticism and theatricality composed of heroic tales set in exotic places amid stormy seas or in the context of dangerous conflicts. Like Anderson, Sutherland chose as his literary favourite Scott's romantic Waverley novels. In later years, he represented himself not as an orthodox, favoured P&O servant, but as a daring figure whose initiatives and independence placed him frequently 'under a cloud' of mistrust from his London-based masters but from which he always appeared to emerge vindicated.[34]

Sutherland's appointment to the management staff in March 1868 was not coincidental. The Company faced twin problems, the first in relation to compound engines and the second in connection with a through passage to India and beyond. Both threatened the financial stability of the Line. The Board, smarting under the legacy of the Humphrys design, had returned to tried-and-tested simple engines. The Line took delivery of the large, oscillating-engined screw steamer *Mongolia* from Scott in 1865. Between 1866 and 1868 they also purchased seven screw steamers already under construction to Scottish builders' own accounts at Dumbarton (Dennys) and Granton (Keys) during the late-1860s economic crisis. In the same period, the Suez Canal neared completion.

[32] P&O Minutes (30 June, 29 August, 20 October, 22, 29 December 1865 and 27 March 1866); Sutherland (1903: 102–8); Jones [1938]: vol. 2, 156–65).

[33] Kendall, Typescript (30 November, 12 December 1863), P&O Collection.

[34] Sutherland (1903: 108); Jones ([1938]: vol. 2, 180).

If taken up by rivals, the change would leave P&O with a useless complex system of railways, hotels, farms, coaling arrangements, warehouses and repair facilities in Egypt. Moreover, the P&O system had been built up over thirty years on the basis, not of a seamless passage to the east, but of travel divided into three connected stages, two by sea and one at least by land.[35]

Concomitant with these external threats to P&O's stability, head office was itself at risk of internal strife over compounding in the later 1860s, just as Sutherland took up his management post in London. According to Bain's highly critical account, the only director in support of the double-cylinder engine was George Carleton L'Estrange who had commanded the confidence of Irish proprietors since his appointment in 1866 and who played a major role in persuading Benjamin Disraeli as Chancellor of the Exchequer to support renewal of the mail contract in 1867. Of the MDs, Bayley was a strong opponent. In the winter of 1867–8 Bain, under written instructions from the Board and almost certainly in the company of Bayley, was on a tour of inspection of the seven vessels under construction in the north of Britain. Evidently aware of Elder's and Scott's growing reputation for marine engine economy, Bain desired to visit Elder's Glasgow works but, as he later confessed, to have done so would have been 'an evasion of direct written instructions'.[36]

Unwilling to tangle with Bayley, Bain invited Elder to visit him at Granton on the Firth of Forth where he was routinely inspecting P&O's *Travancore*, one of the steamers recently purchased and fitting out with Humphrys & Tennant simple engines between December 1867 and May 1868. Nearby, however, lay a steamer with Elder's compound engines. As a result of Elder's method of meeting all doubters on their own ground (Chapter 16), Bain found himself satisfied 'of their great value in economizing fuel, by a personal examination of these engines, ... [here] conducted under the eye of the eminent Engineer ... who virtually brought them to their present state of perfection'.[37]

With no direct access to the P&O Board, Bain found an ally in L'Estrange. As long as the MDs remained highly sceptical of the new engines, Elder could not be invited to P&O's Leadenhall Street Office and in any case would not have attended if they objected to his presence. According to Bain's account, the 'difficulty was overcome only by a visit [of Board members] to Mr Elder at his hotel'. Keeping out of sight meant that Bain escaped the fate of L'Estrange who took the 'grave blame ... for what was deemed an offence'. But as a result of this surreptitious meeting with Elder, a Board Room decision in favour of the introduction of compound engines now received sufficient support to override managerial opponents. The likelihood is that the meeting took place in the first half

[35] Rabson and O'Donoghue (1988: 67–73) (steamers); Jones ([1938]: vol. 2, 166–7) (Suez Canal).
[36] Bain (1873a: 19) (Elder); Harcourt (2006: 162, 191–224) (mail contract).
[37] Bain (1873a: 19); Rabson and O'Donoghue (1988: 73) (*Travancore*).

Figure 18.1. P&O's new-generation steamship *Khedive* (1871) from Caird's of Greenock

of 1868, by which time Sutherland had joined the management staff alongside Bain. Proud of his maritime experience as shipmaster and of his close connections to the late chairman, Bain was not to take kindly to the ambitious former clerk, however well-schooled in office management and politics.[38]

With Elder's yard engaged in PSNC and RMSP contracts, P&O revived their connection with Caird's at Greenock (Chapter 14). In May 1869 P&O received a tender from Caird offering to construct a large steamer for £100,000. Shortly afterward the Board resolved that 'the vessel be fitted with the compound high and low pressure engines in conformity with the recommendations of Messrs Caird & Co., concurred in by Mr Lamb'. That autumn, P&O accepted tenders from Caird for two more large steamers at a price of £107,000 each. Launched on 8 April 1870 and on a passage through the Suez Canal by late October, the *Australia* became the Line's first large compound-engined steamer constructed after the Humphrys era. With a total of eight similar steamers delivered up to the end of 1872 by Caird and Denny alone, P&O's Board Report in November that year characterized the recent era of compounding as having 'caused almost a revolution in Steam Navigation' (Figure 18.1). The Company also contracted with Denny in 1872 for the staged re-engining of four members of the 'old' fleet.[39]

[38] Bain (1873a: 10, 19).

[39] P&O Minutes (7 May, 15 June and 9 July 1869, 8 April and 21 October 1870, and 12 April 1872); Bain (1873a: 19) (quoting the Board Report of 29 November 1872).

Bain regarded himself as the only nautical member of the management staff 'with a great responsibility in dealing with the question of *ships*'. As exemplified by his account of the meetings with Elder, the Board itself contained only two nautical members, one Admiral William Hall and the other the long-serving Captain Engledue. From Bain's perspective, Messrs Allan, Bayley and now Sutherland were former clerks, not captains. And the rest of the Board consisted primarily of mercantile gentlemen with limited nautical experience. But it was the ambitious Sutherland that Bain found especially annoying. 'The building of the *Australia* was delayed for a considerable time in consequence of a controversy between myself and Mr Sutherland', Bain wrote in 1873. 'In fact, it is but right to state here, that nearly all my controversies on nautical subjects, especially regarding reconstruction, and those arising out of my anxiety for increased cargo capacity in new ships, have been with that gentleman'. In particular, he claimed that when Sutherland won the day on an important point, special arrangements with the builder 'for the sole purpose of correcting his error entail[ed] very considerable expense on the Company'.[40]

In January 1872, after serving P&O since the beginning, Allan asked to be relieved of his principal duties. In response to a request from the Board for advice as to a successor, he suggested that Sutherland be nominated as a director and, if elected, appointed an MD in June that year. Disgruntled about Sutherland's elevation, Bain felt himself left playing a supporting role to two former clerks. At a deeper level, however, Bain seemed convinced that the Line's precarious financial position had been brought about by a Board with limited nautical skills. Seeking a Committee of Investigation into the management of the Company, Bain was accused by the Board of 'an act of grave insubordination' for threatening to make public the Company's affairs.[41]

The controversy reached its peak when Bain, having resigned from the management staff and fought with varying degrees of success for access to relevant Company documents, published two pamphlets addressed directly to proprietors. The documents detailed a wide-ranging set of management misdemeanours. Bain's credibility rested not only on his association with Anderson and his past record as a competent P&O master, but especially on his insider's knowledge of the fleet. Central to Bain's attacks on P&O's management was its lack of nautical expertise, leading to the construction of unsuitable steamers and a resistance to the adoption of a credible new marine engineering solution to concerns over competition, canal and economy.[42]

At the AGM in December 1873 the presiding chairman was P. D. Hadow, a barrister who had been a director since 1849 and who hailed from a merchant

[40] Bain (1873a: 10, 21–2).
[41] P&O Minutes (28 April, 12, 23 May and 16 June 1871 (Bain) and 26, 30 January 1872 (Allan)).
[42] Bain (1873a and 1873b).

family in Bombay with involvement in the opium trade.[43] Hadow opened his remarks to the assembled shareholders with the statement that for the first time since the establishment of the Company 'your Directors have been charged with neglect, with mal-administration ... and with so far neglecting your interests that they have wasted millions, it is alleged, of your money in the short space of seven years'. Hadow's tactic was to mock Bain's personal credibility. 'Who is this great individual who takes upon himself to bring these charges against the Board?' the Chairman asked without dignifying him by name or title. 'An ex-employee of this Company – a gentleman whose interest here is limited to £50 [in shares]'. In a speech interspersed with the occasional solitary 'laugh', Hadow referred to the 'bulky pamphlets, ill-written teeming with misrepresentations ... and with malicious insinuations ... Shareholders are often credulous; but it is taxing their credibility to an enormous extent to believe that one tittle of this is true'.[44]

Most proprietors, anxious about a looming loss of confidence, supported dismissal of the claims. Some, referring to *Captain* Bain by both name and title, felt that 'he had been accused somewhat of not being a gentleman' and deserved to be heard. Bain vainly sought to obtain a ballot of all proprietors and not just those present at the AGM. Sutherland then addressed the meeting for two hours with the skills of a consummate and cultured rhetorician. He hailed 'the magnitude of the company's operations, the magnificence of its fleet, and the efficiency of its service ... [and] the harmony and esprit de corps with which this great service has been conducted for thirty years ... (cheers)'. Whatever the shortcomings of the MDs in other respects, he proclaimed, 'they have always held their position as English gentlemen, and endeavoured to sustain the dignity of this company, and maintain not only the confidence of the Proprietors, but of the public at large (question and order, order)'.[45] The son of an Aberdeen painter had learnt something about self-confidence from the English gentlemen of Britain's eastern colonies.

Assuming the role of moral guardian, Sutherland then asked if Captain Bain were setting himself up as 'the only intellectual and moral character in the company – the alpha and omega of all official virtue? (laughter and cheers)'. After addressing Bain's points in considerable detail, he declared that Bain had never 'displayed any of those qualities which the Directors look for in order to promote gentlemen to high office in the company'. Given his fortuitous attachment to the Company's London office, 'some feeling of modesty and gratitude would not have been unbecoming [on Bain's part] (cheers)'. As he reached his conclusion, Sutherland likened his opponent to the Rev Sydney Smith's

[43] Harcourt (2006: 161–2).
[44] 'Minutes of proceedings at the Annual General Meeting', P&O Collection.
[45] 'Minutes of proceedings at the Annual General Meeting', P&O Collection.

Figure 18.2. P&O's Thomas Sutherland, managing director and later chairman

characterization of Lord John Russell – as someone who believed himself capable of building St Peter's in Rome, constructing an entire battle fleet and acting as a surgeon with the skills to remove a kidney stone but undeterred by the prospects of a cathedral collapse, a fleet foundering and the patient dying.[46]

Finally, Sutherland downgraded Bain to artisanal status. Bain appeared 'to be one of those extraordinary characters [one of whom] ... is described in Shakespeare's "Midsummer Night's Dream", who was called "Bottom"', he told his audience. Bottom, called upon to act a part in a play, 'was not an actor, he was a weaver (a laugh); he had such an idea of his own gigantic abilities that he thought he could not only act the character assigned to him, but every other in the play (a laugh)'.[47] Predictably, Bain's resolutions received barely nine votes against an unquantified show of hands. He persisted in his campaign for some time, but Sutherland's character assassination ensured that Captain Bain would henceforth be relegated to (at most) supercilious references in P&O histories.[48]

Sutherland emerged unassailable from under this threatening cloud to lead the Company as chairman from 1880 until his retirement in 1914 (Figure 18.2).

[46] 'Minutes of proceedings at the Annual General Meeting', P&O Collection.
[47] 'Minutes of proceedings at the Annual General Meeting', P&O Collection.
[48] For example, Jones ([1938]: vol. 2, 168–9) ('Bain was a vain, foolish man'); Cable (1937: 175–7) ('[Bain] had held minor commands of two of the smallest ships ... in the fleet').

During that long reign he found special satisfaction in his friendships with James Caird and with the Greenock shipbuilding firm that Caird's sons took over after their father's death in 1888. In the same period, Sutherland served as Liberal (later Liberal Unionist) MP for Greenock between 1884 and 1900. Beginning with the *Australia*, P&O took delivery of some eighty large steamships, many of them serving as the crack vessels of the fleet, from the Caird's yard between 1870 and the takeover of the firm by Harland & Wolff during the Great War.[49]

When PSNC demonstrated the power of Elder's compound steam engines on their new Liverpool to Valparaiso ocean line, RMSP finally moved to order their initial trio of similar vessels from the Glasgow yard. Setting aside its troubled experience with double-cylinder engines, P&O re-established its links with Greenock shipbuilder James Caird and, under the formidable regime of Thomas Sutherland, maintained that link for more than four decades.

[49] Jones ([1938]: vol. 2, 173–4). Rabson and O'Donoghue (1988: 77–83), summarizes the post-1870 history.

Epilogue: 'The Sovereignty of the Seas'

The Maritime System Builders

> This could have occurred nowhere but in England ... the sea entering into the life of most men, and the men knowing something or everything about the sea, in the way of amusement, of travel, or of bread-winning ... [Seated around the table with others], the lawyer – a fine crusted Tory, High Churchman, the best of old fellows, the soul of honour – had been chief officer in the P&O service in the good old days when mail-boats were square-rigged at least on two masts, and used to come down the China Sea before a fair monsoon with stun'-sails set alow and aloft. We all began life in the merchant service.
>
> *Conrad's narrator sets the scene for 'Marlow's tale' told in* Youth *(1899)*[1]

Published at the very end of the nineteenth century, Conrad's short novel *Youth* follows very closely the author's personal experience in the early 1880s of a voyage from the Tyne to Bangkok as second mate aboard the ageing wooden barque *Palestine*. As with so many similar sailing vessels the coal laden vessel never reached her destination. Fighting a losing battle against fire in the cargo, the officers and crew abandoned their home in waters south-east of Singapore.[2] With characteristic irony, Conrad's opening scene of former seafarers exchanging reminiscences of their youth includes the above description of the days of P&O mail steamers running down the South China Sea from Hong Kong to Singapore before a 'fair monsoon' with not only the standard square-sails, but also the clipper-like studding sails set on extensions to the yardarms. The juxtaposition of coal and wind as complementary and yet contesting powers of nature would not have been lost on early readers of the story.

At the time when Lindsay's early volumes of his *History* appeared in print, total British steam tonnage had accelerated from about 190,000 tons in 1850 to 2 million tons in 1874. Committed to a divinely sanctioned faith in universal progress of which the advance of steam power was a decisive demonstration, Lindsay did not highlight the evidence that British sailing ship tonnage had

[1] Conrad (1984 [1899]: 131).

[2] Villiers (2006: 44–51).

continued to rise over the same period from 4 million to almost 5.5 million tons.[3] At the beginning of the final quarter of the nineteenth century, then, sailing ships accounted for nearly two-thirds of the British merchant fleet in terms of tonnage and far more in terms of ship numbers. Imperial, and indeed global, trade was overwhelmingly being driven not by coal but by wind.

Delivering his presidential address to the BAAS's Mathematical and Physical Section at York in 1881, Professor William Thomson (Chapter 16) focused 'On the sources of energy in nature available to man for the production of mechanical effect [useful work]'. Of the sources derived from the sun's heat, he identified wind and coal stores as the most significant for motive power. '[E]ven in the present days of steam ascendancy', he declared, 'old-fashioned Wind still supplies a large part of the energy used by man'. His claim rested on the evidence of the British shipping register showing that of some 40,000 vessels listed, about 30,000 were sailing vessels. He further observed that the 10,000 steamers on the register drew a very considerable proportion of the 'whole horse-power taken from coal annually in the whole world at the present time'. But far from celebrating an inevitable triumph of coal over wind, or wistfully regretting a decline of sail due to the advance of steam, he instead highlighted 'the fact of a lamentable decadence of wind-power'. Echoing Jevons' analysis in *The Coal Question* (1865) in its global implications (Chapter 17), Thomson accepted that the 'subterranean coal-stores of the world are becoming exhausted surely, and not slowly, and the price of coal is upward bound ... on the whole'. Thus, he concluded, in due time 'wind-motors in some form will again be in the ascendant, and that wind will do man's mechanical work on land at least in proportion comparable to its present doing of work at sea'.[4]

If the winds of heaven resisted quantitative prediction,[5] coal too appeared to defy rational calculation. 'Coal is a thing which it is impossible to measure exactly and in supplying it either you give a fraction too much or if you are an astute merchant you give too little', P&O's Thomas Sutherland insisted. The elusive character his Company's prime source of motive power, he recalled, had almost cut short his own upward career trajectory (Chapter 18). During Britain's protracted conflicts with China, P&O's Hong Kong station supplied the Admiralty with coal over a number of years up to 1860. At the close of the arrangement, Sutherland explained, the Company found that its stock was several thousand tons short of the recorded tally. The unnamed superintendent in charge of the coaling station declared in his defence that 'he had not sold any beyond what he had accounted for'. Much annoyed by the discrepancy but unwilling to censure their senior official in the colony, the P&O Board made

[3] Lindsay (1874–6: vol. 3, 618–19).
[4] Thomson (1889–94: vol. 2, 440–2); Jevons (1866).
[5] Smith (2013c) (Conrad's *Typhoon*).

the then-junior Sutherland its scapegoat, informing him sternly that he would not be returning to Hong Kong after his leave. Eventually persuaded that the anonymous superintendent was 'a perfectly honest man but not an astute one' who had unwittingly 'giving a fraction too much' over time, the Company reinstated Sutherland.[6]

Coal was the most visible manifestation of a steamship's action. Long before a steamer became perceptible, black smoke on the horizon marked its position and long after its departure from view, trails of dark clouds swirled in its wake. Aboard ship, not even the most privileged passengers could escape the all-pervading coal dust and soot. In Company offices and scientific societies, engineering workshops and shipyards, the lives of human agents were devoted to the commodity and its elusive qualities. Through experimental tests, owners came to recognize the great inequalities in the calorific value of coal from different geological regions and even between specific pits (Chapter 9). The variations mattered, as when the Holt brothers warned Captain Middleton before his first voyage to China of the 50 per cent reduction in the work value of Labuan (Borneo) coal compared to that of Welsh steam coal.[7]

The virtues of coal, and how to realize them, long remained matters for dispute. Through their attempts to measure the indicated horsepower of their engine designs in practical working, marine engineers divided over the reliability of indicator measurements and the trustworthiness of the instruments and their designers. McNaught's authority, for example, derived from his track record as a first-rate practical engineer with tangible results to his credit. He therefore expected his clients to take for granted the authority vested in his indicator as a tool that yielded value-free, objective knowledge of steamship performance. It was an expectation with practical consequences for Randolph, Elder's first PSNC contracts (Chapter 16).[8]

The more astute marine engineers, however, recognized that questions of the measurement of engine economy were frequently addressed without an appreciation of the interconnectedness of the steamship as a system. After James Napier read a paper to the IESS in 1864 on superheated steam tested aboard east coast vessels, John Elder took the view that 'a good deal of the acceleration in speed and saving of fuel had arisen from *other* causes as well as from superheating'. He politely observed that had the audience received information on 'the precise alterations in the [paddle] wheels and boilers', it would have aided the arrival 'at a more correct conclusion'. In one case, for example, a 600-ton steamer had been fitted initially with very small diameter wheels which, when the vessel was deeply laden, would detract from its propulsion. He therefore

[6] Quoted in Jones (1935–38: vol. 2, 158–9).
[7] Alfred and Philip Holt to Captain Isaac Middleton, 14 April 1866, AHP, LRO.
[8] I thank Simon Schaffer for the last point.

specifically questioned how much of the increased speed and economy was due to the new and larger wheels, how much to the enhanced heating surface and how much to the superheating? He 'believed that all three had favourably affected the result'. He furthermore suggested that the quality of the coals would affect the calculations, especially as coals on the east coast were well known to be superior to those on the west. Shortening of the length of the fire grate would also produce more effective combustion.[9] Concomitantly, doubts as to indicators resurfaced when Professor Thomson raised with James Napier the question of what precisely these instruments measured (work done by the steam on the piston) and how their results related to the question of the vessel's 'real economy' in terms of the work done *by* the steamer in traversing the seas (Chapter 16).

Screw ships such as P&O's *Bengal* appeared to have the edge over paddle-steamers such as RMSP's *Orinoco* in terms of fuel economy (Chapter 14). But that economy often relied on their much lower-powered engines combined with sailing qualities uninhibited by massive sponsons, the *Shannon*'s impressive Atlantic crossing under sail alone notwithstanding (Chapter 18). Mail lines also had to take into account their commitment to the strict schedules demanded by their Government contracts. But even Cunard's mainline deployment of powerful, heavy-consuming paddlers until the 1860s was not without sail assistance.

Passengers, moreover, continued to dislike the screw steamers' unerring propensity to uncomfortable rolling (see Introduction). As Murray reminded the INA in 1865, it was 'well known that the paddle steamers of the Peninsular and Oriental Company are much more popular than the screws, and that Indian passengers will sometimes wait months for the chance of a passage by one of them'. The paddles, he averred 'have a great steadying effect upon the ship'. Thus while in light breezes the screw would be more economical, 'with heavy seas and head winds the superiority of the paddles is undoubted. ... If these [screw] vessels were put upon an Atlantic or Channel passage, they could not possibly keep time'.[10] It is nevertheless clear from the historical evidence that sails set aboard screw steamers had a variety of functions that included a significant steadying effect in a beam sea, as well as a habitual reduction in coal consumption during an ocean voyage. Although on occasions deployed in the event of a major engine failure, a fractured shaft or a lost propeller, sails were not intended to complete a mail voyage too long to be achieved by means of steam alone.[11]

Randolph and Elder rested promises for the superior character of their engine on winning any 'trial-by-space', both in terms of distance run and internal space

[9] Napier (1864: 91–2) (Elder's remarks). My italics.
[10] Murray (1865: 163).
[11] Compare Mendonca (2013: 1730–5).

saved. Both the commercial press and the more specialist engineering magazines were thus soon pointing to that 50 per cent reduction in fuel consumption over the 'best' simple engines to about 3 lb weight of coal per horsepower per hour. At the same time, the practical measurements to underwrite those sorts of figures remained not only difficult, but often highly contentious (Chapter 16). Steam power nevertheless offered the potential to reorder geographical space in Victorian commercial and public cultures profoundly imbued with rhetoric of 'experiment', 'progress', 'advancement' and 'improvement'. But, to use Henri Lefebvre's conceptualization, the new technologies would not survive as mere projects, ideas or visions.[12] They had, in any 'trial-by-space', to inscribe themselves in space, make themselves durable, and resist not only rival system builders, but also (in Captain Andrew Henderson's subversive words) to struggle against those winds and seas 'whose violence and fury so often overturn theories and wreck mechanical science' (Chapter 15).

Our period closes with Prime Minister Disraeli's famous acquisition of the Suez Canal in 1874. As a principal connecting link in the geographical systems of Empire, the Canal was now under the control of the country whose steamship fleets contributed most to its revenues. Eight years earlier in a re-election speech, Disraeli sought to evoke in the minds of his electors the vision that merchant steamers, now the most conspicuous and trusted servants of Empire, linked the British Isles to the ports of North America, the Caribbean, South America, Asia and Australia. And, he might have added, even the greatest rivers of the world, from the Amazon to the Yangtze, were now served by steamers constructed on the Clyde, the Tyne, the Mersey and the Thames. 'England is no longer a mere European Power', he declared with feeling. It had become rather 'the metropolis of a great maritime empire, extending to the bounds of the farthest oceans'. It was a view already expressed on the other side of the Atlantic. In the words of the *NAR*,

> [w]hen these three great lines [Cunard, P&O and RMSP] were organized, Great Britain had ... more nearly than ever before attained to the sovereignty of the seas; for the Atlantic, the Mediterranean ... and the waters of the East had then been taken possession of for the transmission of her mails and for the accommodation of her commerce, as though they had all been included within her rightful domain.[13]

[12] Lefebvre (1991: 416–17).
[13] [Anon.] (1864: 487–8) (*NAR*); Lindsay (1874–6: vol. 3, 558–9); Harcourt (2006: 214) (Disraeli).

Bibliography

Principal Archival Sources

Liverpool

LRO: Liverpool Record Office
  AHP: Alfred Holt (and family) Papers
  PP: Peacock Papers
MMM: Merseyside Maritime Museum
  OA: Ocean Archive (Ocean Steamship Co.)
  PSNC: Pacific Steam Navigation Co. Archive
LUL: Sidney Jones Library, University of Liverpool
  CP: Cunard Papers

Glasgow

GUA: Glasgow University Archives
  NP: Napier Papers
  SP: John Scott Papers
GMT: Glasgow Museum of Transport (Riverside Museum)
  NC: Napier Collection
ML: Mitchell Library
  IC: Inverclyde Collection
ULG: Special Collections, Glasgow University Library
  KC: Kelvin Collection

London

NMM: Caird Library, National Maritime Museum, Greenwich
  P&O: Peninsular & Oriental Steam Navigation Co. Collection
  RMSP: Royal Mail Steam Packet Co./Royal Mail Line Collection
  WSL: William Schaw Lindsay Collection
UCL: University College London Archives
  RMSP: Royal Mail Steam Packet Co./Royal Mail Line Collection

Newcastle

TWA: Tyne & Wear Archives
HL: Hawthorn Leslie Collection

Edinburgh

ULE: New College Library, University of Edinburgh
TC: Thomas Chalmers correspondence

Halifax, NS

NSA: Nova Scotia Archives
CP: Cunard Papers

Printed Sources

Parliamentary Reports

Hansard's Parliamentary Debates
SC(1831): *Report from the Select Committee on Steam Navigation.*
SC(1834a): *Report from the Select Committee on Steam Navigation to India.*
SC(1834b): *Report from the Select Committee on Steam Navigation.*
PO(1836): *The Fourth Report of the Commissioners Appointed to Inquire into the Management of the Post-Office Department.*
HCO(1840): *West India Mails Contract* (5 June 1840).
SC(1846): *Report from the Select Committee on Halifax and Boston Mails.*
SC(1849): *Report from the Select Committee on Contract Packet Service.*
HCO(1851): *Iron Steam Ships.*
AR(1852): *Report of the Officers Appointed to Conduct the Official Inquiry into the loss of the 'Amazon' (under the Steam Navigation Act, 1851).*
SC(1860): *First Report from the Select Committee on Packet and Telegraphic Contracts.*

Newspapers and Periodicals

Acadian Recorder
American Journal of Science and Arts (*Silliman's Journal*)
Annals of Science

BAAS Reports
Bristol Mirror
BJHS
Bulletin (Liverpool Nautical Research Society)
Coriolis: Interdisciplinary Journal of Maritime Studies
Cultural and Social History
Daily News
Edinburgh Magazine
Engineer, The
Engineering
Glasgow Herald
Harper's New Monthly Magazine
International Journal of Maritime History
Illustrated London News
Journal for Maritime Research
Journal of Transport History
Liverpool Albion
Liverpool Courier
Liverpool Journal
Liverpool Mercury
Londonderry Sentinel
Marine Engineer/Marine Engineer and Naval Architect, The
Mariner's Mirror, The
Mechanics Magazine (MM)
Morning Chronicle
Nautical Magazine (NM)
New York Daily Tribune
Newcastle Daily Chronicle
Newcastle Leader
North American Review (*NAR*)
North British Review
Novascotian
Penny Magazine
Porcupine, The
Proceedings of the Institution of Mechanical Engineers
Proceedings of the [Liverpool] Literary and Philosophical Society
Railway Magazine (RM) (Herapath's *Journal*)
Scotsman, The
Shipbuilder and Marine Engine Builder
Shipping & Mercantile Gazette
Technology and Culture
Times, The

Transactions of the Institution of Engineers and Shipbuilders in Scotland (TIESS)
Transactions of the Institution of Naval Architects (TINA)
Transactions of the Liverpool Polytechnic Society
Victorian Studies

Books, Pamphlets and Periodical Articles

Abbott, Jacob. 1851. 'The Novelty Works'. *Harper's New Monthly Magazine* 2: 721–34.

Abbott, John. 1852. 'Ocean life'. *Harper's New Monthly Magazine* 5: 61–6.

[Alberdi, J. B.]. 1920–22. 'Founder of the P.S.N.C. Life of William the Conqueror'. *Sea Breezes, 2* (1920–1): 330–2, 367–9; *3*(1921–2): 19–20, 57–60, 98–9, 135–6, 170–1, 212–13, 254–5, 295–7, 326–8. Translated for Sea Breezes (PSNC's magazine) from J. B. Alberdi, *La vida y los trabajos industriales de William Wheelwright en la America del Sud* (Paris: Libreria de Garnier Hermanos, 1876).

Albion, Robert Greenhalgh. 1938. *Square-Riggers on Schedule: The New York Sailing Packets to England, France and the Cotton Ports.* Princeton, NJ: Princeton University Press.

1970. *The Rise of New York Port [1815–1860].* Newton Abbot: David & Charles. First published 1939.

Alborn, Timothy L. 1998. *Conceiving Companies. Joint-Stock Politics in Victorian England.* London and New York, NY: Routledge.

Anderson, Arthur. 1840. *Steam Communication with India. A Letter to the Directors of the Projected East Indian Steam Navigation Company, Containing a Practical Exposition of the Prospects of that Proposed Undertaking, and of the Real State of the Question of Steam Communications with India.* London: Smith, Elder.

[Anon.]. 1821. 'Iron boat'. *American Journal of Science 3*: 371–2.

[Anon.]. 1822. 'An iron steam boat'. *American Journal of Science* 5: 396–7.

[Anon.]. 1825. *Narrative of the Loss of the Comet Steam-packet ... on Friday 21st October 1825.* Edinburgh: Hunter; London: Duncan.

[Anon.]. 1834. 'Iron steam-boats'. *Nautical Magazine 3*: 626–7.

[Anon.]. 1837a. 'Steam Navigation at the Late Meeting of the British Association'. *Nautical Magazine 6*: 691–9.

[Anon.]. 1837b. [Logs of the *Atalanta* and *Berenice*]. *Nautical Magazine 6*: 327, 498–507, 596–606, 646–54, 745–52.

[Anon.] 1837c. 'Steam navigation in the Pacific'. *Nautical Magazine 6*: 24–31, 97–100, 255–6.

[Anon.]. 1838. 'Expenditure of coals in the Atalanta and Berenice'. *Nautical Magazine 7*: 93–6.

[Anon.]. 1840. 'Naval chronicle [remarks on Wye Williams *et al.*]'. *Nautical Magazine 9*: 850–2.

[Anon.] 1842 'West India Mail Steam Packet Company'. *Nautical Magazine* 11: 419–20.

[Anon.]. 1843a. 'A day at the Clyde steam-boat works'. *Penny Magazine.* Supplement to September number, 377–84.

[Anon.]. 1843b. 'Charles Dickens. Notes for general circulation'. *North American Review 56*: 212–39.

[Anon.]. 1844. *Trial of Pedro de Zulueta, Jun., on a Charge of Slave Trading, with an Address to the Merchants, Manufacturers and Traders of Great Britain by Pedro de Zulueta, Jun., Esq., and Documents Illustrative of the Case*. London: Wood.

[Anon.]. 1846. 'The Great Britain'. *Nautical Magazine 15*: 616.

[Anon.]. 1847. 'Dr Chalmers'. *North British Review 7*: 560–74.

[Anon.]. 1849. 'The Modern Baal; or, the Railway God'. *The Scottish Christian Journal: Conducted by Ministers and Members of the United Presbyterian Church 1*: 156–8.

[Anon.]. 1850. 'Steam-bridge of the Atlantic'. *Harper's New Monthly Magazine 1*: 411–14 (from Chambers's Edinburgh Journal).

[Anon.]. 1851. 'Atlantic packet steamers'. *Nautical Magazine 20*: 99–100.

[Anon.]. 1852a. *Narrative of the Loss of the Amazon, Steam-vessel, on Sunday Morning, Jan. 4, 1852*. London: Richard Bentley.

[Anon.]. 1852b. 'Destruction of the Royal Mail new steamer "Amazon," and the loss of 115 lives'. *Nautical Magazine 21*: 89–108.

[Anon.]. 1852c. 'Sunderland ships; or, an historical sketch of the rise and progress of shipbuilding on the Wear'. *Nautical Magazine 21*: 581–93.

[Anon.]. 1854. 'The Screw Steamer "Brandon"'. *Nautical Magazine 23*: 507–09.

[Anon.]. 1860. 'Humphrys' marine steam engines'. *The Engineer 9*: 73.

[Anon.]. 1864. 'Ocean steam navigation'. *North American Review 99*: 483–522.

[Anon.]. 1866. 'The Late Mr David Elder'. *Engineering 1*: 103.

[Anon.]. 1867. 'The late Mr Edward Humphrys'. *Engineering 3*: 528.

[Anon.]. 1871. [Obituary of Captain W.H. Walker]. *Nautical Magazine 40*: 196.

[Anon.] 1875. 'Memoir of the late John Grantham'. *Transactions of the Institution of Naval Architects 16*: 270–2.

[Anon.]. 1876. *Memorial of Old College Church (Blackfriars'). Glasgow*. Glasgow: Thomas Murray.

[Anon.]. 1886. *Memoirs and Portraits of One Hundred Glasgow Men*. 2 vols. Glasgow: James Maclehose.

[Anon.]. 1889–90. 'The late Robert Duncan, shipbuilder, Port Glasgow'. *Marine Engineer 11*: 192–3.

[Anon.]. 1891. *The Elder Park Govan. An Account of the Gift of the Elder Park and of the Erection and Unveiling of the Statue of John Elder*. Glasgow: James Maclehose.

[Anon.]. 1892–93. 'Obituary of Laurence Hill [shipbuilder]'. *Transactions of the Institution of Engineers and Shipbuilders in Scotland 36*: 319–20.

[Anon.]. 1905–06. 'Obituary. Robert Mansel, Shipbuilder'. *Marine Engineer 27*: 452.

[Anon.]. 1909. *Some Account of the Works of Palmers Shipbuilding & Iron Company Limited*. 4th edn. Newcastle: Franklin.

[Anon.]. 1913–14. 'James Howden'. *Marine Engineer 36*: 165.

[Anon.]. 1914. *The Edinburgh Academy Register. A Record of All Those Who Have Entered the School since Its Foundation in 1824*. Edinburgh: Constable.

Armonsky, Ruth and Susie Cox. 2012. *P&O. Across the Oceans Across the Years. A Pictorial Voyage*. Woodbridge: Antique Collectors' Club.

Atkins, Thomas. 1852. *The Claims of Seaman Advocated and the Duties of Seaman Enforced, a Sermon, on the Occasion of the Loss of the Amazon, R.M.S.S.* Southampton: G. L. Marshall.

Babcock, F. Lawrence. 1931. *Spanning the Atlantic*. New York, NY: Alfred A. Knopf.

Bain, George. 1873a. *Letter and Statement for the Proprietors of the Peninsular & Oriental Steam Navigation Co.* London: Yates & Alexander (printer).

1873b. *Investigation. Letter and Statement for the Proprietors of the Peninsular & Oriental Steam Navigation Co.* London: Yates & Alexander (printer).

Baird, Christina. 2007. *Liverpool China Traders.* Bern: Peter Lang.

Banbury, Philip. 1971. *Shipbuilders of the Thames and Medway.* Newton Abbot: David & Charles.

Bayly, C. A. 1989. *Imperial Meridian. The British Empire and the World, 1780–1830.* London: Longman.

Bell, David. 1912. *David Napier Engineer 1790–1869. An Autobiographical Sketch.* Glasgow: Maclehose.

Bellamy, Martin. 2012. 'A ludicrous travesty? James R. Napier and the *Lancefield*'. *Mariner's Mirror 98*:2: 161–77.

Black, R. D. Collison (ed.). 1973. *Papers and Correspondence of William Stanley Jevons. Volume II. Correspondence 1850–1862.* Clifton, NJ: Kelly.

[Blood, William]. 1852a. *'The Amazon.' A Sermon Preached at St Andrew's Church, Plymouth.* London: Aylott and Jones.

1852b. *A Description of the Destruction, by Fire, of 'The Amazon', with Reflections, as Delivered by the Rev. William Blood, One of the Survivors, in All Saints Church, Southampton.* London: Hall.

1852c. *The Gospel the Power of God: A Sermon, Preached on the Destruction, by Fire, of 'The Amazon'.* London and Bristol: Hamilton et al.

Bonsor, N. R. P. 1975–80. *North Atlantic Seaway.* 5 vols. Newton Abbot: David & Charles; Jersey: Brookside.

Booth, Henry. 1844–46. 'On the prospects of steam navigation &c'. *Transactions of the Liverpool Polytechnic Society* 2: 24–31.

Bourne, John. 1852. *A Treatise on the Screw Propeller, with Various Suggestions of Improvement.* London: Longman.

Bowen, F. C. nd. *A Century of Atlantic Travel 1830–1930.* London: Sampson Low.

Boyce, Gordon H. 1995. *Information, Mediation and Institutional Development. The Rise of Large-Scale Enterprise in British Shipping, 1870–1919.* Manchester and New York, NY: Manchester University Press.

Brooke, John H. 1977: 'Natural theology and the plurality of worlds: Observations on the Brewster-Whewell debate'. *Annals of Science 34*: 221–86.

1991: *Science and Religion. Some Historical Perspectives.* Cambridge: Cambridge University Press.

Brown, Alexander C. 1961. *Women and Children Last. The Sinking of the Liner Arctic with the Loss of More than 300.* London: Muller.

Brown, Callum G. 1997. *Religion and Society in Scotland since 1707.* Edinburgh: Edinburgh University Press.

Bruce, A. B. 1888. *The Life of William Denny, Shipbuilder, Dumbarton.* London: Hodder & Stoughton.

Buchanan, Angus. 2002. *Brunel. The Life and Times of Isambard Kingdom Brunel.* London: Hambledon & London.

Burgess, D. R. 2016. *Engines of Empire. Steamships and the Victorian Imagination.* Stanford: Stanford University Press.

Burns, John. 1828. *Principles of Christian Philosophy.* 2nd edn. London: Longman.

Bushell, T. A. 1939. *'Royal Mail' A Centenary History of the Royal Mail Line 1839–1939*. London: Trade & Travel Publications.
Cable, Boyd. 1937. *A Hundred Year History of the P. & O. Peninsular and Oriental Steam Navigation Company 1837–1937*. London: Nicholson and Watson.
Cain, P. J. and A. G. Hopkins. 2002. *British Imperialism, 1688–2000*. First published 1993. Harlow: Pearson.
Caird, John. 1899. *The Fundamental Ideas of Christianity*. 2 vols. Glasgow: Maclehose.
[Cameron, John B.] 1856. *The Wreck of the West India Steam-Ship Tweed. By an Eye Witness*. London: SPCK.
Cantrell, John. 2002. 'Henry Maudslay'. In John Cantrell & Gillian Cookson (eds), *Henry Maudslay & the Pioneers of the Machine Age*, pp. 18–38. Brimscombe Port Stroud: Tempus.
Cantrell, John and Gillian Cookson (eds). 2002. *Henry Maudslay & the Pioneers of the Machine Age*. Brimscombe Port Stroud: Tempus.
Cardwell, Donald and R. L. Hills. 'Thermodynamics and practical engineering in the nineteenth century'. *History of Technology 1*: 1–20.
Carey, Hilary M. 2011. *God's Empire. Religion and Colonialism in the British World, c.1801–1908*. Cambridge: Cambridge University Press.
Carnot, Sadi. 1986. *Reflexions on the Motive Power of Fire. A Critical Edition with the Surviving Scientific Manuscripts*. Trans. and ed. Robert Fox. Manchester and New York, NY: Manchester University Press. First published 1824.
Chadwick, F. E. *et al.* 1891. *Ocean Steamships. A Popular Account of Their Construction, Development, Management and Appliances*. Six co-authors. New York, NY: Charles Scribner's Sons.
Chalmers, Thomas. 1820. *Application of Christianity to the Commerce and Ordinary Affairs of Life. In a Series of Discourses*. Glasgow: Chalmers & Collins.
1836–42. *The Works of Thomas Chalmers*. 25 vols. Glasgow: Collins.
Chandler, George. 1960. *Liverpool Shipping. A Short History*. London: Phoenix House.
Chisholm, J. A. (ed.). 1909. *The Speeches and Public Letters of Joseph Howe*. 2 vols. Halifax: Chronicle Publishing.
Chittick, V. L. O. 1924. *Thomas Chandler Haliburton ('Sam Slick') A Study in Provincial Toryism*. New York, NY: Columbia University Press.
Clark, Michael. 2012. 'William Schaw Lindsay: The largest shipowner in the world?' *History Scotland 12*: 44–9.
Clarke, David J. 2008. 'The development of a pioneering steamship line: William Wheelwright and the origins of the Pacific Steam Navigation Company'. *International Journal of Maritime History 20*: 221–50.
Clarke, J. F. [1979]. *Power on Land & Sea. 160 Years of Industrial Enterprise on Tyneside. A History of R.&W. Hawthorn Leslie & Co. Ltd. Engineers and Shipbuilder*. Newcastle: Hawthorn Leslie.
Cleland, James. 1837. *Statistical Facts Descriptive of the Former and Present State of Glasgow*. Glasgow: Bell & Bain
1840. *Statistical Facts Descriptive of the Former and Present State of Glasgow*. Glasgow: Bell & Bain.
Collins, Timothy. 2002. *Transatlantic Triumph & Heroic Failure. The Galway Line*. Cork: Collins Press.
Conrad, Joseph. 1906. *Mirror of the Sea*. London: Dent.

1950. *The Rescue. A Romance of the Shallows*. Harmondsworth: Penguin.

1984. *Sea Stories*. London: Granada. First published 1899.

Corlett, Ewan. 1980. *The Iron Ship. The History and Significance of Brunel's Great Britain*. First published 1975. Bradford-on-Avon: Moonraker Press.

Cottrell, O. L. 1992. 'Liverpool shipowners, the Mediterranean and the transition from sail to steam during the mid-nineteenth century'. In L. R., Fischer (ed.), *From Wheelhouse to Counting House: Essays in Maritime and Economic History in Honour of Professor Peter Neville Davies*, pp. 153–202. St John's: International Maritime History Association.

Courtney, Stephen and Crosbie Smith. 2013. '"Or vast fiery cross, on the banner of morn": Reading the Royal Mail Steam Packet Company's shipwrecks'. *Journal for Maritime Research 15*: 183–203.

Cutler, Carl C. 1984. *Greyhounds of the Sea. The Story of the American Clipper Ship*. 3rd edn. Wellingborough: Patrick Stephens. First published 1930.

Daunton, Martin J. 1977. *Coal Metropolis Cardiff 1870–1914*. Leicester: Leicester University Press.

1985. *Royal Mail. The Post Office Since 1840*. London: Athlone.

Dawson, Charles. 2009. 'Fossets'. *The Bulletin (LNRS) 53*: 35–41.

2012. 'John Grantham, pioneer marine engineer and naval architect'. *The Bulletin (LNRS) 56*: 22–7.

Delap, Lucy. 2006. '"Thus does man prove his fitness to be the master of things:" Shipwrecks, chivalry and masculinities in nineteenth- and twentieth-century Britain'. *Cultural and Social History 3*: 45–74.

Dickens, Charles. 1842. *American Notes for General Circulation*. New York, NY: Harper.

Duckworth, C. L. D. and G. E. Langmuir. 1977. *Clyde and Other Coastal Steamers*. 2nd (revised) edn. First published 1939. Prescot: T. Stephenson.

Dugan, James. 1953. *The Great Iron Ship*. London: Hamish Hamilton.

Duncan, J. H. 2002. *A Brief History of the Church in the Parish of Row*. Dumbarton: np.

Elder, John. 1859. 'Engines of the Callao, Lima, and Bogota steamships'. *The Engineer 8*: 291.

Emmerson, George S. 1977. *John Scott Russell. A Great Victorian Engineer and Naval Architect*. London: John Murray.

nd. *The Greatest Iron Ship. S.S. Great Eastern*. Newton Abbot: David & Charles.

Falkus, Malcolm. 1990. *The Blue Funnel Legend. A History of the Ocean Steam Ship Company 1865–1973*. Basingstoke: Macmillan.

Feirreiro, Larrie D. 2007. *Ships and Science. The Birth of Naval Architecture in the Scientific Revolution, 1600–1800*. Cambridge, MA and London: MIT Press.

Fifer, J. Valerie. 1998. *William Wheelwright. Steamship and Railroad Pioneer. Early Yankee Enterprise in the Development of South America*. Newburyport, MA: Historical Society of Old Newbury.

Fletcher, R. A. 1910. *Steam-Ships. The Story of Their Development to the Present Day*, London: Sidgwick & Jackson.

Forbes, R. B. 1849. 'The loss of the Charles Bartlett'. *Nautical Magazine 18*: 436–9.

Forrester, Robert E. 2014. *British Mail Steamers to South America, 1851–1965. A History of the Royal Mail Steam Packet Company and Royal Mail Lines*. Farnham: Ashgate.

Fox, Robert. 1986. 'Introduction'. In Robert Fox (ed.), *Sadi Carnot. Reflexions on the Motive Power of Fire. A Critical Edition*, pp. 1–57. Manchester: Manchester University Press.

Fox, Stephen. 2003. *The Ocean Railway. Isambard Kingdom Brunel, Samuel Cunard and the Revolutionary World of the Great Atlantic Steamships*. London: Harper Collins.

Fry, Henry. 1896. *The History of North Atlantic Steam Navigation with Some Account of Early Ships and Shipowners*. London: Sampson Low.

Gannett, Ezra S. 1840. *The Arrival of the Britannia. A Sermon Delivered in the Federal Street Meeting-House, in Boston, July 19, 1840*. Boston: Dowe.

Gillin, Edward. 2012. '"Diligent in business, serving the Lord": John Burns, evangelicalism and Cunard's culture of speed, 1878–1901'. *Journal for Maritime Research 14*: 15–30.

2015. 'Prophets of progress. Authority in the scientific projections and religious realizations of the *Great Eastern* steamship'. *Technology and Culture 56*: 928–56.

Grant, Kay. 1967. *Samuel Cunard. Pioneer of the Atlantic Steamship*. London; New York, NY and Toronto: Abelard-Schuman.

Grantham, John. 1842. *Iron as a Material for Ship-Building; Being a Communication to the Polytechnic Society of Liverpool*. London, Liverpool, Glasgow and Dublin: Simpkin, Marshall et al.

1866. 'Some particulars of the iron screw yacht "Vesta"'. *Transactions of the Institution of Naval Architects 7*: 55–6.

Green, Edwin and Michael Moss. 1982. *A Business of National Importance. The Royal Mail Shipping Group, 1902–1937*. London and New York, NY: Methuen.

Griffiths, Denis, Andrew Lambert and Fred Walker. 1999. *Brunel's Ships*. London: NMM.

[Haliburton, Thomas]. 1840. *The Letter-Bag of the Great Western; or, Life in a Steamer*. Philadelphia: Lea & Blanchard.

Hamblin, Katy. 2011. '"*Men of Brain and Brawn and Guts": The Professionalization of Marine Engineering in Britain, France and Germany 1830–1914*'. University of Exeter PhD.

Hannan, John. 1866. 'Description of an improved steam engine indicator'. *Transactions of the Institution of Engineers in Scotland 9*: 75–80.

Harcourt, Freda. 1992. 'Charles Wye Williams and Irish steam shipping'. *Journal of Transport History 13*: 141–62.

2006. *Flagships of Imperialism. The P&O Company and the Politics of Empire from Its Origins to 1867*. Manchester: Manchester University Press.

Harvey, W. J. and P. J. Telford. 2002. *The Clyde Shipping Company Glasgow 1815–2000*. Np: World Ship Society.

Haws, Duncan. 1978. *Merchant Fleets. The Ships of the P&O, Orient and Blue Anchor Lines*. Cambridge: Patrick Stephens.

1982. *Merchant Fleets. Royal Mail Line & Nelson Line*. Crowborough: TCL.

1986. *Merchant Fleets. Pacific Steam Navigation Company (P.S.N.C.)*. Burwash: TCL.

Headrick, Daniel R. 1981. *The Tools of Empire. Technology and European Imperialism in the Nineteenth Century*. New York, NY and Oxford: Oxford University Press.

Heaton, P. M. 1986. *Lamport and Holt*. Newport, Gwent: Starling Press.

Henderson, Andrew. 1854. 'On ocean steamers and clipper ships'. *BAAS Report 24*: 152–6.

Hills, Richard L. 1989. *Power from Steam. A History of the Stationary Steam Engine*. Cambridge: Cambridge University Press.

Hilton, Boyd. 1991. *The Age of Atonement. The Influence of Evangelicalism on Social and Economic Thought 1785–1865*. First published 1988. Oxford: Oxford University Press.

Hodder, Edwin. 1890. *Sir George Burns, Bart. His Times and Friends*. London: Hodder & Stoughton.

Hollett, David. 2008. *More Precious than Gold. The Story of the Peruvian Guano Trade*. Cranbury, NJ: Associated University Presses.

Hollis, R. B. 1852. *Sudden Destruction: A Discourse on the Loss of the Amazon*. London: Partridge and Oakey.

[Holt, George]. 1995. *A Brief Memoir of George Holt, Esquire of Liverpool*. Reprint of private 1861 edition Cambridge: Baily.

Holt, Alfred. 1877–78. 'Review of the progress of steam shipping during the last quarter of a century [and discussion]'. *Minutes of the Proceedings of the Institution of Civil Engineers 51*: 2–135.

Holt, Alfred. 1911. *Fragmentary Autobiography of Alfred Holt*. Privately printed.

Holt, Anne. 1938. *Walking Together. A Study in Liverpool Nonconformity, 1688–1938*. London: Allen & Unwin.

Hope, Ronald. 1990. *A New History of British Shipping*. London: John Murray.

Howarth, David and Stephen Howarth. 1986. *The Story of P&O*. London: Weidenfeld & Nicholson.

Howland, S. S. 1843. *Steamboat Disasters and Railroad Accidents in the United States*. Worcester, MA: Lazell.

Hughes, John. 1906. *Liverpool Banks and Bankers, 1760–1837*. Liverpool: Young.

Hughes, Thomas P. 1983. *Networks of Power. Electrification in Western Society, 1880–1930*. Baltimore, MD: Johns Hopkins University Press.

1987. 'The evolution of large technological systems'. In Wiebe E. Bijker, Thomas P. Hughes and Trevor Pinch (eds), *The Social Construction of Technological Systems. New Directions in the Sociology and History of Technology*, pp. 51–82. Cambridge, MA and London: MIT Press.

1994. 'Technological momentum', In Merritt Roe Smith and Leo Marx (eds), *Does Technology Drive History? The Dilemma of Technological Determinism*, pp. 100–13. Cambridge, MA and London: MIT Press.

2004. *Human-Built World. How to Think About Technology and Culture*. Chicago, IL and London: University of Chicago Press.

Hyde, F. E. 1956. *Blue Funnel: A History of Alfred Holt & Company 1865–1914*. Liverpool: Liverpool University Press.

1967. *Shipping Enterprise and Management*. Liverpool: Liverpool University Press.

1973. *The Far Eastern Trade, 1860–1914*. Liverpool: Liverpool University Press.

1975. *Cunard and the North Atlantic*. London and Basingstoke: Macmillan.

Ince, Laurence. 2002. 'Maudslay, Sons & Field, 1831–1904'. In John Cantrell & Gillian Cookson (eds), *Henry Maudslay & the Pioneers of the Machine Age*, pp. 166–84. Brimscombe Port Stroud: Tempus.

Irish, Bill. 2001. *Shipbuilding in Waterford 1820–1882. A Historical, Technical and Pictorial Study*. Bray: Wordwell.

Jevons, W. Stanley. 1865. *The Coal Question. An Inquiry Concerning the Progress of the Nation, and the Probable Exhaustion of our Coal Mines*. London: Macmillan.

John, A. H. 1959. *A Liverpool Merchant House. Being the History of Alfred Booth and Company 1863–1958*. London: Allen & Unwin.

Jones, Clement. 1935–38. *Pioneer Shipowners*. 2 vols. Liverpool: Journal of Commerce.

Kemble, Fanny. 1878. *Record of a Girlhood*. 3 vols. London: Bentley.

Kennedy, John. 1903. *The History of Steam Navigation*. Liverpool: Charles Birchall.

Keynes, R. D. (ed.). 1988. *Charles Darwin's Beagle Diary*. Cambridge: Cambridge University Press.

Kingsley, Charles. 1887. *At Last. A Christmas in the West Indies*. London: Macmillan. First published 1871.

Knight, Alan. 1999. 'Britain and Latin America'. In Andrew Porter (ed.), *The Oxford History of the British Empire. Volume III. The Nineteenth Century*, pp. 122–45. Oxford and New York, NY: Oxford University Press.

Kurihara, Ken. 2011. '"The voice of God upon the waters:" Sermons on steamboat disasters in antebellum America'. *Coriolis: Interdisciplinary Journal of Maritime Studies* 2: 1–16.

Laird, Macgregor and R. A. K. Oldfield. 1837. *Narrative of an Expedition into the Interior of Africa, by the River Niger, in the Steam-Vessels Quorra and Alburkah, in 1832, 1833, and 1834*. 2 vols. London: Bentley.

Lambert, Andrew. 1991. *The Last Sailing Battlefleet. Maintaining Naval Mastery 1815–1850*. London: Conway.

1999. 'Brunel, the Navy and the screw propeller'. In Griffiths *et al.* (eds), *Brunel's Ships*, pp. 27–52. London: NMM.

2011. 'John Scott Russell –Ships, science and scandal in the age of transition'. *International Journal for the History of Engineering 81*: 100–18.

Lamport, William J. 1849. 'The education of the mercantile classes'. *Proceedings of the [Liverpool] Literary and Philosophical Society* 5: 72–3.

Langley, John G. 2006. *Steam Lion. A Biography of Samuel Cunard*. Shelter Island, NY: Brick Tower.

[Lardner, Dionysius]. 1837. 'Atlantic steam navigation'. *Edinburgh Review 65*: 118–46.

Laudan, L. L. 1970. 'Thomas Reid and the Newtonian turn of British methodological thought'. In R. E. Butts and J. W. Davis (eds), *The Methodological Heritage of Newton*, pp. 103–31. Oxford: Oxford University Press.

Lawson, Will. 1927. *Pacific Steamers*. Glasgow: Brown, Son & Ferguson.

Le Fleming, H. M. 1961. *Ships of the Blue Funnel Line*. Southampton: Adlard Coles.

Lefebvre, Henri. 1991. *The Production of Space*. Trans. Donald Nicholson-Smith. Oxford: Blackwell. First published 1974.

Leggett, Don. 2013. 'William Froude, John Henry Newman and scientific practice in the culture of Victorian doubt'. *English Historical Review 128: 571–95.*

2015. *Shaping the Royal Navy. Technology, Authority and Naval Architecture, c.1830–1906*. Manchester: Manchester University Press.

Leggett, Don and Richard, Dunn (eds). 2012. *Re-inventing the Ship. Science, Technology and the Maritime World 1800–1918*. Farnham: Ashgate.

Lindsay, W. S. 1874–76. *History of Merchant Shipping and Ancient Commerce*. 4 vols. London: Sampson Low.

Lubbock, Basil. 1946. *The China Clippers*. Glasgow: Brown, Son & Ferguson.

1948. *The Colonial Clippers*. Glasgow: Brown, Son & Ferguson.

1950. *The Blackwall Frigates*. 2nd edn. Glasgow: Brown, Son & Ferguson. First edn 1924.

McBride, Ian. 1998. *Scripture Politics: Ulster Presbyterians and Irish Radicalism in the Late Eighteenth Century*. Oxford: Oxford University Press.

McGarry, E. John. 2006. *Ploughing the South Sea. A History of Merchant Shipping on the West Coast of South America*. Bloomington, IN and Milton Keynes: Author House.

MacKinnon, [Lauchlan]. 1853. 'English and American ocean steamers'. *Harper's New Monthly Magazine 7*: 205–08.

MacLeod, Christine. 2007. *Heroes of Invention. Technology, Liberalism and British Identity, 1750–1914*. Cambridge: Cambridge University Press.

Macleod, Donald. 1876. *Memoir of Norman Macleod, D.D.* 2 vols. London: Daldy, Isbister.

1883. *A Nonogenarian's Reminiscences of Garelochside & Helensburgh and the People Who Dwelt Thereon and Therein*. Helensburgh, Dumbarton and Glasgow: Morison et al.

MacMechan, A. 1931. 'Halifax in trade'. *Canadian Geographical Journal 3*: 151–4.

[McQueen], James. 1819–20. 'Monthly register: Commercial report'. *Edinburgh Magazine 6*: 589–95.

1838a. *A General Plan for a Mail Communication by Steam between Great Britain and the Eastern and Western Parts of the World; Also, to Canton and Sydney, Westward by the Pacific; to Which Are Added Geographical Notices of the Isthmus of Panama, Nicaragua, & c. With Charts*. London: Fellowes.

1838b. *West Indian Mail Communication: In a Letter to Francis Baring, Esq. M.P. Secretary to the Treasury*. London: Fellowes.

McRonald, Malcolm. 2005–07. *The Irish Boats*. 3 vols. Brimscombe Port Stroud: Tempus.

2009. 'Fawcett, Preston & Co. – Marine engines'. *The Bulletin (LNRS) 54*: 27–32.

Maginnis, Arthur J. 1892. *The Atlantic Ferry. Its Ships, Men, and Working*. London: Whittaker.

Mansel, Robert. 1882. 'On some points in the history and applications of the theory of thermodynamics'. *Transactions of the Institution of Engineers and Shipbuilders in Scotland 25*: 85–132.

Marriner, Sheila. 1961. *Rathbones of Liverpool 1845–73*. Liverpool: Liverpool University Press.

Marriner, Sheila and F. E. Hyde. 1967. *The Senior. John Samuel Swire 1825–98. Management in Far Eastern Trades*. Liverpool: Liverpool University Press.

Marsden, Ben. 1998a. '"A most important trespass": Lewis Gordon and the Glasgow chair of civil engineering and mechanics 1840–1855'. In Crosbie Smith and Jon Agar (eds), *Making Space for Science. Territorial Themes in the Shaping of Knowledge*, pp. 87–117. Basingstoke: Macmillan.

1998b. 'Blowing hot and cold: Reports and retorts on the status of the air-engine as success or failure'. *History of Science 36*: 373–420.

2006. 'Superseding steam: The Napier and Rankine hot-air engine'. *Transactions of the Newcomen Society 76*: 1–22.

2008. 'The administration of the "engineering science" of naval architecture at the British Association for the Advancement of Science, 1831–1872'. *Yearbook of European Administrative History 20*: 67–94.

Marsden, Ben and Crosbie Smith. 2005. *Engineering Empires. A Cultural History of Technology in Nineteenth-Century Britain*. Basingstoke: Palgrave Macmillan.

Marshall, F. C. 1881. 'On the progress and development of the marine engine'. *Proceedings of the Institution of Mechanical Engineers* 32: 449–509.

Maugham, William C. 1896. *Annals of Garelochside*. Paisley: Gardner.

Melvill, Henry. 1846. *Sermons Preached on Public Occasions*. London: Rivington.

Mendonca, Sandro. 2013. 'The "sailing ship" effect: Reassessing history as a source of insight on technical change'. *Research Policy 42*: 1724–38.

'Mercator.' 1839a. 'Atlantic steam navigation'. *Nautical Magazine 8*: 39–44.

'Mercator.' 1839b. 'On the construction of British merchant shipping'. *Nautical Magazine 8*: 179–92.

'Mercator.' 1840a. 'Atlantic steam navigation. [Letter] To the Directors of the Royal Mail Steam Packet Company'. *Nautical Magazine 9*: 114–18, 429–39; 191–2 (Roberts' reply).

'Mercator.' 1840b. 'Atlantic steamers'. *Nautical Magazine 9*: 272–3.

'Mercator.' 1840c. 'Atlantic steam navigation. "Theory v. Practice"'. *Nautical Magazine 9*: 789–92.

Middlemass, Norman L. 2002. *Merchant Fleets. Blue Funnel Line*. Newcastle: Shield Publications.

Miller, David Philip. 2012. 'Testing power and trust: The steam indicator, the "Reynolds controversy," and the relations of engineering science and practice in late nineteenth-century Britain'. *History of Science 50*: 212–50.

Miller, Russell *et al.* 1980. *The East Indiamen*. Amsterdam: Time Life.

Milne, Graeme J. 2000. *Trade and Traders in Mid-Victorian Liverpool. Mercantile Business and the Making of a World Port*. Liverpool: Liverpool University Press.

Morrell, Jack. 1974. 'Reflections on the history of Scottish science'. *History of Science 12*: 81–94.

Morrell, Jack and Arnold Thackray. 1981. *Gentlemen of Science. Early Years of the British Association for the Advancement of Science*. Oxford: Oxford University Press.

Murray, Robert. 1860. 'On various means and appliances for economising fuel in steam-ships'. *Transactions of the Institution of Naval Architects 1*: 171–83.

1865. 'Some recent experiences in marine engineering'. *Transactions of the Institution of Naval Architects 6*: 158–69.

1867. 'Further experiences in marine engineering'. *Transactions of the Institution of Naval Architects 8*: 155–67.

Murray, Andrew and Robert Murray. 1861. *The Theory and Practice of Ship-Building [Andrew]; Steam-Ships [Robert]*. Combined in a single volume from entries in the *Encyclopaedia Britannica*.

Napier, James R. 1864. 'On the effects of superheated steam and oscillating paddle wheels on the speed and economy of steamers (from data supplied chiefly by Mr William Beardmore)'. *Transactions of the Institution of Engineers in Scotland 7*: 86–102.

1866. 'Memoir of the Late Mr David Elder'. *Transactions of the Institution of Engineers in Scotland* 9: 92–105.

Napier, James. 1904. *Life of Robert Napier of West Shandon*. Edinburgh and London: William Blackwood.

Nasmyth, James. 2010. *James Nasmyth, Engineer. An Autobiography*. Ed. Samuel Smiles. Facsimile edn. Cambridge: Cambridge University Press. First published 1883.

'Nauticus.' 1841. 'Atlantic Steam Navigation'. *Nautical Magazine 10*: 258–60.

Newall, Peter. 1999. *Union-Castle. A Fleet History*. London: Carmania Press.

2004. *Orient Line. A Fleet History*. Preston: Ships in Focus.

Nicol, Stuart. 2001. *Macqueen's Legacy*. 2 vols. Brimscombe Port Stroud: Tempus.

Nicolson, John. 1914. *Arthur Anderson. A Founder of the P and O Company*. Paisley: Gardner.

Oakley, Francis. 1961. 'Christian theology and the Newtonian science'. *Church History 30*: 433–57.

Osborne, Brian D. 1995. *The Ingenious Mr Bell*. Glendaruel: Argyll.

Osborne, John. [1844]. *Guide to the Madeiras, Azores, British and Foreign West Indies, Mexico, and Northern South America*. London: Simpkin & Marshall.

Palmer, Sarah. 1982. '"The most indefatigable activity": The General Steam Navigation Company 1824–50'. *Journal of Transport History 3*: 1–22.

Parry, Edward. 1858. *Memoirs of Rear-Admiral Sir W. Edward Parry*. 4th edn. London: Longmans. First published 1857.

Paton, Andrew. 1853. *The Sunday Steamer: Remonstrance of the Established Presbytery of Glasgow with the Answers of the Owners of the Steamer Emperor*. Glasgow: Love.

[Peacock, Thomas Love]. 1835. 'On steam navigation to India'. *Edinburgh Review 60*: 445–82.

Peacock, George. 1837. 'On propelling steam vessels without the aid of steam'. *Nautical Magazine 6*: 730–2.

1838. 'Raising sunken vessels'. *Nautical Magazine 7*: 776–8.

1839. 'Pictou Roads and Harbour, Nova Scotia'. *Nautical Magazine* 8: 146–8.

1840a. 'Rogers' Anchor'. *Nautical Magazine 9*: 447–8.

1840b. 'Naval tactics'. *Nautical Magazine 9*: 599–600.

1842. 'Pacific Steam Navigation Company's steam-vessel Chili'. *Nautical Magazine 11*: 413–15.

[Peacock, George (ed.)]. 1859. *Official Correspondence, Certificates of Service, and Testimonials of Mr George Peacock, F.R.G.S. Formerly a Master in the Royal Navy of 1835*. Privately printed: Exeter.

Perrey, Abraham. 1826. *Memorial of the Loss of the Comet Steam Packet: In Three Parts. I. A Narrative of That Event II. An Address to Youth. III. A General Improvement*. Glasgow: [sold by] Maurice Gale.

Pickstone, John. 1993. 'Ways of knowing: Towards a historical sociology of science, technology and medicine'. *The British Journal for the History of Science 26*: 433–58.

Pole, William (ed.). 1970. *The Life of Sir William Fairbairn, Bart*. Newton Abbot: David & Charles. First published 1877.

Pollard, Sidney and Paul Robertson. 1979. *The British Shipbuilding Industry 1870–1914*. Cambridge, MA and London: Harvard University Press.

Pond, E. Leroy. (ed.). 1927 [1971]. *Junius Smith. A Biography of the Father of the Atlantic Liner*. Facsimile edn. New York, NY: Books for Libraries Press.

[Pray, I. C.]. 1855. *James Gordon Bennett and his Times*. New York, NY: Stringer & Townsend.

Rabson, Stephen and Kevin O'Donoghue. *P&O. A Fleet History*. Kendall: World Ship Society.

Rankine, W. J. Macquorn. 1855. 'On the means of realizing the advantages of the air-engine'. *Edinburgh New Philosophical Journal 1* (new series): 1–32.

1871. *A Memoir of John Elder. Engineer and Shipbuilder*. Edinburgh and London: William Blackwood.

Reid, Thomas. 1846–63. 'Essays on the active powers of man'. In Sir William Hamilton (ed.), *The Works of Thomas Reid, D.D., vol.* 2, pp. 509–697. 2 vols. Edinburgh: Maclachlan & Stewart.

Riddell, John F. 1979. *Clyde Navigation. A History of the Development and Deepening of the River Clyde*. Edinburgh: Donald.

Ridgely-Nevitt, Cedric. 1981. *American Steamships on the Atlantic*. Newark, DE: University of Delaware Press.

Rieger, Bernard. 2005. *Technology and the Culture of Modernity in Britain and Germany, 1890–1945*. Cambridge: Cambridge University Press.

Rigg, Arthur. 1870. 'On the compound marine steam engine'. *Transactions of the Institution of Naval Architects 11*: 136–53.

Ritchie, L. A. 1985. 'The Floating Church of Loch Sunart'. *Records of the Scottish Church History Society 22*: 159–73.

Robb, Johnston Fraser. 1993. *Scotts of Greenock. Shipbuilders and Engineers 1820–1920. A Family Enterprise*. 2 vols. University of Glasgow Ph.D.

[Robb, Johnston Fraser]. 2009. *Truly 'Clyde Built'. The Scott Family Enterprise. Scotts' of Greenock*. Compiled by William Kane. Glasgow: Brown, Son & Ferguson.

Roberts, G. W. 2009. 'Magnetism and chronometers: The research of the Reverend George Fisher'. *The British Journal for the History of Science 42*: 57–72.

Rowan, Frederick J. 1880. 'On the introduction of the compound engine, and the economical advantage of high pressure steam; with a description of the system introduced by the late Mr J.M. Rowan'. *Transactions of the Institution of Engineers and Shipbuilders in Scotland 23*: 51–97

Rowland, K. T. 1970. *Steam at Sea. A History of Steam Navigation*. Newton Abbot: David & Charles.

Roy, James A. 1935. *Joseph Howe. A Study in Achievement and Frustration*. Toronto: Macmillan.

Russell, John Scott. 1841. *On the Nature, Properties, and Applications of Steam and on Steam Navigation*. Edinburgh: A&C Black.

1861. 'On the late Mr John Wood and Mr Charles Wood, naval architects, of Port Glasgow'. *Transactions of the Institution of Naval Architects* 2: 141–8.

1863. *Very Large Ships, Their Advantages and Defects: A Lecture Delivered at the Athenaeum, Bristol. April 5th, 1863*. London and Bristol: Longman.

Salamo, Lin and H. E. Smith (eds). 1997. *Mark Twain's Letters Volume 5 1872–1873*. Berkeley, CA; Los Angeles, CA and London: University of California Press.

Scarlett, Peter C. 1838. *South America and the Pacific*. 2 vols. London: Henry Colburn.

Schaffer, Simon. 2007. '"The charter'd Thames": Naval architecture and experimental spaces in Georgian Britain'. In Lissa Roberts, Simon Schaffer and Peter Dear (eds), *The Mindful Hand: Inquiry and Invention from the Late Renaissance to Early Industrialization*, pp. 279–305. Amsterdam: Koninkliijke Nedelandse Akademie van Wetenschappen.

Schivelbusch, Wolfgang. 1986. *The Railway Journey. The Industrialization of Time and Space in the 19th Century*. New edn. Leamington Spa: Berg. First published 1977 as *Geschichte der Eisenbahnreise*.

[Scotts]. 1906. *Two Centuries of Shipbuilding by the Scotts at Greenock*. London: Offices of 'Engineering'.

Searle, G. R. 1998. *Morality and the Market in Victorian Britain*. Oxford: Oxford University Press.

Secord, James A. 2000. *Victorian Sensation. The Extraordinary Publication, Reception, and Secret Authorship of 'Vestiges of the Natural History of Creation'*. Chicago, IL and London: University of Chicago Press.

Shapin, Steven. 1989. 'The invisible technician'. *American Scientist 77*: 554–63.

1994. *A Social History of Truth. Civility and Science in Seventeenth-Century England*. Chicago, IL and London: University of Chicago Press.

Shields, John. 1949. *Clyde Built. A History of Shipbuilding on the Clyde*. Glasgow: William Maclellan.

[Silliman, Benjamin]. 1831. 'Safety of steam boats'. *American Journal of Science 19*: 143–6.

Slater, Isaac (ed.). 1856. *National Commercial Directory of Ireland*. Manchester and London: Isaac Slater

Smith, Junius. 1839. 'Letters on Atlantic steam navigation'. *The American Journal of Science and Arts 35*: 160–7.

Smith, T. T. Vernon. 1857. *The Past, Present and Future of Atlantic Ocean Steam Navigation. Read before the Fredericton Athenaeum, June 15, 1857*. Fredericton: The Fredericton Athenaeum.

Smith, Crosbie. 1979. 'From design to dissolution: Thomas Chalmers' debt to John Robison'. *The British Journal for the History of Science 12*: 59–70.

1998a. *The Science of Energy. A Cultural History of Energy Physics in Victorian Britain*. Chicago, IL and London: University of Chicago Press.

1998b. '"Nowhere but in a great town:" William Thomson's spiral of classroom credibility'. In Crosbie Smith and Jon Agar (eds), *Making Space for Science. Territorial Themes in the Shaping of Knowledge*, pp. 118–46. Basingstoke: Macmillan.

2004. 'Meikleham, William'. In Bernard, Lightman (ed.), *The Dictionary of Nineteenth-Century British Scientists*, vol. 3, pp. 1390–2. 4 vols. Bristol: Thoemmes Continuum.

2011a. '"A most terrific passage": Putting faith into Atlantic steam navigation'. In Robert, Lee (ed.), *Commerce and Culture. Nineteenth-Century Business Elites*, pp. 285–315. Farnham: Ashgate.

2011b. 'The "crinoline" of our steam engineers': Re-inventing the marine compound engine 1850–1885'. In David Livingstone and CharlesWithers (eds), *Geographies of Nineteenth-Century Science*, pp. 229–54. Chicago, IL and London: University of Chicago Press.

2012. '"This great national undertaking:" John Scott Russell, the Master Shipwrights and the Royal Mail Steam Packet Company'. In Don Leggett and Richard Dunn (eds), *Re-inventing the Ship. Science, Technology and the Maritime World 1800–1918*, pp. 25–52. Farnham: Ashgate.

2013a. 'Engineering energy: Constructing a new physics for Victorian Britain'. In Jed Z. Buchwald and Robert Fox (eds). *The Oxford Handbook of the History of Physics*, pp. 508–32. Oxford: Oxford University Press.

2013b. '"We never make mistakes": Constructing the Empire of the Pacific Steam Navigation Company'. In Miles Taylor (ed.), *The Victorian Empire and Britain's Maritime World, 1837–1901. The Sea and Global History*, pp. 82–112. Basingstoke: Palgrave Macmillan.

2013c. '"I have in mind a study of a Scotch seaman:" Witnessing power in Joseph Conrad's early literature of the sea'. In Ben Marsden, Hazel Hutchison and Ralph O'Connor (eds), *Uncommon Contexts: Encounters between Science and Literature, 1800–1914*, pp. 145–66, 223–7. London: Pickering & Chatto.

2014. 'Witnessing power: John Elder and the making of the marine compound engine 1850–1858'. *Technology and Culture 55*: 76–106.

Smith, Crosbie and M. Norton Wise. 1989. *Energy and Empire. A Biographical Study of Lord Kelvin*. Cambridge: Cambridge University Press.

Smith, Crosbie and Anne Scott. 2007. '"Trust in Providence:" Building confidence into the Cunard Line of Steamers'. *Technology and Culture 48*: 471–96.

Smith, Crosbie, Ian Higginson and Phillip Wolstenholme. 2003a. '"Avoiding equally extravagance and parsimony". The moral economy of the ocean steamship'. *Technology and Culture 44*: 443–69.

2003b. '"Imitations of God's own works:" Making trustworthy the ocean steamship'. *History of Science 41*: 379–426.

Somner, Graeme. 2000. *The Aberdeen Steam Navigation Company Ltd*. Gravesend: World Ship Society.

Stammers, Michael K. 1978. *The Passage Makers*. Brighton: Teredo.

Statham-Drew, Pamela. 2003. *James Stirling. Admiral and Founding Governor of Western Australia*. Crawley, WA: University of Western Australia Press.

Stewart, Randall (ed.). 1962. *The English Notebooks by Nathaniel Hawthorne*. New York, NY: Russell & Russell.

Stirling, James. 'Stirling's air-engine'. *Mechanics Magazine 45*: 559–66.

Storey, Graham *et al.* 1974. *The Letters of Charles Dickens Volume Three 1842–1843*. Oxford: Oxford University Press.

et al. (eds). 1999. *The Letters of Charles Dickens Volume Eleven 1865–1867*. Oxford: Oxford University Press.

et al. (eds). 2002. *The Letters of Charles Dickens Volume Twelve 1868–1870*. Oxford: Oxford University Press.

Strang, John. 1850. 'On the progress of Glasgow, in population, wealth, manufactures, &c'. *BAAS Report 20*: 162–9.

Sutherland, Thomas. 1903. 'Stray notes'. *Alma Mater. Aberdeen University Magazine 20*: 102–08.

Taylor, James. 1976. *Ellermans. A Wealth of Shipping*. London: Wilton House Gentry.

Taylor, James. 2006. *Creating Capitalism. Joint-Stock Enterprise in British Politics and Culture, 1800–1870*. Woodbridge: Boydell Press.

Thiesen, William H. 2006. *Industrializing American Shipbuilding. The Transformation of Ship Design and Construction, 1820–1920*. Gainesville, FL: University Press of Florida.

Thom, John H. 1832. *Sermons and Occasional Services, Selected from the Papers of the Late Rev. John Hincks, with a Memoir of the Author by J.H. Thom.* London: Longman.

1836. *The Doctrine of Waste: A Discourse Delivered in Prince's Street Chapel, Cork, 8th May 1836.* Cork: King.

Thomson, James. 1849. 'Theoretical considerations on the effects of pressure in lowering the freezing point of water'. *Transactions of the Royal Society of Edinburgh 16*: 575–80.

Thomson, William. 1847. 'Notice of Stirling's air-engine'. *Proceedings of the Glasgow Philosophical Society 2*: 169–70.

1889–94. 'On the sources of energy in nature available to man for the production of mechanical effect'. In W. Thomson (ed.), *Popular Lectures and Addresses, vol. 2*, pp. 433–50. 3 vols. London and New York, NY: Macmillan.

Thorne, R. G. (ed.). 1986. *The History of Parliament. The House of Commons 1790–1820 Volume 4*. London: David & Charles.

Trollope, Anthony. 1985. *The West Indies and Spanish Main*. Gloucester: Alan Sutton. First published 1859.

Tyler, David Budlong. 1939. *Steam Conquers the Atlantic*. New York, NY and London: Appleton-Century.

Villiers, Peter. 2006. *Joseph Conrad. Master Mariner*. Rendlesham, Suffolk: Seafarer Books.

Walton, John K. and Alistair Wilcox (eds). 1991. *Low Life and Moral Improvement in Mid-Victorian England. Liverpool through the Journalism of Hugh Shimmin.* Leicester: Leicester University Press.

Warde, Paul. 2007. *Energy Consumption in England and Wales 1560–2000*. Rome: Instituto di Studi sulle Societe del Mediterraneo.

Wardle, Arthur C. 1940. *Steam Conquers the Pacific. A Record of Maritime Achievement 1840–1940*. London: Hodder & Stoughton.

Webb, R. K. 1978. 'John Hamilton Thom: Intellect and conscience in Liverpool'. In P. T. Phillips (ed.), *The View from the Pulpit: Victorian Ministers and Society*, pp. 211–43. Toronto: Macmillan.

[Wheelwright, William]. 1838. *Statements and Documents Relative to the Establishment of Steam Navigation in the Pacific, with Copies of the Decrees of the Governments of Peru, Bolivia, and Chile, Granting Exclusive Privileges to the Undertaking* (including PSNC Prospectus). London: Whiting.

Willox, John. 1865. *Steam Fleets of Liverpool*. Liverpool: Peat.

Winter, Alison. 1994. '"Compasses All Awry": the iron ship and the ambiguities of cultural authority in Victorian Britain'. *Victorian Studies 38*: 69–98.

Wise, M. Norton. 1989–90. 'Work and waste: political economy and natural philosophy in nineteenth-century Britain (I)-(III)'. With the collaboration of Crosbie Smith. *History of Science 27*: 263–301; 391–449; *28*: 221–61.

Woodman, Richard. 2009. *Masters under God: Makers of Empire: 1816–1884*. Brimscombe Port Stroud: The History Press.

Wrigley, E. A. 2010. *Energy and the Industrial Revolution*. Cambridge: Cambridge University Press.

Index

Printed in the United States
by Baker & Taylor Publisher Services

Printed in the United States
by Baker & Taylor Publisher Services